Embedded Robotics

Thomas Bräunl

EMBEDDED ROBOTICS

From Mobile Robots
to Autonomous Vehicles
with Raspberry Pi
and Arduino

Fourth Edition

 Springer

Thomas Bräunl
School of Engineering
The University of Western Australia
Perth, WA, Australia

ISBN 978-981-16-0803-2   ISBN 978-981-16-0804-9   (eBook)
https://doi.org/10.1007/978-981-16-0804-9

1st–3rd editions: © Springer-Verlag Berlin Heidelberg 2003, 2006, 2008
4th edition: © The Editor(s) (if applicable) and The Author(s), under exclusive license to Springer Nature Singapore Pte Ltd. 2022

This Springer imprint is published by the registered company Springer Nature Singapore Pte Ltd.
The registered company address is: 152 Beach Road, #21-01/04 Gateway East, Singapore 189721, Singapore

PREFACE

This book gives a practical, in-depth introduction to embedded systems and autonomous robots, using the popular Raspberry Pi and Arduino microcontrollers. We demonstrate how to build a variety of mobile robots using a combination of these controllers, together with powerful sensors and actuators. Numerous application examples are given for mobile robots, which can be tested either on the real robots or on our freely available simulation system *EyeSim*.

This book combines teaching and research materials and can be used for courses in embedded systems as well as in robotics and automation. We see laboratories as an essential teaching and learning method in this area and encourage everybody to reprogram and rediscover the algorithms and systems presented in this book.

Although we like simulations for many applications and treat them in quite some depth in several places in this book, we do believe that students should also be exposed to real hardware in both areas, embedded systems and robotics. This will deepen their understanding of the subject area and, of course, create a lot more fun, especially when experimenting with small mobile robots.

We started this robotics endeavor over twenty years ago, when we first interfaced a digital image sensor to our own processor board. The *EyeBot-1* was based on a Motorola 68332 and our own operating system *RoBIOS*. The controller was soon followed by a variety of driving, walking, flying, swimming and diving robots that we called the EyeBot family. More powerful (and more expensive) controller architectures followed, until the availability of cheap, powerful boards like the Raspberry Pi and the Arduino Uno let us reconsider our approach. We redesigned our robot family around these popular controllers, but maintained the look and feel of our easy-to-use operating software.

Even on the simulation side, we have gone new ways. Our original stand-alone robot simulation has now been replaced by our new *EyeSim VR* package, which uses the Unity game environment with a much more realistic physics engine. The companion book *Robot Adventures in Python and C* (Springer 2020) concentrates on the software aspect of mobile robots and extends the hardware/software approach of this book.

For any embedded application, the processor power (and cost) needs to match to the given problem. For low-level control of a mobile robot with two DC motors and a few simple sensors, an 8-bit controller such as the Arduino might be sufficient. However, if we want to do image processing or learning, we need a more powerful controller like the 32-bit Raspberry Pi.

The EyeBot family consists of mobile robots with all sorts of propulsion systems. We and numerous other universities use these robots and simulation systems for laboratory experiments in embedded systems as part of the computer engineering, electrical engineering and mechatronics curriculum.

Acknowledgements

A number of colleagues and students contributed to the chapters, software systems and robotics projects presented in this book.

Working on the RoBIOS robot operating system were Remi Keat (base system), Marcus Pham (high-level control), Franco Hidalgo (low-level control) and Klaus Schmitt (driving routines).

The EyeBot controller board layout was done by Ivan Neubronner, and the EyeSim VR simulator was implemented by Travis Povey (Unity), Joel Frewin (robot models and applications), Michael Finn (terrain, underwater, swarms) and Alexander Arnold (VR).

The following colleagues and former students contributed to this book: Adrian Boeing on the evolution of walking gaits, Mohamed Bourgou on car detection and tracking, Christoph Braunschädel on PID control graphs, Louis Gonzalez and Michael Drtil on AUVs, James Ng on Bug and Brushfire algorithms, David Venkitachalam on genetic algorithms, Joshua Petitt on DC motors, Bernhard Zeisl on lane detection, Alistair Sutherland on balancing robots, Jordan King on traffic sign recognition and Nicholas Burleigh on deep learning for autonomous driving.

Additional Materials

All system software discussed in this book, the RoBIOS operating system, C/C++ compilers for Windows, MacOS and Linux, system tools, image processing tools, simulation system, and a large collection of example programs are available free from:

```
http://robotics.ee.uwa.edu.au/eyebot/
http://robotics.ee.uwa.edu.au/eyesim/
```

Software specific to the Raspberry Pi and Arduino (Nano) controllers can be downloaded from:

```
http://robotics.ee.uwa.edu.au/rasp/
http://robotics.ee.uwa.edu.au/nano/
```

As further reading, please refer to our companion book on programming of mobile robots: *Robot Adventures in Python and C*. This book makes heavy use of the EyeSim simulator, so it can be used for practical experiments without the need for a physical robot.

Perth, Australia

Thomas Bräunl

January 2022

CONTENTS

Contents

Contents

Part II: Robot Hardware

Part III: Robot Software

Contents

APPENDICES

PART I
EMBEDDED SYSTEMS

ROBOTS AND CONTROLLERS

Robotics has come a long way. For mobile robots particularly, a trend is happening similar to that seen for computer systems: the transition from mainframe computing—via workstations and PCs—to handheld devices. In the past, mobile robots were controlled by large and expensive computer systems that could not be carried and had to be linked via cable or wireless modems. Today, we can build small mobile robots with numerous actuators and sensors that are controlled by inexpensive, small and lightweight embedded computer systems that are carried on-board the robot.

There has been a tremendous increase of interest in mobile robots. Robots are not just interesting toys or inspirations for science fiction stories or movies (famously the stories by Asimov[1])—they are the perfect tool for engineering education. Mobile robots are used today at almost all universities in undergraduate and graduate courses in computer science/computer engineering, information technology, cybernetics, electrical engineering, mechanical engineering, automation and mechatronics.

What are the advantages of using mobile robot systems as opposed to traditional ways of education, such as mathematical models or computer simulation?

First of all, a robot is a tangible, self-contained piece of real-world hardware. Students can relate to a robot much better than to a formula or even a piece of software. Tasks to be solved involving a robot are of a practical nature and directly make more sense to students than algorithms without context.

Secondly, all problems involving real-world hardware, such as a robot, are in many ways harder than solving a theoretical problem. The *perfect world*, which often is the realm of pure software systems, does not exist here. Any actuator can only be positioned to a certain degree of accuracy, and all sensors have intrinsic reading errors and certain limitations. Therefore, a working robot program will be much more than just a logical solution coded in software. It

---

[1] I. Asimov. *I Robot*, Doubleday, New York NY, 1950.

T. Bräunl, *Embedded Robotics*,
https://doi.org/10.1007/978-981-16-0804-9_1

needs to be a robust system that takes into account and overcomes inaccuracies and imperfections. In summary: it requires a valid engineering approach to a specific, industry-relevant problem.

Third and finally, building and programming of mobile robots is enjoyable and an inspiration to students. The fact that there is a moving system whose behavior can be specified by a piece of software is a challenge. The fun factor can even be increased by introducing robot competitions, where two robot teams compete in solving a complex task (see Bräunl[2]).

1.1 Mobile Robots

Over the years, we have developed a large number of mobile robots in the Automation and Robotics Lab at The University of Western Australia (UWA). We have built wheeled, tracked, legged, flying and underwater robots. We call these robots the *EyeBot family* of mobile robots (Figure 1.1).

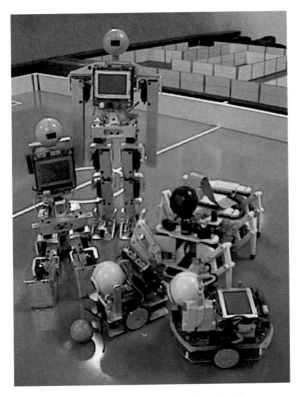

Figure 1.1: Members of the EyeBot robot family

[2]T. Bräunl. *Research Relevance of Mobile Robot Competitions*, IEEE Robotics and Automation Magazine, Dec. 1999, pp. 32–37 (6).

1.1 Mobile Robots

The simplest types of mobile robots are wheeled robots, as shown in Figure 1.2. Wheeled robots possess one or more *driven wheels* (drawn solid in the figure) and have optional passive *caster wheels* for stability (drawn hollow). They may also have *steered wheels* (drawn inside a circle to show their rotation axis). Most designs require a total of two motors for driving and steering of a mobile robot.

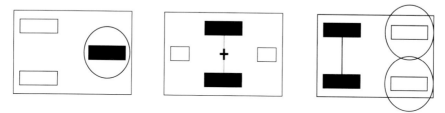

Figure 1.2: Wheeled robots

The design on the left of Figure 1.2 has a single driven wheel that is also steered. It requires two motors, one for driving the wheel and one for turning. The advantage of this design is that the driving and turning actions are completely decoupled by using two different motors. Therefore, the control software for driving straight or curves will be very simple. A disadvantage of this design is that the robot cannot turn on the spot, since the driven wheel is not located at its center.

The robot design in the middle of Figure 1.2 is called *differential drive* and is one of the most commonly used mobile robot designs. The combination of two driven wheels allows the robot to be driven straight, in a curve, or to turn on the spot. The execution of driving commands, for example a curve of a given radius, requires the generation of corresponding wheel speeds for the left and the right wheels, which has to be done by software. Another advantage of this design is that motors and wheels are in fixed positions and do not need to be turned as in the previous design. This simplifies the robot mechanics design considerably.

Finally, on the right of Figure 1.2 is the so-called *Ackermann Steering*, which is the standard drive and steering system of a rear-driven passenger car. We have one motor for driving both rear wheels via a differential gear box and one motor for steering of both front wheels.

It is interesting to note that all of these different mobile robot designs require two motors in total for driving and steering.

A special case of a wheeled robot is the omni-directional *Mecanum drive* robot in Figure 1.3, left. It uses four driven wheels with a special wheel design and will be discussed in more detail later.

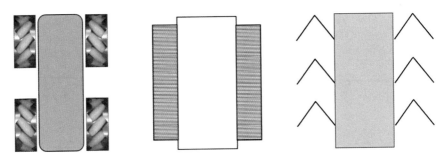

Figure 1.3: Omni-directional, tracked and walking robots

One disadvantage of all wheeled robots is that they require a paved road or some sort of flat surface to drive on. Tracked robots (see Figure 1.3, middle) are more flexible and can maneuver over rough terrain. However, they cannot navigate as accurately as a wheeled robot. Tracked robots also need two motors, one for each track on the left and right side.

Legged robots (see Figure 1.3, right) are the final category of land-based mobile robots. Like tracked robots, they can navigate over rough terrain, but can also climb up and down stairs. There are many different designs for legged robots, depending on their number of legs. The general rule is, the more legs, the easier to balance. For example, the six-legged robot shown in the figure can be operated in such a way that three legs are always on the ground while three legs are in the air. The robot will be stable, resting on a tripod formed by the three legs currently on the ground—provided its center of mass falls in the triangle described by these three legs. The less legs a robot has, the more complex it gets to balance and walk. For example, a robot with only four legs needs to be carefully controlled to avoid falling over. A biped (two-legged) robot cannot play the same trick with a supporting triangle, since that requires at least three legs. So, other techniques for balancing need to be employed, as we will discuss in more detail later. Legged robots usually require two or more motors (*dof—degrees of freedom*) per leg, so a typical six-legged robot has at least twelve motors. Many biped robot designs have five or more motors per leg, which results in a rather large number of degrees of freedom and also in considerably high weight and cost.

An interesting conceptual abstraction of actuators, sensors and robot control are the vehicles described by Braitenberg.[3] In one example, we have a simple interaction between motors and light sensors of a differential drive robot. If a light sensor is activated by a light source, it will proportionally increase the speed of the motor it is linked to.

---

[3]V. Braitenberg, *Vehicles – Experiments in Synthetic Psychology*, MIT Press, Cambridge MA, 1984.

1.1 Mobile Robots

In Figure 1.4, our robot has two light sensors, one on the front left and one on the front right. The left light sensor is linked to the left motor and the right sensor to the right motor. If a light source appears in front of the robot, it will start driving toward it, because both sensors will activate both motors. However, what happens if the robot gets closer to the light source and goes slightly off course? In this case, one of the sensors will be closer to the light source (the left sensor in the figure), and therefore, the corresponding motor (the left motor) will spin faster than the other. This will result in a curve trajectory of the robot vehicle, and it will miss the light source.

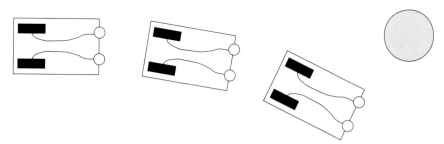

Figure 1.4: Braitenberg vehicle avoiding light (phototroph)

Figure 1.5 shows a very similar scenario of a Braitenberg vehicle. However, here we have linked the left sensor to the right motor and vice versa. If we conduct the same experiment as before, again the robot will start driving when encountering a light source. But when it gets closer and goes slightly off course (veering to the right in the figure), the left sensor will now receive more light and therefore accelerate the right motor. This will result in a left curve, so the robot is brought back on track to find the light source.

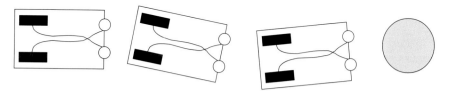

Figure 1.5: Braitenberg vehicle searching light (photovore)

Braitenberg vehicles are only an abstraction with limited use for building real robots. However, a number of control concepts can easily be demonstrated by using them.

1.2   Embedded Controllers

Throughout this book, we use two low-cost embedded microcontroller types for controlling robots—an Arduino controller, either for simple tasks or as the low-level controller in a dual-controller configuration, and a Raspberry Pi controller for complex and high-level tasks (Figure 1.6). While the Raspberry Pi has sufficient processing power to do complex tasks, such as image processing and sensor data evaluation, it is not ideal for generating actuator output signals. The Raspberry Pi was not designed to directly drive a robot, so it misses crucial hardware components, such as motor drivers or pulse counting input registers. In our design, only the camera and LCD plus any high-level USB sensors are directly connected to the Raspberry Pi. All actuators and low-level sensors are connected to a dedicated I/O board that communicates to the Raspberry Pi (or any other computer system) via USB.

Figure 1.6: Raspberry Pi and Arduino Nano clone

Our EyeBot7 I/O board is based on an Atmel Atmega128A1U processor with similar performance to an Arduino Mega. It can drive up to four motors and 14 servos and provide interfaces to a large number of sensors (see Figure 1.7).

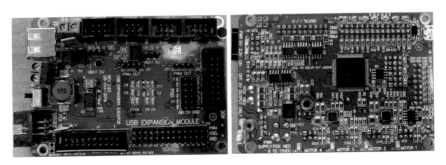

Figure 1.7: EyeBot/IO board, top and bottom

1.2 Embedded Controllers

The EyeBot-7 I/O board has:

- USB data port for communication
- Power regulator for 7.2 V input
- Micro-USB port for supplying regulated 5 V to Raspberry Pi
- 4 DC motor drivers
- 4 dual-encoder inputs
- 14 servo/PWM output
- 16 digital input/outputs
- 8 analog inputs including battery level sensor
- 6 PSD sensor inputs (infrared distance sensors)
- Microphone for audio input

The I/O board runs a continuous control program without any operating system that waits for input from a high-level controller on its USB port and then executes this command. Its schematic diagram is given in Figure 1.8. A simple command language is used for communication. Below are some examples; see the Appendix for details.

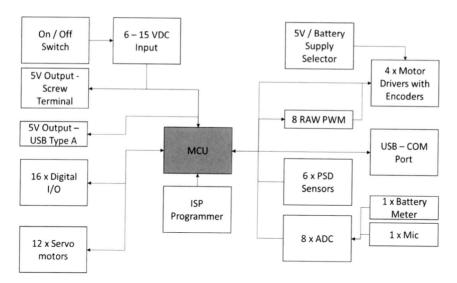

Figure 1.8: EyeBot-7 I/O structure

Command:

m 1 50   set **m**otor 1 to 50% speed.
e 1   read matching **e**ncoder 1.
s 2 200   set **s**ervo 2 to position 200.
p 3   read and return value from **P**SD distance sensor no. 3
a 10   read and return value from **a**nalog input pin no. 10.

1.3 Robot Design

Figure 1.9 shows the system design with a Raspberry Pi and an Atmel-based low-level controller, specialized for input/output, which provides the link between sensors and actuators to the high-level controller. The Raspberry links directly to the camera and the optional touch-screen LCD, as well as to any high-level sensors that require a USB or Ethernet LAN interface, e.g., satellite-based global positioning system (GPS), inertial measurement unit, internal sensor (IMU) or light detection and ranging, laser-based (Lidar). All motors with encoders, servos and all simple sensors are connected to the I/O board, which communicates to the Raspberry Pi via a bidirectional USB link. The advantage of this approach is that the high-level controller is freed from time-consuming low-level control. Also, since the Raspberry Pi usually does not run a real-time operating system, it cannot always maintain the correct time interval, which is essential for motor and servo control. Instead, this will be done by the low-level controller, which does not have an operating system itself, but can make use of various built-in hardware timers to guarantee real-time signals.

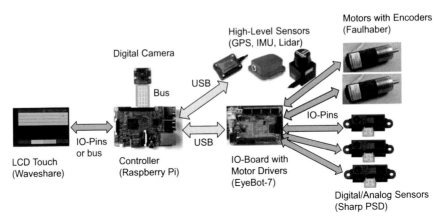

Figure 1.9: Mobile robot design with Raspberry Pi and Atmel/Arduino

The alternative to using the EyeBot-7 I/O board, or any secondary processor board for that matter, is to use separate extension boards for each of the required functionalities. These are:

- Voltage controller board
  required, e.g., for converting 7.2 V battery supply to 5 V, 3 A for Raspberry Pi and high-level sensors.

- Dual (or quad) motor driver board
  required to drive DC motors.

- Encoder interface board
  required to properly control motor speeds. Without this, a differential drive robot cannot drive straight.

- Servo controller board
  required to generate PWM output signals for model servos, such as the steering in a remote-controlled car. However, these signals can also be generated done in software if a lower PWM accuracy is acceptable.

- Analog sensor interface board
  required to interface to analog sensors, such as distance sensors.

- Digital sensor interface board
  required for sensors that return their data as a timed sequence of digital values. Simple binary sensors (on/off, 1/0) can be directly connected to an I/O pin of the Raspberry Pi.

Somewhat simpler is the approach without a dedicated I/O board that is shown in Figure 1.10. Camera, LCD and any high-level sensors are directly connected to the Raspberry Pi as before. However, motors are linked through an external motor driver to an output pin of the Raspberry Pi, while low-level sensors cannot be connected at all. If a model car with built-in motor driver and servo for steering is being used instead of a robot driving hardware, then Raspberry-I/O lines can be used directly to drive the model car. As the Raspberry Pi does not have any hardware timers, care has to be taken that the software pulse generation for the steering command does not result in a jitter of the servo.

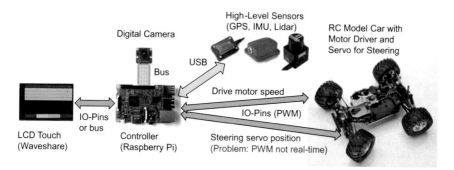

Figure 1.10: Mobile robot design with Raspberry Pi only

Equally simple, but also a lot less powerful is using an Arduino controller by itself for robot control, as shown in Figure 1.11. Although LCDs, camera modules and some high-level sensors are available, the compute power of the Arduino is not really capable to make the best use of them. Connecting motors requires an external motor driver board (more on this in the chapter on actuators), and encoders would need another specialized hardware addition. Only low-level sensors can be directly connected to the Arduino pins.

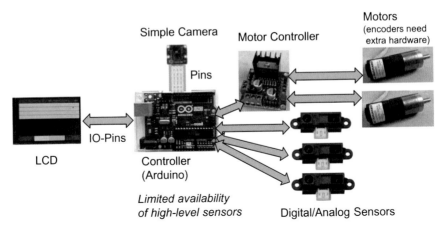

Figure 1.11: Mobile robot design with Arduino only

Figure 1.12 shows two of our driving robot designs, which use a stack of hardware boards (from top to bottom):

- Touch-screen LCD, optional, but very helpful
- Raspberry Pi controller, linking to I/O board and camera
- Atmel I/O board, connecting to motors and sensors.

Figure 1.12: Robots S4 and EyeCart with stacked LCD, Raspberry Pi and EyeBot-7 board

1.4 Operating System

Embedded systems can have anything between a complex operating system, such as Linux (used in the Raspberry Pi) or just the "naked" application program by itself with no operating system whatsoever (as in the Arduino).

1.4 Operating System

It all depends on the intended application area and the performance of the controller.

For the Raspberry Pi, we wanted to keep the *look and feel* and the convenience of our previous operating system Robot Basic Input Output System (*RoBIOS*), but we also needed the Linux system for complex sensor interfacing and communication, so we ended up implementing a similar approach to what Apple is using in their MacOS systems. The Raspberry hardware is running Linux, but we provide additional libraries for hardware control plus a monitor program for easy access to robot functionality and application programs.

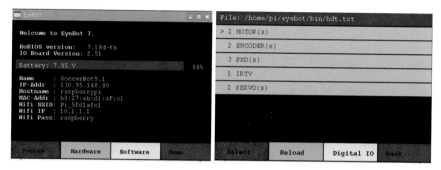

Figure 1.13: RoBIOS screens

When the Raspberry Pi boots up, it automatically starts RoBIOS and will display the start-up screen on the LCD (Figure 1.13). By pressing simple touch buttons, the user can test all relevant actuators and sensors of the robot, as well as starting example programs or own applications programs, whose I/O is being directed to the touch-screen LCD. Figure 1.14 shows the testing of a connected motor and the selection of one of the supplied demo programs.

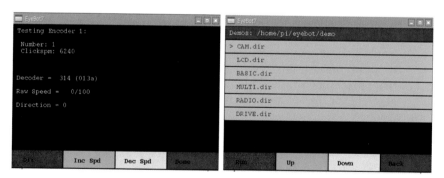

Figure 1.14: RoBIOS motor testing and program selection

Figure 1.15 shows the amended structure of the RoBIOS-7 system. A *hardware description table* (HDT) provides the link between the general RoBIOS monitor program and the specific hardware configuration of the robot.

It is a special text file that contains information for the Raspberry Pi about which sensors and actuators are connected to the controller and at which I/O pins.

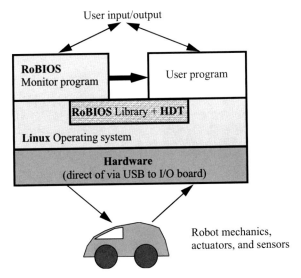

Figure 1.15: RoBIOS structure

1.5 Simulation

In this book, we will work with real robots as well as with our simulation system EyeSim. We should stress that we put the emphasis on creating a very realistic approximation of real robot behavior. Robot programs can be directly transferred from the simulation to the real robot without having to change a single line of source code. There are no unrealistic assumptions, no virtual (wishful) sensors, no "cheating." We even apply realistic error settings to the robot's sensors and actuators. A comprehensive companion book on robot algorithms and simulation is available in *Robot Adventures in Python and C*, Bräunl.[4]

EyeSim-VR[5] (Figure 1.16) runs natively on MacOS, Windows or Linux and also supports Virtual Reality (VR) devices. It was implemented by Travis Povey and Joel Frewin in 2018 and extended by Michael Finn and Alexander Arnold in 2019.

---

[4]T. Bräunl, *Robot Adventures in Python and C*, Springer-Verlag, Heidelberg, Berlin, 2020.

[5]T. Bräunl, *EyeSim VR – EyeBot Mobile Robot Simulator*, online: https://robotics.ee.uwa.edu. au/eyesim/.

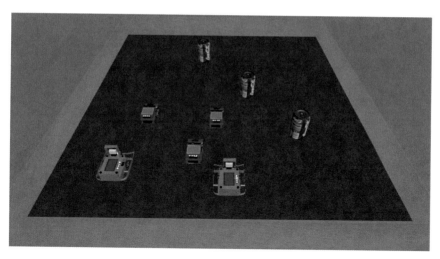

Figure 1.16: EyeSim-VR simulation of a robot scene

1.6   Tasks

1. Buy the embedded controller of your choice, e.g., Arduino or Raspberry Pi, and then spend some time with it. Install the required tools on your PC and get familiar with them. Learn how to transfer programs and data to and from the controller. Learn how to compile and run a simple program.

2. Install the EyeSim simulator onto your system, find out how to run it and play with it:
    https://robotics.ee.uwa.edu.au/eyesim/
    Install the pack of example programs from:
    https://robotics.ee.uwa.edu.au/eyesim/ftp/EyeSim-Examples.zip
    Compile and run the example programs. Understand how they work, modify them and create your own.

CENTRAL PROCESSING UNIT

The CPU (central processing unit) is the heart of every embedded system and every personal computer. It comprises the ALU (arithmetic logic unit), responsible for the number crunching, and the CU (control unit), responsible for instruction sequencing and branching. Modern microprocessors and microcontrollers provide, on a single chip, the CPU and a varying degree of additional components, such as counters, timing co-processors, watchdogs, SRAM (static RAM) and Flash-ROM (electrically erasable ROM).

Hardware can be described on several different levels, from low-level transistor-level to high-level hardware description languages (HDLs). The so-called register-transfer level is somewhat in-between, describing CPU components and their interaction on a relatively high level. We will use this level in this chapter to introduce gradually more complex components, which we will then use to construct a complete CPU. With the simulation system Retro,[1,2] we will be able to actually program, run, and test our CPUs.

One of the best analogies for a CPU is a mechanical clockwork (Figure 2.1). A large number of components interact with each other, following the rhythm of one central oscillator, where each part has to move exactly at the right time.

---

[1]Chansavat, B., Bräunl, T. *Retro User Manual*, Internal Report UWA/CIIPS, Mobile Robot Lab, 1999, pp. (15), web: http://robotics.ee.uwa.edu.au/retro/ftp/doc/UserManual.PDF.

[2]Bräunl, T. *Register-Transfer Level Simulation*, Proc. of the Eighth Intl. Symposium on Modeling, Analysis and Simulation of Computer and Telecommunication Systems, MASCOTS 2000, San Francisco CA, Aug./Sep. 2000, pp. 392–396 (5).

© The Author(s), under exclusive license to Springer Nature Singapore Pte Ltd. 2022
T. Bräunl, *Embedded Robotics*,
https://doi.org/10.1007/978-981-16-0804-9_2

Figure 2.1: Working like clockwork

2.1 Logic Gates

On the lowest level of digital logic, we have logic gates AND, OR, NOT (Figure 2.2). The functionality of each of these three basic gates can be fully described by a truth table (Table 2.1), which defines the logic output value for every possible combination of logic input values. Each logic component has a certain delay time (time it takes from a change of input until the updated output is being produced), which limits its maximum operating frequency.

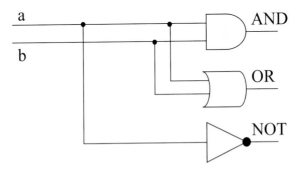

Figure 2.2: AND, OR, NOT gates

Input a, b	Output a AND b	Output a OR b	Output NOT a
0, 0	0	0	1
0, 1	0	1	1
1, 0	0	1	0
1, 1	1	1	0

Table 2.1: Truth table

Gates are built by using electronically activated switches. These are transistors in today's technology, while relays and vacuum tubes have been used in the past. However, for the understanding of the mechanisms inside a CPU, we do not need to know any further low-level details.

The layer of abstraction above gate level is formed by so-called combinatorial logic circuits. These do not have any timing components, so everything can be explained as a combination of AND, OR, NOT gates.

In the following we will denote negated signals with an apostrophe in the text, i.e. a' for NOT(a), and as a dot at a gate's input or output port in diagrams (see Figure 2.3).

2.1.1 Encoder and Decoder

A decoder can be seen as a translator device of a given binary input number. A decoder with n input lines has 2^n output lines. Only the output line corresponding to the binary value of the input line will be set to '1', all other output lines will be set to '0'. This can be described by the formula:

$$Y_i = \begin{cases} 1 & \text{if } i = X \\ 0 & \text{else} \end{cases}$$

Only the output matching the binary input pattern is set to '1'. So, if e.g. $n = 2$ and input X is a binary 2, meaning $X_1 = 1$ and $X_0 = 0$, then output line Y_2 will be '1', while Y_0, Y_1, and Y_3 will be '0'.

Figure 2.3 shows a simple decoder example with two input lines and consequently four output lines. Its implementation with combinatorial logic requires four AND gates and four NOT gates. Decoders are being used as building blocks for memory modules (ROM and RAM) as well as for multiplexers and demultiplexers.

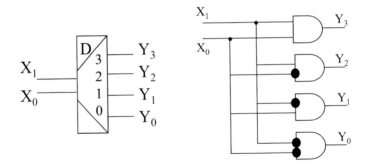

Figure 2.3: Decoder symbol and implementation

Encoders perform the opposite function of a decoder. They work under the assumption that only a single one of their input lines is active at any time. Their output lines will then represent the input line number as a binary number. Consequently, encoders with n output lines have 2^n input lines. Figure 2.4 shows the implementation of an encoder using only two OR gates. Note that X_0 is not connected to anything, as the output lines will default to zero if none of the other X lines are active. Figure 2.5 shows the interaction between an encoder and a decoder unit, reconstructing the original signal lines.

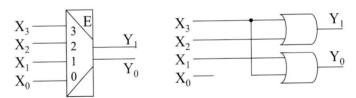

Figure 2.4: Encoder symbol and implementation

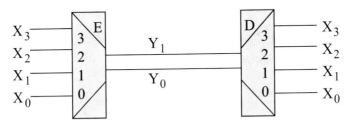

Figure 2.5: Encoder and Decoder

2.1.2 Multiplexer and Demultiplexer

The next level of abstraction are multiplexers and demultiplexers. A multiplexer routes exactly one of its inputs (X_0, ..., X_{n-1}) through to its output Y, depending on the selection lines S. Each input X_i and output Y have the same width (number of lines), so they can either be a single line as in Figure 2.6 or can all be e.g. 8 bits wide.

2.1 Logic Gates

The width (number of lines) of selection line S depends on the number of multiplexer inputs n, which is always a power of 2:

$n = 2^k$ with k being the width of S.

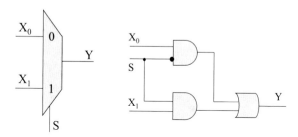

Figure 2.6: Multiplexer 2-way and implementation

In the example in Figure 2.6, we only have two inputs, so we only need a single selection line to distinguish between them. In this simple case, we can write the logic equation for a multiplexer as:

$$Y = S \cdot X_1 + S' \cdot X_0$$

The equivalent circuit built from AND, OR, NOT gates is shown on the right-hand-side of Figure 2.6.

When building a larger multiplexer, such as the 4-way multiplexer in Figure 2.7, using a decoder circuit makes the implementation a lot easier (Figure 2.7, right). For each case, the input position matching the selection lines is routed through, which can be written in short as:

$$Y = X_S$$

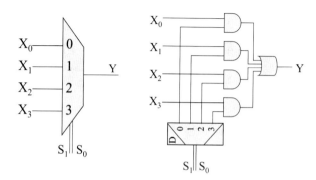

Figure 2.7: Multiplexer 4-way and implementation

A demultiplexer has the opposite functionality to a multiplexer. Here we connect a single input X to one of several outputs Y_1 ... Y_n, depending on the status of the select line S. In fact, if multiplexers and demultiplexers were built like a mechanical pipe system, they would be the same thing—just turning it around would turn a multiplexer into a demultiplexer and vice versa. Unfortunately, this does not work in the electronics world, as it is not easy to exchange inputs and outputs. Most electronic circuits have a 'direction', as becomes clear from the demultiplexer's equivalent circuit made out of AND and NOT gates in Figures 2.8 and 2.9.

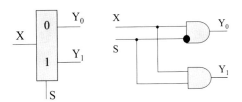

Figure 2.8: Demultiplexer 2-way and implementation

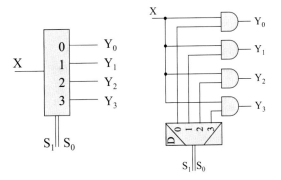

Figure 2.9: Demultiplexer 4-way and implementation

The logic formula for a general demultiplexer is very similar to a decoder, however, remember that input X and outputs Y_i can be wider than a single line:

$$Y_i = \begin{cases} X & \text{if } i = s \\ 0 & \text{else} \end{cases}$$

2.1.3 Adder

The adder is a standard textbook example, which we will only briefly describe. The first step is building a half adder that can add two bits input (X, Y) and produces a single-bit output plus a carry bit. It can be constructed by using an XOR and an AND gate (Figure 2.10).

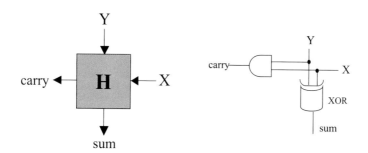

Figure 2.10: Half-Adder symbol (2-bit) and implementation

Two half adders and an OR gate are being used to build a full adder cell. The full adder adds two input bits plus an input carry and produces a single-bit sum plus an output carry (Figure 2.11). It will later be used in a bit slice manner to build adders with arbitrarily long word inputs, e.g. 8 bits wide.

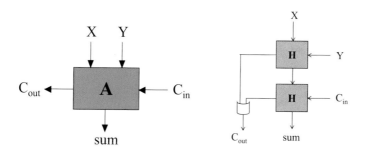

Figure 2.11: Full-Adder symbol (3-bit) and implementation

2.2 Function Units

Function units are essentially higher-level combinatorial logic circuits. This means each one of them can be represented by a set of AND, OR, NOT gates, but using the higher-level building blocks from the previous section will help to simplify and better understand their functionality.

2.2.1 Adding

The adder for two n-bit numbers is the first function unit we introduce here (Figure 2.12). Note that we draw fat lines to indicate that an input or output consists of multiple lines (in same cases showing the width as a numeric number next to the fat line).

Internally, an adder is built by using n full-adder components, each taking one input bit each from X and Y. Note that the adder's propagation delay is n times the propagation delay of a bit-slice full-adder component, so the carry bits have sufficient time to percolate through from right to left.

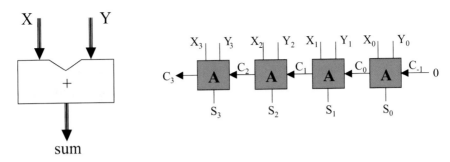

Figure 2.12: Adder function unit and implementation

Incrementing a counter by one is a standard operation for which it would be useful to have a function unit available, ready to use. Figure 2.13 shows the definition of an incrementor function unit with a single n-bit number as input and a single n-bit output. The incrementor can easily be implemented by using the adder for two n-bit numbers and hard-wiring the left input to hexadecimal value 01_{16}. (base 16)—equivalent to binary value $0000\ 0001_2$ (base 2). By "hard-wiring" we mean to connect all input bits where the binary word has a '0' to electric ground, and to connect the only bit where the binary word has a '1' to the supply voltage (possibly using a pull-up resistor).

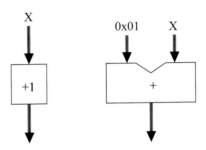

Figure 2.13: Incrementor function unit and implementation

2.2 Function Units

A function unit for multiplying the input number by two is another example where we have to be careful with reusing function units that are too complex for the task (Figure 2.14, left). Although, we could implement "multiply by two" with a single adder (Figure 2.14, middle), the operation is equivalent with a *shift left* operation, and this we can realize with a simple reordering of the wires. No active components are required for this solution (Figure 2.14, right).

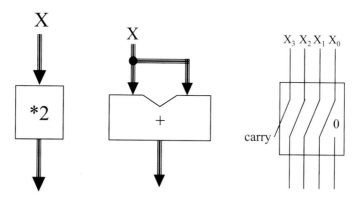

Figure 2.14: Multiply by two and implementations

2.2.2 Logic Functions

The one's complement of a single input is simply the inverse of all its bits. We can implement this function unit by using n NOT gates, one for each bit in the word (Figure 2.15).

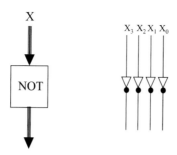

Figure 2.15: One's complement and implementation

Having function units for AND and OR is useful, and their implementation is equally simple, since each bit can be calculated independent of the other bits. The implementation in Figure 2.16 uses n AND gates, each connected to the corresponding input bits from X and Y.

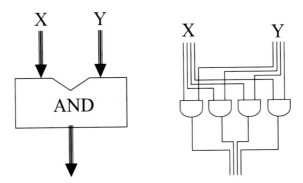

Figure 2.16: AND of two operands

The two's complement returns the negated value of an input number (Figure 2.17). We can implement this function unit by combining two of the function units we have constructed before, the one's complement (NOT) and the incrementor, executed one after the other.

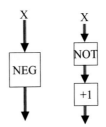

Figure 2.17: Two's complement and implementation

2.2.3 Subtracting

The subtractor shown in Figure 2.18, left, is another important function unit. We can implement it with the help of the previously defined function units for adding and negation (Figure 2.18, middle).

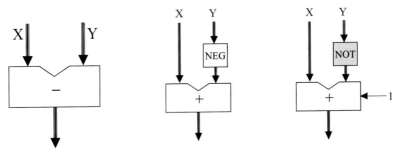

Figure 2.18: Subtractor, simple implementation and optimized implementation

Although this solution will work, it takes up an unnecessarily large area of chip real estate, since the NEG function contains an adder in the incrementor, in addition to the main adder with two inputs. Instead of adding the two's complement of Y:

$$Z = X - Y$$

$$Z = X + (\text{NOT } Y + 1)$$

we can also add the one's complement of Y and then increment the result later:

$$Z = (X + \text{NOT } Y) + 1$$

This final increment can be done by using the $carry_0$ input of the adder, so it requires no additional circuitry (see Figure 2.18, right).

2.2.4 Comparisons

Comparing numbers is a fundamental component in software as in hardware design. Each comparator takes one or two n-bit word as input and has only a single output line (yes or no, 1 or 0). We'll start here by comparing a single number to zero, then look at comparing two numbers with each other for equality as well as for ranking (greater than or less than).

Comparing a number to zero is the simplest comparison. Since in a zero-word all bits are equal to '0', we can implement the zero-comparator by using a single NOR gate that connects to all input bits (Figure 2.19).

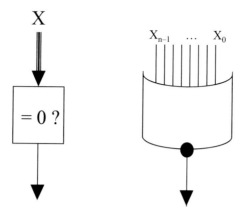

Figure 2.19: Function unit and implementation for comparing with zero

Next we want to check two numbers for equality, so we define a function unit with two n-bit inputs and a single output (yes or no, see Figure 2.20). We could implement this function unit by using the previously defined function units for subtraction and check for equality to zero (Figure 2.20, middle). However, while this would be correct in a mathematical sense, it would be a very poor choice of implementation, both in terms of hardware components required and in the required delay time (computation time). Checking two n-bit numbers for equality can be much easier achieved by using n EQUIV gates (negated XORs) for a bit-wise equality check and a single AND gate to combine the results (Figure 2.20, right).

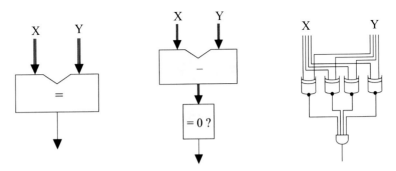

Figure 2.20: Equality of two operands and implementations

Performing comparisons with integer values can be quite tricky, especially when there is a mix of unsigned and signed numbers in a system. Figure 2.21 shows a comparator that checks whether a single signed input number is less than zero (remember that an unsigned number is by definition always positive). In two's complement representation the highest bit of a signed number determines whether the number is negative or positive. So, the implementation in Figure 2.21 takes advantage of this fact and therefore does not require any active components.

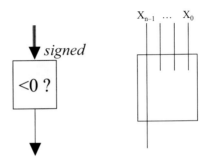

Figure 2.21: Signed comparison and implementation

For comparing two numbers for equality, we showed a simple solution using combinatorial gates. However, when comparing whether one input number is less or greater than the other, we cannot get away with this simple implementation. For a general number comparison, we have to conduct a subtraction and then subsequently check whether the result (as a signed number) is less than zero (Figure 2.22, right).

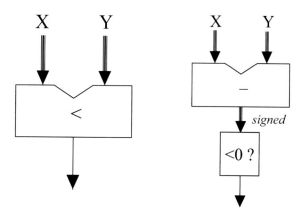

Figure 2.22: Comparison of two operands and implementation

The list of function units shown in this section is not meant to be complete. More function units can be designed and implemented using the methods shown here, whenever a specific function is considered useful for a design. The good thing about this additional level of abstraction is that we can now forget about the (hopefully efficient) implementation of each function unit and can concentrate on how to use function units in order to build more complex structures.

2.3 Registers and Memory

So far, we have been using combinatorial logic exclusively, so a combination of AND, OR and NOT gates, without any feedback loop, clock or system state. This will change when we want to store data in a register or in memory.

The smallest unit of information is one *bit* (short for *binary digit*), which is the information that can be held by a single latch (level-triggered) or a single flip-flop (edge-triggered). The RS (reset/set) latch in Figure 2.23, left, has inputs for setting and resetting the latch (each input acts on a logic '1' signal, which we call *active-high*). The latch's single-bit contents will always be displayed at output Q, while Q' displays the negated output.

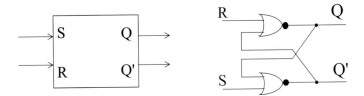

Figure 2.23: RS latch and implementation

The latch implementation in in Figure 2.23, right, shows that we only need two NOR gates to construct it (two NAND gates are also possible). The output of each becomes the input of the other and thereby *traps* a single bit in this circuit. If the latch is in state zero, activating S (with a 1) will change its state and output to one, which will stay stable until R gets activated, and vice versa. The RS latch introduces the concept of a *state* to our circuits. S and R will set the state to 1 or 0, respectively, which will be kept indefinitely until either input gets activated again—or power is being turned off.

Drawbacks of the RS latch are that the data inputs (S and R) are two separate lines instead of a single data line and that the inputs are *level triggered*, instead of being *edge triggered* (either rising edge/low-to-high or falling edge/high-to-low).

On a level triggered latch any change on the input lines will cause an instantaneous change of the flip-flop contents and its output Q. However, we would like to be able to decouple the input data (ideally a single data line) from an activation line (ideally edge triggered).

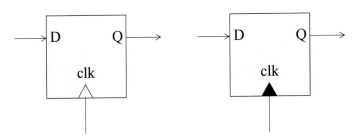

Figure 2.24: D flip-flop, rising and falling edge-triggered

The D-type flip-flop shown in Figure 2.24, left, has a single data input line D and one output line Q. On the rising edge of the clock input *clk*, the current input of D is copied into the flip-flop and will from then on be available on its output Q. There is also an equivalent version of the D flip-flop that switches on the falling edge of the clock signal in Figure 2.24, right. We draw this version with a solid clock arrow instead of a hollow one, resembling the negation dot of a NOT gate.

2.3 Registers and Memory

The D flip-flop implementation is best explained in two steps:

1. Extend a level-triggered RS latch to a latch with inputs D (*data*) and En (*enable*)

2. Construct an edge-triggered D flip-flop from two level-triggered D latches

For the first step (Figure 2.25, left) we add a NOT and two AND gates to the RS latch. Only when En is set to '1', D will activate either S or R, depending on its value, '1' for S (set) or '0' for R (reset).

For the second step (Figure 2.25, right), we use the master/slave combination of two D latches in series. The output of the master latch is the input for the slave latch, but both latches are enabled with opposite clock levels (the master is enabled on a '0' clock level, while the slave is activated on a '1'). This interlocking design accomplishes the transition from level-triggered latches to edge-triggered flip-flops.

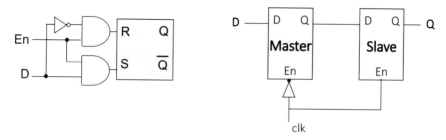

Figure 2.25: D flip-flop implementation (rising edge)

A register is now simply a bank of D flip-flops with all their clock lines linked together (Figure 2.26). That way, we can store a full data word with a single control line (clock signal). We use a box with digits in the register symbol to denote its current contents—sort of like a window to its memory contents.

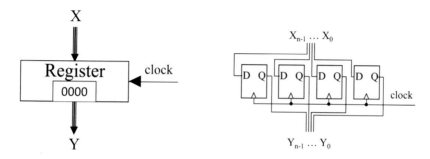

Figure 2.26: Register (4-bit) and implementation

The final components are the memory modules RAM (random access memory—read and write access) and ROM (read only memory) as shown in Figure 2.27. Memory modules come in various sizes, so they will have different numbers of address lines (determining the number of memory cells) and various numbers of data lines (determining the size of each memory cell). A typical memory chip might have 20 address lines, which let it access 2^{20} different memory cells. If this memory module has eight data lines (8 bits = 1 Byte), then the whole module has 1'048'576 Bytes, which equals 1 Megabyte (1 MB).

Both ROM and RAM modules in our notation have chip select lines (CS', active low) and output enable lines (OE', active low), which are required if a system has multiple memory modules or if other devices need to write to the data bus. Only the RAM module has an additional Read/Write' line (read when set to 1, write when set to 0) that allows data to be written back to the RAM module.

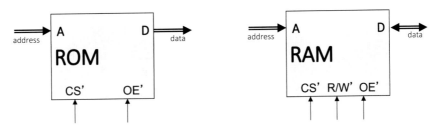

Figure 2.27: Memory modules ROM and RAM

Note that because of the complexity of memory modules, their typical delay times are significantly larger than those of simple gates or function units, which again limits the maximum CPU clock speed. At this level of abstraction, we do not distinguish between different types of ROM (e.g. mask-ROM vs. flash-ROM, etc.) and RAM (e.g. SRAM vs. DRAM, etc.). It does not matter for our purposes here.

2.4 Retro

Before we proceed with the major CPU blocks, we introduce the Retro,[3,4] hardware design and simulation system. Retro is a tool for visual circuit design at register-transfer level, which gives students a much better understanding of

[3]Chansavat, B., Bräunl, T. *Retro User Manual, Internal Report* UWA/CIIPS, Mobile Robot Lab, 1999, pp. (15), web: http://robotics.ee.uwa.edu.au/retro/ftp/doc/UserManual.PDF.

[4]Bräunl, T. *Register-Transfer Level Simulation*, Proc. of the Eighth Intl. Symposium on Modeling, Analysis and Simulation of Computer and Telecommunication Systems, MASCOTS 2000, San Francisco CA, Aug./Sep. 2000, pp. 392–396 (5).

how to construct a complex digital system and how a computer system works in detail.

Retro supplies a number of basic components and function units (as discussed in the preceding sections) that can be selected from a palette and placed on a canvas where they will be interconnected. Components can be linked by either a single signal line, or by a bus of variable size (e.g. 8, 16, 32 lines). All palette components are grouped into libraries that can be loaded into the system, making Retro extendable with new component types. Retro can run in several demo modes, displaying signal levels as colors and data in hex displays. Similar to a debugger, the simulator can be run in single step mode and its execution can be halted at any time. Retro is implemented in Java and can run in most operating systems, including Windows, MacOS and Linux.

Figure 2.28 shows a sample Retro setup with the component library palette on the left and execution control buttons (VCR-style control buttons) on the top.

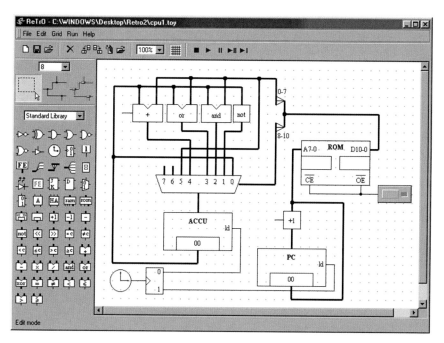

Figure 2.28: Retro simulator with library component palette

All synchronous circuits require a central clock, which is a component from the palette. The clock speed can be set in relation to the components' latencies and to the simulated time passing. Since most synchronous circuits require a number of timing signals derived from the central clock, the standard palette also includes a high-level pulse generator (Figure 2.29, left). The pulse generator has a variable number of outputs, for each of which a repetitive timing pattern can be specified.

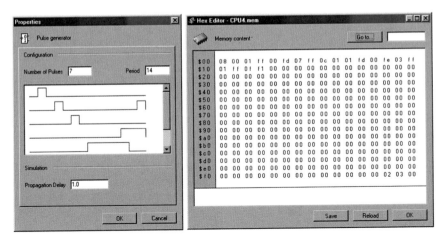

Figure 2.29: Pulse generator component and memory contents tool

The palette component for memory modules ROM and RAM are more complex than other components. They allow detailed propagation delay settings for various memory aspects and also include a tool for displaying and changing memory contents in a window or for saving to a file. Since memory data is stored in a separate data file and not together with the circuit design data, the same hardware can be used with several different programs for individual experiments (Figure 2.29, right).

Retro was originally implemented by Chansavat[5] at UWA with later additions by numerous other students. Retro was inspired by Wirth's textbook on Digital Circuit Design.[6]

2.5 Arithmetic Logic Unit

The first major component of any CPU is the ALU (arithmetic logic unit). It is the number cruncher of a CPU, supplying basic arithmetic operations such as addition and subtraction (in more advanced ALUs also multiplication and division) and logic operations such as AND, OR, NOT for data words of a specific width. So in fact, one can imagine the ALU as a small pocket calculator inside the CPU.

One of the most important decisions to make when designing an ALU is how many registers to use and how many operands to receive from memory per instruction. For our first ALU we will use the simplest possible case: only one register and just one operand per instruction. This is called a one-address machine (assuming the operand is in fact an address—more about this later).

[5]B. Chasavat, T. Bräunl, *Retro – Register-Transfer-Object Hardware Simulator,* online at: http://robotics.ee.uwa.edu.au/retro/.

[6]N. Wirth, *Digital Circuit Design*, Springer-Verlag, Heidelberg, 1995.

Since here each instruction has only one operand, we need to use some intermediate steps for even simple operations, such as adding of two numbers. In the first step we load the first operand into the register (which we will call the *accumulator* from now on). In the second step, we add the second operand to the accumulator.

ALUs which can perform this operation in a single step are called two-address machines. Each of their instructions can supply two operands (e.g. a + b) and the result will be stored in the accumulator. Three-address machines provide an address for the result as well (e.g. c = a + b), so there is no need for a central accumulator in such a system. And, just for completeness, there are also zero-address machines where all operands and results are pushed and popped from a stack.

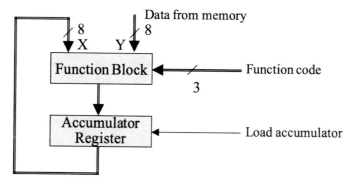

Figure 2.30: ALU structure

Figure 2.30 shows the basic ALU structure for a one-address machine. Only one operand (eight bits wide) at a time comes in from memory, so each operation (3 bits wide) is between the accumulator (i.e. the result of the previous operation) and the operand. Also, we have made no provisions for writing a data value back to memory.

We already know what a register is, so the remaining secret of ALU-1 is the central function block. Figure 2.31 reveals this black box. We are using one large multiplexer that is being switched by the function code (also called *opcode*, or *machine code*). The 3-bit function code gives us a total of $2^3 = 8$ different instructions and each of them is defined by the respective multiplexer input.

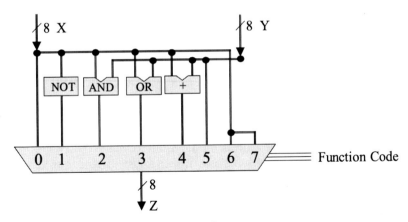

Figure 2.31: ALU function block

- Opcode 0: **NOP**
  Routes the *left* operand through.
  Since the output is linked to the accumulator, this instruction will not change the accumulator contents. This is known as a NOP (short for *no operation*).
- Opcode 1: **NOT**
  Inverts the left operand, so effectively inverts the accumulator.
  No memory data is used for this instruction.
- Opcodes 2–4: **AND, OR, ADD**
  These opcodes perform logic AND, OR and arithmetic addition, respectively, between left and right operand (i.e. between accumulator and memory operand).
- Opcode 5: **LOAD**
  Routes the *right* operand through. So the accumulator will be loaded with the memory operand.
- Opcodes 6–7: **NOP**
  Operation is identical to opcode 0, so from the ALU point of view they are also NOPs.

It might seem like a waste of resources to calculate all possible results (NOT, AND, OR, ADD) for every single instruction, and then discard all but one. However, since we need all of these operations at some stage in a program, there is no possible savings in terms of chip space or execution time. There may be a possible energy consumption issue, but we do not look at this now.

We can now summarize the function of these eight opcodes in table form as machine code with mnemonic abbreviations (Table 2.2).

No	Opcode (bin.)	Operation
0	000	Z = X
1	001	Z = NOT X
2	010	Z = X AND Y
3	011	Z = X OR Y
4	100	Z = X + Y
5	101	Z = Y
6	110	Z = X
7	111	Z = X

Table 2.2: Operations for ALU-1

2.6 Control Unit

The CU (control unit) is the second part of each CPU, enabling step-by-step program execution and branching. The central register used in the CU is the program counter. The program counter addresses the memory in order to load opcodes and operands (*immediate* data or memory *addresses*) from the memory chip.

Figure 2.32 shows a first, very simple CU structure. The program counter is incremented by one in each step and its output is used for addressing the memory unit. This means, every instruction (opcode + operand) will be a single word. Each program on this CU will be executed line after line with no exceptions; there are no provisions for branching forward or backward.

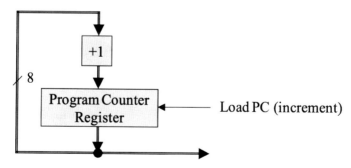

Figure 2.32: CU structure

2.7 Central Processing Unit

In order to build a fully functional CPU, we link together an ALU and a CU with a memory module. In this section we will introduce a number of CPU designs, starting with the simplest possible design, then successively adding more features when building more complex CPUs.

2.7.1 CPU-1—Minimal Design

To build the first complete CPU, we use the ALU and CU from the previous two sections, linked by a ROM module to build CPU-1 (Figure 2.33).

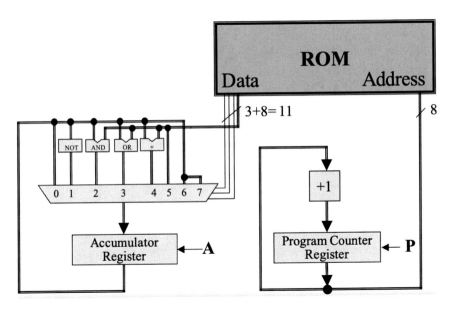

Figure 2.33: CPU-1 design

As has been established before, this CPU design does not allow for any branching and only immediate operands (constant values) are used in a single memory word of 11 bits that combines opcode and operand. Figure 2.34 shows the identical CPU-1 design in the Retro system (with the exception of unused/disconnected opcodes 6 and 7). The function block shows how all internal details and the load signals for accumulator and program counter are wired up to a pulse generator, driven by the central clock.

2.7 Central Processing Unit

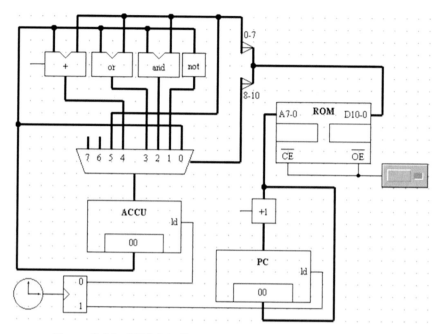

Figure 2.34: CPU-1 in Retro

As can be verified from the multiplexer configuration, ALU-1 supports eight opcodes, of which only the first six are being used: NOP, NOT, AND, OR, ADD and LOAD, for opcodes 0, ..., 5. On the CU-1 side, the program counter (PC) always addresses the memory module and its output is fed back via an incrementor. This means program steps are always executed consecutively, branches or jumps are not possible.

The memory module uses an unusual 11-bit data format, which further simplifies the design, because operator (3-bit opcode) and immediate operand (8-bit data) can be encoded in a single instruction and no additional registers are required to store them. The splitting of the two is simply done by dividing the data bus wires coming from the memory module.

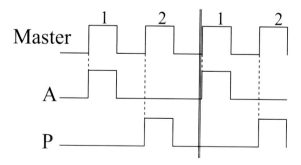

Figure 2.35: Timing diagram for CPU-1

The timing requirements for CPU-1 are minimal. Since only two registers need to be triggered, all we need are two alternating signals, derived from a master clock. First the accumulator gets triggered, then the program counter is incremented (see Figure 2.35). Table 2.3 summarizes the available instructions for CPU-1 and lists their specific accumulator and program counter operations.

Opcode	Description	Mnemonic
0	acc ← acc pc ← pc + 1	NOP
1	acc ← NOT acc pc ← pc + 1	NOT
2	acc ← acc AND constant pc ← pc + 1	AND const
3	acc ←acc OR constant pc ← pc + 1	OR const
4	acc ← acc + constant pc ← pc + 1	ADD const
5	acc ← constant pc ← pc + 1	LOAD const
6	not used	
7	not used	

Table 2.3: CPU-1 opcodes

We can now look at the software side, the programming of CPU-1. We do this by writing opcodes and data directly into the ROM. The simple program shown in Program 2.1 adds the two numbers 1 and 2. With the first instruction we load constant 1 into the accumulator (code: 501). In the second step we add constant 2 (code: 402).

Program 2.1 CPU-1 Addition Program

Address	Opcode	Operand	Comment
00	5	01	LOAD 1
01	4	02	ADD 2
02	0	00	NOP
...	...	...	...
FF	0	00	NOP

This program also shows some of the deficiencies of CPU-1's minimal design:

a. Operands can only be constant values (immediate operands). Memory addresses cannot be specified as operands.

b. Results cannot be stored in memory.

c. There is no way to *stop* the CPU or at least bring to a dynamic halt. This means after executing a large number of NOP instructions, the PC will eventually come back to address 00 and repeat the program, overwriting the result.

2.7.2 CPU-2—Double Byte Instructions and Branching

CPU-1 gave a first impression of CPU design, stressing the importance of timing and interaction between hardware and software. For this second design, CPU-2, we would like to address the major deficiencies of CPU-1, which are the lack of branching and the lack of memory data access (read or write).

For CPU-2 we choose an 8-bit opcode followed by an 8-bit memory address and an 8-bit wide RAM/data bus configuration. This design choice requires two subsequent memory accesses for each instruction. CPU-2 requires two additional registers, a code register and an address register for storing opcode and address, respectively, which are being loaded subsequently from memory. Figure 2.36 shows the CPU-2 schematics (top) and the Retro implementation (bottom).

The instruction execution sequence, defined by the timing diagram (also called microprogramming) now requires several steps:

1. Load first byte of instruction (opcode) and store it in the code register.

2. Increment program counter by 1.

3. Load second byte of instruction (address) and store it in the address register.

4. Use the address value for addressing the memory and retrieving the actual data value, which is then passed on to the ALU.

5. Update the accumulator with the ALU calculation result.

6. (a) If required by the opcode no. 6, write accumulator data back to memory.
    (b) If required by opcode no. 7, use address parameter as PC increment.

7. Increment program counter for the second time (+1 for arithmetic instructions or offset for branching).

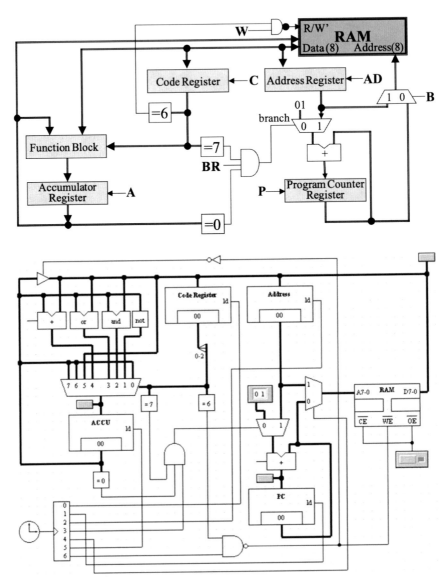

Figure 2.36: CPU-2 schematics and Retro implementation

Figure 2.37 shows the timing diagram required for CPU-2. It is important to note that the program counter will now be incremented twice for each instruction.

2.7 Central Processing Unit

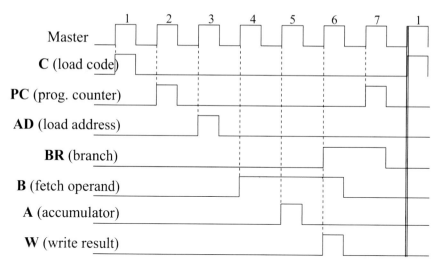

Figure 2.37: Timing diagram for CPU-2

CPU-2 uses the same ALU as CPU-1, but also makes use of the previously unused opcodes 6 and 7. As can be seen in Figure 2.36, opcode 6 is individually decoded (box '=6') and used to write the accumulator result back to the memory. Of course, this can only work if the timing diagram has made provisions for it. In Figure 2.37, we can see that one impulse of each master cycle is reserved to activate signal *W*, which is used to switch the RAM from read to write and also to open the tri-state gate allowing data to flow from the accumulator back towards the RAM. The memory address for writing comes from the address register.

The final major addition is a conditional branch for opcode 7 (see box '=7'). The condition is true if the accumulator is equal to zero. Therefore, a branch will only occur if:

- opcode 7 is used, *and*
- the accumulator is equal to zero, *and*
- timing signal BR is present.

Signal BR overlaps (and therefore replaces) the second increment of the program counter in the timing diagram (Figure 2.37), which again demonstrates the importance of proper timing design. Table 2.4 now shows the complete set of opcodes for CPU-2.

Interaction between CPU-2's components can be seen best by following the execution of an instruction, which takes seven cycles of the master clock (Figure 2.37). Assuming the PC is initialized with zero, it will address the first byte in memory (mem[0]), which will be put on the data bus. The data bus is linked to the code register, the address register, and (over a multiplexer) to the ALU, but only one of them will take the data at a time. Since the first cycle activates signal C (line 0 from the pulse generator), this triggers *load* on the code register, so mem[0] will be copied into it.

Opcode	Description	Mnemonic
0	acc ← acc pc ← pc + 2	NOP
1	acc ← NOT acc pc ← pc + 2	NOT
2	acc ← acc AND memory pc ← pc + 2	AND
3	acc ← acc OR memory pc ← pc + 2	OR
4	acc ← acc + memory pc ← pc + 2	ADD
5	acc ← memory pc ← pc + 2	LOAD
6	Memory ← acc pc ← pc + 2	STORE
7	(* acc unchanged *) if acc = 0 then pc ← pc + address else pc ← pc + 2	BEQ

Table 2.4: CPU-2 opcodes

In the second cycle, *PC load* will be triggered (line 1). The PC's input is an *adder* with the left-hand side being the constant 1 (via a multiplexer) and the right-hand side being the PC's previous value (0 in the beginning). So as long as the multiplexer is not being switched over, a pulse on *PC load* will increment it by one. With the PC now holding value 1, the second memory byte (mem[1]) gets on the data bus, and at timing cycle 3 (line AD) it will be copied into the address register.

Cycle 4 will activate signal B (line 4), which switches the RAM addressing from the program counter to the current address register contents. This is needed since each instruction in CPU-2 has an address operand instead of an immediate (constant) operand in CPU-1 (see also opcodes in Table 2.4). With the address register now being connected to the RAM, the data bus will get the memory contents to which the address register points to. This value will be selected as input for the ALU.

With the ALU's left-hand side input being the accumulator's old value and the right-hand side input being set to the data bus (memory operand), the code register selects the desired operation over the large ALU multiplexer on top of the accumulator. Cycle 5 then activates the accumulator's load signal (line 5), to copy the operation's result.

Cycle 6 will active signal W (line 6). In case the instruction opcode is 6 (STORE memory, see Table 2.4), the RAM's output enable line will be activated (see box '= 6' and NAND gate) and the accumulator's current value is written back to the RAM at the address specified by the address register. In case the current instruction is not a STORE, the RAM's write enable and the tri-state gate will not be activated, so this cycle will have no effect.

Cycle 6 also activates signal BR (line 3), in order to not to waste another cycle. This flips CU-2's adder input from constant '+1' to the address register contents, but only if the instruction's opcode is 7 (BEQ, branch if equal) and also if the current accumulator contents is equal to zero (see box '=0').

Finally, on cycle 7, the program counter is updated a second time, either by '+1' (in case of an arithmetic instruction) or by adding the contents of the address register to it (for the BEQ instruction, if the branch condition is met).

This concludes the execution of one full instruction. On the next master clock cycle the system will start over again with cycle 1.

Program 2.2 shows the implementation of an addition program, similar to the one for CPU-1. However, in CPU-2 there are no constant values, instead all operands are addresses. Therefore, we first load the contents of memory cell A1 (instruction: 05 A1), then add the contents of cell A2 (instruction: 04 A2), and finally store the result in cell A3 (instruction: 06 A3).

After that, we would like to bring the program to a dynamic halt. We can do this with the operation BEQ -1 (in hexadecimal: 07 FF). Although each instruction takes two bytes, we must decrement the program counter only by -1, not -2. This is because at the time the data value -1 (FF) is added to the program counter, it has only been incremented once so far, not twice.

Program 2.2 CPU-2 Addition Program

Address	Code	Data	Comment
00	05	A1	LOAD mem [A1]
02	04	A2	ADD mem [A2]
04	06	A3	STORE mem [A3]
06	05	A0	LOAD mem [A0] "0"
08	07	FF	BEQ -1

We also have to consider that the branching instruction is conditional, so for the unconditional branch needed here, we have to make sure the accumulator is equal to 0. Since we no longer have constants (immediate values) that we can load, we need to execute a LOAD memory instruction from an address that we know has value zero before we can actually execute the branch instruction. In the example program, memory cell A0 has been initialized to 0 before program start.

2.7.3 CPU-3—Addresses and Constants

The design of CPU-3 (Figure 2.38) is extending CPU-2 by adding:

- Load operation for constants (immediate values)
- Unconditional branch operation

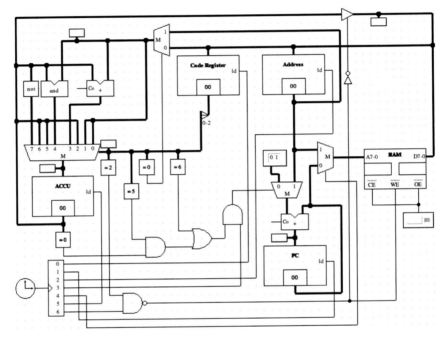

Figure 2.38: CPU-3 design

CPU-3 still only uses 8 different opcodes in total (three bits), so we reduced the functionality of ALU-3 in this design. An additional multiplexer, controlled by opcode 0 (see box '=0'), allows switching between feeding ALU-3 with memory output (memory operand, direct addressing) and address register contents (constant, immediate operand). However, this trick only works for the LOAD operation. Since the opcodes for ADD and AND are not equal to 0, these instructions will still receive data from memory.

The second change is to include both, conditional branching (opcode 5) and unconditional branching (opcode 6). Note that the STORE instruction in CPU-3 has been changed to opcode 2. A branch will now be executed either if opcode is 6 (see box '=6') or if opcode is 5 (see box '=5') and the accumulator is equal to 0 (see Figure 2.38).

2.7.4 CPU-4—Symmetrical Design

While CPU-3 addressed some of the deficiencies of CPU-2, its design is more an 'ad hoc' solution. The redesigned CPU-4 shown in Figure 2.39 shows a much clearer, symmetrical design.

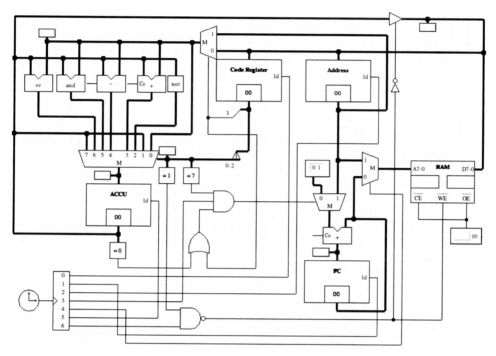

Figure 2.39: CPU-4 design

We are now using one additional bit for opcodes (4 bits), resulting in $2^4 = 16$ instructions in total. The design is symmetrical in the way that the highest bit (bit 3) of the opcode is used to switch between constants (immediate operands) and memory operands, therefore two versions of each instruction exist. Instructions with opcodes 0–7 use memory data (direct address), while instructions with opcodes 8–15 use constants (immediate data) instead. Bit 3 of the opcode is also used to distinguish between conditional branch (opcode 7) and unconditional branch (branch always, opcode 15). Bits 4–7 of each opcode byte are not used in CPU-4, but could be utilized for extensions. All opcodes for CPU-4 are listed in Table 2.5.

As can be seen in Figure 2.39, bit no. 3 is split from the code register output and used to switch the multiplexer between immediate (constant) and direct (memory) operands for every opcode. This symmetric design is characteristic for good CPU designs and highly appreciated by assembly programmers and compiler code generators.

A similar solution for including conditional and unconditional branches as in CPU-3 has been implemented. Here, opcodes 7 and 15 have been selected, as they correspond to each other with respect to opcode bit no. 3 and always use an immediate address parameter, never a value from memory.

As example program we selected the multiplication of two numbers by repeated addition (see Program 2.3). The program expects the two operands in memory cells FD_{16} and FE_{16} and will place the result in FF_{16}.

Opcode	Description	Mnemonic
0	acc ← memory pc ← pc + 2	LOAD mem
1	memory ← acc pc ← pc + 2	STORE mem
2	acc ← NOT acc pc ← pc + 2	NOT
3	acc ← acc + memory pc ← pc + 2	ADD mem
4	acc ← acc – memory pc ← pc + 2	SUB mem
5	acc ← acc AND memory pc ← pc + 2	AND mem
6	acc ← acc OR memory pc ← pc + 2	OR mem
7	(* acc unchanged *) if acc = 0 then pc ← pc+address else pc ← pc+2	BEQ mem
8	acc ← constant pc ← pc + 2	LOAD const
9	not used	
10	not used	
11	acc ← acc + constant pc ← pc + 2	ADD const
12	acc ← acc – constant pc ← pc + 2	SUB const
13	acc ← acc AND constant pc ← pc + 2	AND const
14	acc ← acc OR constant pc ← pc + 2	OR const
15	(* acc unchanged *) pc ← pc + address	BRA addr

Table 2.5: CPU-4 opcodes

Program 2.3 CPU-4 Program for Multiplying Two Numbers in Memory

Addr	Code Data	Mnemonic	Comment
00	08 00	LOAD #0	Clear result memory cell ($FF)
02	01 FF	STORE FF	
04	00 FD	LOAD FD	Load first operand ($FD).
06	07 FF	BEQ -1	···done if 0 (BEQ −1 is equiv. to dynamic halt)
08	0C 01	SUB #1	Subtract 1 from first operand
0A	01 FD	STORE FD	
0C	00 FE	LOAD FE	Load second operand ($FE) and add to result
0E	03 FF	ADD FF	
10	01 FF	STORE FF	
12	0F F1	BRA -15	Branch to loop (address 4)

First, the result cell is cleared. Then, at the beginning of the loop, the first operand is loaded. If it is equal to zero, the program will terminate. For halting execution, we use a branch relative to the current address −1. This will engage the CPU in an endless loop and effectively halt execution.

If the result is not yet zero, 1 is subtracted from the first operand (which is then updated in memory) and the second operand is loaded and subsequently added to the final result. The loop code ends with an unconditional branch statement to the beginning of the loop.

2.8   Structured Design

Retro can also be used for more complex CPU designs, such as a parallel processing system using an SIMD (single instruction, multiple data) structure. For this, we use the structured design feature of Retro, added by T. Forrest[7], which allows us to design new components with inputs and outputs and then use them in a more complex hardware structure. Figure 2.40 shows such a design for a SIMD system with 5 × 5 processors.

---

[7]T. Forrest, *SIMD Microprocessor for Image Processing*, Bachelor Thesis, supervised by T. Bräunl, UWA 2014, online: https://robotics.ee.uwa.edu.au/theses/2014-Retro-Forrest.pdf.

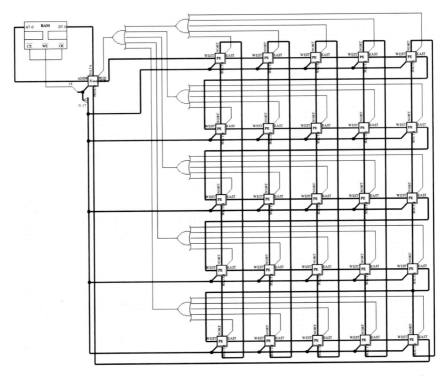

Figure 2.40: Structured SIMD system with 5 × 5 processing elements by T. Forrest, UWA

In this design, the *sequencer* (main array processor) and the PEs (*processing elements* or slave processors) are new components, drawn as boxes with double borders. These components need their own description files in Retro, which are shown in Figure 2.41 for the PE with its characteristic links to North, South, East and West, and in Figure 2.42 for the sequencer. Each PE is a simplified CPU, which only contains an ALU plus data communication to the four ports. The sequencer is the only full CPU with its own (scalar) ALU and a CU that controls the whole processor system. It is linked into the PE grid like another array element, but is also gets a status line input OR-connected over all PEs. This is needed, among others, to determine when a data-parallel loop has been finished.

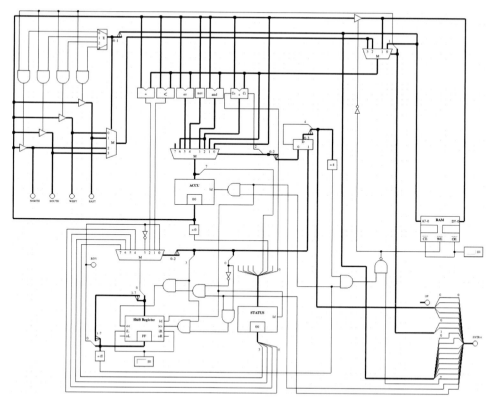

Fig. 2.41 SIMD PE by T. Forrest, UWA

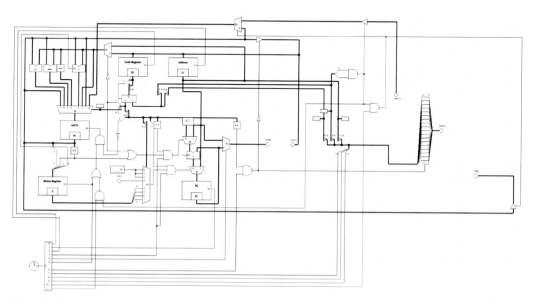

Fig. 2.42 SIMD sequencer by T. Forrest, UWA

2.9   Tasks

1. Download Retro and example hardware structures from:
    https://robotics.ee.uwa.edu.au/retro/
    Java binary program:
    https://robotics.ee.uwa.edu.au/retro/ftp/Version43/Retro_V4.3.jar
    Examples:
    https://robotics.ee.uwa.edu.au/retro/ftp/Version43/examples.zi p

2. Create your own instruction set.

3. Build an ALU and a CU to match this instruction set.

4. Combine ALU and CU to a complete CPU in Retro. You can do this either from scratch or by modifying one of the existing CPUs.

5. Program your CPU for the task to calculate $\Sigma(1,n)$ for the number n in memory.

6. Program your CPU for the task of multiplying two numbers from memory locations.

7. Build a parallel SIMD system for an image array (1 PE per pixel) to calculate the Sobel edge detector (see chapter on image processing).

ARDUINO

From the previous chapter, we now completely understand how a CPU works—in fact we could build one from simple logic gates—and I have made my students do exactly this, it is a great project! But for practical and budget reasons, it makes more sense to either buy a CPU chip and build your own controller—like we have done for many years with the EyeBot-1 to EyeBot-6 generations (Figure 3.1)—or to buy a complete controller such as the Arduino Uno/Mega/Nano or one of the Raspberry Pi versions. As these are mass-produced in the millions, it is simply not possible to build any small production run in a similar price bracket. So we stopped building our own controllers and switched over to using Arduino for low-level tasks and Raspberry Pi for high-level robotics applications.

In this chapter, we will start with the Arduino hardware and software. The following Web link contains some useful tools and documents:

https://robotics.ee.uwa.edu.au/nano/

T. Bräunl, *Embedded Robotics*,
https://doi.org/10.1007/978-981-16-0804-9_3

Figure 3.1: EyeBot controller generations

3.1 Arduino Hardware

The Arduino family of controllers was developed in 2005 by the Interaction Design Institute Ivrea in Italy. Hardware and software are open source, which added to their success and the availability of low-cost versions. The Nano controller has a small footprint and is ideal for small projects as it only costs a few dollars. For our course in embedded systems, we give one controller to every student at the start of semester for experimenting. Larger versions such as the Arduino Uno and Arduino Mega can take add-on boards or *shields*, but for simple projects, the Nano is all we need.

The Arduino, like many other controller boards such as the AVR Butterfly, uses an Atmel AVR 8-bit microcontroller. For the Nano version, it is an Atmel ATmega328P. Programming can be easily done in C using Atmel Studio or in a simplified C version using the Arduino app.

The Arduino has a small built-in ROM memory and cannot be extended. Powering up is simply by connecting its USB socket to a power supply or a PC (see Figure 3.2). Upload and download of programs and data can also be accomplished over the USB link.

3.1 Arduino Hardware

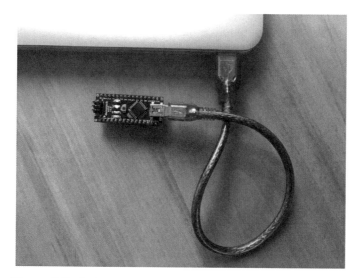

Figure 3.2: Arduino Nano connected to a laptop PC

For writing a program for the Arduino and uploading it to the controller, we use the free Arduino app[1] or integrated development environment (IDE). When starting this on a Mac, Windows or Linux PC, one is greeted by an empty program with two C functions, *setup* and *loop*, as shown in Figure 3.3.

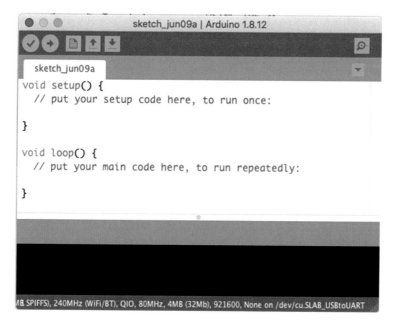

Figure 3.3: Arduino app with empty program

[1] Arduino DIE online at: https://www.arduino.cc.

Function *setup* will run only once at program start (after power-up or after a push on the reset-button), while *loop* will from then on run in an infinite loop. So as a complete C program, the hidden *main* program actually looks like Program 3.1:

Program 3.1: Hidden Main Program of the Arduino

```
int main()
{ setup();
  while (1) loop();
}
```

The term *while (1)* or *while (true)* (when including *stdbool.h*) denotes an infinite loop, which means that subroutine *loop* will be called endlessly over and over again until someone either presses the reset button or disconnects power.

If you are new to programming, you probably do not think of this infinite loop to be a big deal. After all, this makes a lot of sense for any embedded systems: you set up the system in the beginning, then you continue to run the task. Think of your TV remote: after you put in the batteries, its internal program starts by initializing the system. Then it will run an infinite loop that only comprises two steps: (a) wait for you to press a key and then (b) transmit this key's code serially via its infrared LED. And this continues until the batteries run flat or you replace them.

However, if you have completed any programming course before, most likely you have been told that "*whatever you do, make sure your program terminates*"! Clearly, any program we can write on the Arduino now contradicts this rule. None of our programs will ever terminate and as I mentioned before, this is how it has to be. How do these conflicting rules come together?

The rule to have your program terminate only applies to programs on a PC or other devices that have an operating system (OS), such as MacOS, Windows or Linux. In fact, every operating system itself has this infinite loop of Program 3.1 built in, so no application program should introduce another one.

Things are different, however, on a simple microcontroller *without* any operating system, such as the Arduino. There is no fancy OS running in the background that could help you with administrative tasks (but would also generate some significant run-time overhead). All that runs on the microcontroller are the few lines of code that you enter into the Arduino app. This has to include an infinite loop, because if your program was to terminate, where should the program branch to? There is no operating system, no other code and you cannot just halt the CPU's execution, so the program counter might wander into uninitialized program areas or even data areas, executing random bytes as commands with unpredictable and unwanted outcomes.

So in short, if you write a program for a system with an operating system, make sure it terminates—but if you write a program for an embedded system without an OS, make sure it has an infinite loop!

In the Arduino app, the infinite loop is always implicitly there, so you do not even have to write the code for it.

3.2 Arduino Programming

The Arduino app comes with a variety of example programs. One of them is the *blink* program that lets us play with one of the built-in LEDs on top of the controller. In this demo, function *setup* in Program 3.2 has only a single line, calling system function *pinMode* with the address of the pin (here: *LED_BUILTIN*) and the desired initialization value (here: *OUTPUT*).

The *loop* function in Program 3.2 turns the LED on – waits – turns the LED off – waits again, all in an infinite loop. System function *digitalWrite* with the address of the pin (again: *LED_BUILTIN*) and the desired output value (*HIGH* for 1 or *LOW* for 0) is used for actually turning the LED on or off. However, a *delay* statement is required, so we can actually see any effect. If we were to omit the two delay statements, the LED would change state from on to off and back so quickly that we would only see a dim glow. By changing the delay times (specified in milliseconds), we can achieve different blinking patterns.

Program 3.2: Arduino Demo Program "Blink," Modified from Arduino App

```
void setup() {
    pinMode(LED_BUILTIN, OUTPUT);        // init LED in as output
}

void loop()  {
    digitalWrite(LED_BUILTIN, HIGH);     // turn LED on
    delay(500);                          // wait 500ms
    digitalWrite(LED_BUILTIN, LOW);      // turn LED off
    delay(500);                          // wait 500ms
}
```

As the first step, we need to set up the Arduino app to match our Nano controller. From the *Tools* menu select as board *Arduino Nano* and as processor *ATmega328P*. Then under *Port* select the correct COM port or USB port of your PC to which you connected the Nano.

Now to get this program running, click on the tick symbol in the top left of the Arduino app to compile the program into machine code. If all goes well, you get a green message saying *"Done compiling"* with some statistical data (see Figure 3.4).

```
Done compiling.

Sketch uses 924 bytes (3%) of program storage space. Maximum is 30720 bytes.
Global variables use 9 bytes (0%) of dynamic memory, leaving 2039 bytes for local variables.
```

Figure 3.4: Done compiling

The next step is clicking the right arrow symbol (also in the top left of the app) for uploading the program to the controller. The upload should start, indicated by the text "*Uploading ...*" in the Arduino app and by rapid blinking of the Nano's send and receive LEDs. Completion of the successful upload is then acknowledged by the message in Figure 3.5.

```
Done uploading.
Sketch uses 924 bytes (3%) of program storage space. Maximum is 30720 bytes.
Global variables use 9 bytes (0%) of dynamic memory, leaving 2039 bytes for local variab
```

Figure 3.5: Successful upload

If you have selected the wrong communication port or there is any problem with the connection, you will get an error message. Some Arduino clones require additional drivers or an update of the built-in bootloader. Some versions of the Arduino app have the entry "*... (Old Bootloader)*" in the processor menu, which works for older hardware or firmware versions on clones.

After the successful upload, the user program will start to run straight away. So in our case, the blinking pattern will start and run and run and run—until we disconnect power, press reset or overwrite the program. The application program is now permanently stored in the Arduino Nano, so even after disconnecting and reconnecting power, the program will start again and display the blinking pattern.

3.3 Arduino Interfacing

Interfacing a controller to the outside world, to sensors and actuators as well as to displays and input devices is one of the central tasks to build an embedded system. On the Arduino, input/output lines are organized in ports of eight lines each, and each line can work either as an input or an output. This means that we need to initialize each I/O line before we can use it.

Only some of the input/output ports of the Atmel CPU are made available as pins on the Nano board; some larger boards offer more I/O lines. The labels printed onto the Nano board unfortunately do not match the port numbers of the Atmel CPU. Labels A0–A7 ('A' for analog) refer in fact straight to CPU ports C0–C7, and labels D2–D13 ('D' for digital) are linked to various port pins from port B and port D (see Figure 3.6). Try not to use pin D13, as it is internally connected to the onboard LED.

Figure 3.6: Arduino Nano board pins and matching port numbers

We start with a simple input of a binary switch. On the Nano board, we select pin D12 (digital 12), which happens to be linked to the controller's port C0 (see Figure 3.7, left). While the switch is open, a pull-up resistor of 10 kΩ to the 3.3 V pin brings the input to a high state (logic 1). Once the button is pushed, the D12 input gets directly linked to ground, so the pin registers a low state (logic 0).

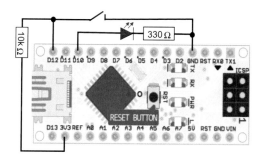

Figure 3.7: Arduino Nano with external input and output; left schematics, right breadboard

For the output, we choose pin D10 (digital 11) and connect it to an LED, which links via a 300 Ω resistor on its negative side to ground. Figure 3.7, right, shows the actual implementation on a breadboard.

Resistor values are coded as color rings, as their surface is too small to print any actual values on. In the simplest case, resistors have four rings, which stand for two digits, a multiplier and a quality indicator (which we can ignore for now). The digit colors are as in Table 3.1.

Table 3.1 Resistor ring colors

Denoting the values of these three rings with (a, b, c), then the resistor's value is given by:

$a\ b \times 10^c$.

For example, a resistor with colors (brown, black, orange) = (1, 0, 3) has the value $10 \times 10^3\ \Omega = 10\ k\Omega$.

A resistor with colors (orange, orange, brown) = (3, 3, 1) has the value $33 \times 10^1\ \Omega = 330\ \Omega$.

While resistors have no polarity, diodes and light emitting diodes (LEDs) have a positive and a negative pin. A regular diode has a marker ring at its negative side, while the two pins of an LED differ in length. The longer of the two pins is the positive pin, see Figure 3.8.

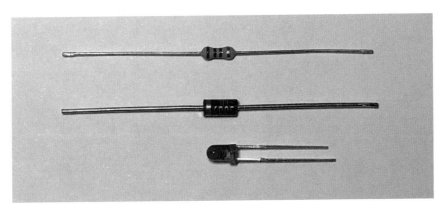

Figure 3.8: 2.2 kΩ resistor (red, red, red, gold), diode (left neg.) and LED (bottom pos.)

There are tables as well as apps and tools that help to quickly identifying resistor values, like the one from Digi-Key[2] shown in Figure 3.9.

---

[2]Digi-Key Electronics, *4 Band Resistor Color Code Calculator*, 2021, online: https://www.digikey.com.au/en/resources/conversion-calculators/conversion-calculator-resistor-color-code.

3.3 Arduino Interfacing

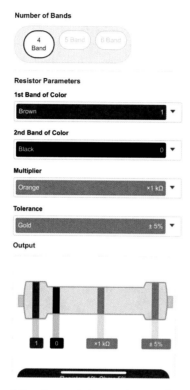

Figure 3.9: Tool for determining resistor values (Digi-Key)

To write the software for this hardware setup, we can extend the *blink* program as shown in Program 3.3. In the *setup* part, we need to initialize both pins. Pin D12 should be an input, while pin D10 should be an output. In the *loop* part, we continuously read the status of the switch and then set the LED output accordingly. There is no need for a delay in this program.

Program 3.3: Reading from and Writing to External Ports

```
void setup() {
  pinMode(12, INPUT);     // pin for switch
  pinMode(10, OUTPUT);    // pin for LED
}
int sw;
void loop() {
  sw = digitalRead(12);   // read switch status
  digitalWrite(10, sw);   // set LED accordingly
}
```

3.4 Arduino Communication

The USB port of the Arduino is not just for providing power and uploading new programs. It can also be used for transmitting data to and from the embedded controller to a host PC. In a stand-alone application, we might want to add some LEDs or a display to the Arduino to show data, but with a PC link we can easily transmit data from the Arduino to the PC (or vice versa) and display it in a window. This is especially helpful during the debugging phase.

In Program 3.4, we added a line of code for initializing the serial transmission in *setup* and one line that transmits the switch's status back to the PC in *loop*.

Program 3.4: Transmitting Data to a PC

```
void setup() {
   pinMode(12, INPUT);      // pin for switch
   pinMode(10, OUTPUT);     // pin for LED
   Serial.begin(115200);    // init. serial port at 115200 bps
}

int sw;

void loop() {
   sw = digitalRead(12);    // read switch status
   digitalWrite(10, sw);    // set LED accordingly
   Serial.print(sw);        // send bit to PC
   delay(200);              // slow things down (wait 200ms)
}
```

Figure 3.10 shows the *Serial Monitor* tool from the Arduino app for this program.

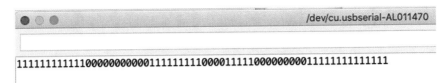

/dev/cu.usbserial-AL011470

111111111111000000000001111111110000111100000000001111111111111

Figure 3.10: Serial monitor window displaying Arduino output on a PC

In Program 3.5, we simplify the setup to remote control the LED from the PC. The desired LED value now no longer comes from the switch on the Arduino, but from data input on the PC. We initialize the LED to off at the start in *setup*. In the main *loop,* we first check if new data is available on the serial port; only then do we read and process it. If the transmitted character is an 'A', we turn on the LED, if it is a 'Z', we turn it off; all other characters are being ignored. Note that depending on how you transmit the characters from your PC, there may be additional stray characters, such as carriage return/line feed (CR/LF).

Program 3.5: Reading Data from a PC

```
void setup() {
  pinMode(10, OUTPUT);      // pin for LED
  digitalWrite(10, 0);      // turn off LED at start
  Serial.begin(115200);     // init. serial port at 115200 bps
}

char ch;

void loop() {
  if (Serial.available()>0) // only if data has been received:
  { ch = Serial.read();        // read the new value from PC
    if (ch == 'A') digitalWrite(10, 1); // turn on LED
    if (ch == 'Z') digitalWrite(10, 0); // turn off LED
  }
}
```

For testing, we can now enter these characters from the *Serial Monitor* in the Arduino app, but they could also come from any other application or another embedded controller, such as another Arduino or a Raspberry Pi.

3.5 Beyond Arduino

This section looks at alternative programming languages for the Arduino and at alternative controller boards in a similar performance and price range to the Arduino.

3.5.1 Python for Arduino

The Arduino can be programmed in Atmel Assembly language or C. Python is not an option, as Python is an interpreted language as opposed to being a compiled language like C. A user program in C gets compiled into machine code and then only the machine code gets transferred to the processor, which it can directly execute. For an interpreted language, a so-called interpreter program is running on the processor, which line-by-line reads the source code (e.g., Python) and executes the commands as they come. Such interpreters require a significant amount of memory space and usually also introduce a certain amount of processing time overhead, so they are not really suitable for small embedded controllers such as the Arduino. However, by using the Arduino's USB communication port, a Python program can run on a host computer, e.g., a laptop or desktop PC connected to the Arduino, and then transmit command by command to the embedded controller. This is not a stand-alone solution, but may work for a number of applications. See *pyserial*[3] for details.

---

[3]Project pyserial documentation, online: https://pythonhosted.org/pyserial/.

An alternative is using MicroPython[4] on a *pyboard*[5] embedded controller. MicroPython is an implementation of Python3 and pyboard is a system using the STM32F405RG (a 32-bit Cortex-M4F ARM microcontroller) plus additional 3-axis acceleration sensors and a real-time clock. MicroPython does not run on the standard Arduino boards.

3.5.2 ArduPilot

Besides the number of Arduino variations, such as Arduino Uno, Arduino Nano and Arduino Mega, there are Arduino-based controllers that have added sensors, software and ready to plug in connectors for controlling small model aircraft, boats or cars. Two of the most prominent such controllers are the Pixhawk and the somewhat lower specced and cheaper ArduCopter (see Figure 3.11). Both are Atmel/Arduino-based controllers with a built-in IMU (inertial measurement unit—see chapter on sensors) and lots of additional software available. The controllers are part of the overall ArduPilot[6] project, which has a large repository of information for the application areas Copter (drones), Plane (fixed wing), Rover (cars), and Sub (submarines).

As the name suggests, ArduCopter was developed for automating model helicopters and drones, but can also be used for all other autonomous projects. The Pixhawk[7] flight controller is similar in principle, but has a significantly higher performance due to its 32-bit ARM Cortex M4 processor.

Figure 3.11: ArduCopter controller

[4]MicroPython, online: https://micropython.org.

[5]MicroPython pyboard feature table, online: https://store.micropython.org/pyb-features.

[6]ArduPilot, online: https://ardupilot.org.

[7]Pixhawk, *The open standards for drone hardware*, online: https://pixhawk.org.

3.5.3 ESP32

There are a number of small single-chip embedded controllers on the market now, many with much better processing power than the Arduino, which can be quite useful for many applications. One such prominent set of processors is the ESP32 family of controllers, developed by *Espressif Systems* in Shanghai, China. These boards contain Wi-Fi and Bluetooth communication and there are even versions with OLED display or digital camera on the same board. Espressif offers libraries for the Arduino app that can be downloaded to allow code generation for the EPS32 boards from the Arduino app, which makes a transition very simple. Figure 3.12 shows an example application for the ESP32-based board TTGO that includes several input buttons and an OLED display.

Figure 3.12: TTGO board based on ESP32 with OLED display

3.6   Tasks

1. Buy an Arduino or EPS32 embedded controller of your choice.

2. Install the Arduino app on your PC and familiarize yourself with compiling and uploading programs via USB to the controller.

3. Explore the example programs provided for the Arduino (or ESP32).

4. Place your controller onto a breadboard and connect an input switch and an LED with the required resistors to it as shown in this chapter. Implement a program that lets the button turn the LED on/off.

5. Connect 8 LEDs to the controller output pins of a single port. Implement a binary counter that advances at a speed of 1 Hz.

6. Connect 8 LEDs to the controller output pins of a single port. Implement a single running light from right (bit 0) to left (bit 7), then from left to right and so on.
    Set the speed to 3 Hz.

7. Connect a temperature sensor to the controller then continuously send the sensor data as serial messages via USB to a connected PC. On the PC write a program that displays the incoming data values as text output as well as a graph curve.

Raspberry Pi

4

The Raspberry Pi is much more complex than the Arduino; it is a full single-board computer, not just a single-chip microcontroller. The first-generation Pi was developed by the Raspberry Pi Foundation in Cambridge, England, in 2011. It uses a Broadcom system-on-chip CPU with USB and Ethernet communication ports, HDMI monitor output, special camera interface, and in the current versions even Wi-Fi and Bluetooth (Figure 4.1). So if you connect monitor, keyboard and mouse, you will have a fully functional PC.

There are several different operating systems available for the Pi, most of them dialects of Linux, such as the *Raspberry Pi OS* (originally called *Raspbian*). All software, operating system and user programs reside on an external SD card (on the first-generation Pi this was a full-size SD card, now micro-SD cards are used). This solution has pros and cons. The advantages are that within seconds, the complete system software can be swapped out by simply replacing one SD card with another and different system versions can be kept on separate SD cards. The disadvantages, however, are that an additional component has to be purchased and that SD card data can easily get corrupted so that the Pi no longer boots up.

4.1 Raspberry Pi Operating System and Setup

In the following, we will only be using the Raspberry Pi OS (Raspbian), which is sufficient for most applications. It should be noted that although this is not a real-time operating system, it is sufficient for most robotics applications. Even complex robotics tasks rarely require real-time operation, as long as all sensor data can be marked with real-time clock readings.

The first step is to download the Pi's operating system from the organization's website[1] and then transfer it to an SD memory card. Unfortunately, the Pi does not use a file system compatible with Windows or MacOS, so you cannot

---

[1]Raspberry Pi Downloads, online at https://www.raspberrypi.org/downloads/.

© The Author(s), under exclusive license to Springer Nature Singapore Pte Ltd. 2022
T. Bräunl, *Embedded Robotics*,
https://doi.org/10.1007/978-981-16-0804-9_4

Figure 4.1: Raspberry Pi 4 board

simply put the SD card into your PC and copy the files over. Instead, you need to use an imager tool, such as *Win32DiskImager* for Windows or *ApplePi-Baker* for Mac or the *Raspberry Pi Imager* for Ubuntu/Linux. Each of these programs allows you to copy a system image file from your hard disk to the SD card or vice versa (see Figure 4.2).

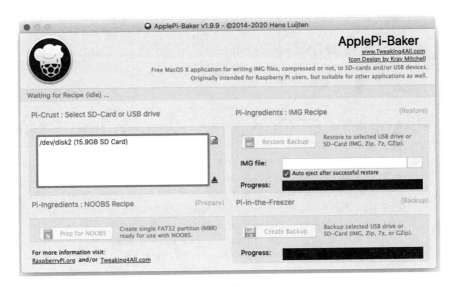

Figure 4.2: ApplePi-Baker for Mac

If you have multiple volumes connected to your PC, be extra careful, as selecting the wrong drive will erase its contents and overwrite it with the selected system image. A typical image file for the Raspberry Pi OS is either 8 GB or 16 GB in size. Transferring such a large file will take quite some time. Even on a high-quality fast SD card, transferring a 16 GB image will take 15 min or more.

We keep Raspberry Pi operating system image files specialized for our EyeBot robots as well as robot basic I/O system (RoBIOS) library files at this server address:

https://robotics.ee.uwa.edu.au/rasp/

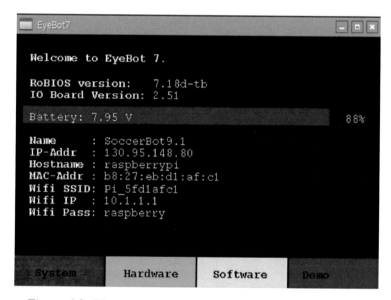

Figure 4.3: Pi startup screen with EyeBot software

RoBIOS will boot with a startup screen as shown in Figure 4.3. The RoBIOS library comprises a number of basic functions that facilitate the writing of robot application programs. A full list of RoBIOS functions is in the appendix of this book. In our EyeBot-Raspbian version we have set a number of useful defaults, which otherwise would have to be set up manually using the Pi's command *raspi-config* as well as other configuration tools. The RoBIOS defaults are:

- Default user is *pi* with password *rasp*
- Auto-login at boot time is enabled
- Remote shell access is enabled
- Raspberry camera is enabled
- DHCP when connected to a router is enabled
- Otherwise a static LAN address is used with IP address **10.0.0.1**

- Wi-Fi hotspot is enabled with SSID **Pi_1234567** (the number is derived from the Pi's unique hardware MAC address) with static IP address **10.1.1.1**
- Video output and touch-screen input for a 3.5 inch Waveshare[2] LCD are enabled.

If *not* using a router, you can directly connect a PC to the Pi via LAN or WLAN. For WLAN, the Pi will act as WLAN hotspot, so just connecting to its SSID is sufficient. For direct LAN setup without a router, you need to set your PC to a fixed IP address in the same range as the Pi's default address (10.0.0.1), so, e.g., 10.0.0.2 with network mask 255.255.255.0 (see Figure 4.4).

Figure 4.4: Static IP setup

You can verify your local IP setting by using the command *ifconfig* (Mac, Linux) or *ipconfig* (Windows):

```
➢ ifconfig
...
inet 10.0.0.2 netmask 0xffffff00 broadcast 10.0.0.255
...
```

---

[2]Waveshare 3.5 TF touch screen for Raspberry Pi, https://www.waveshare.com/wiki/3.5inch_RPi_LCD_(A).

After that, you can verify that the Pi is reachable by using the *ping* command from a terminal/command window. This will display network communication response times from the Pi:

```
➢ ping 10.0.0.1
PING 10.0.0.1 (10.0.0.1): 56 data bytes
64 bytes from 10.0.0.1: icmp_seq=0 ttl=64 time=0.716 ms
64 bytes from 10.0.0.1: icmp_seq=1 ttl=64 time=0.892 ms
...
```

4.2   Raspberry Pi Tools and Programming

Many modern programming languages are available for the Pi. Besides the low-level ARM Assembly language, these include C, C++, Python and Scratch. Because the Raspberry Pi has a much more powerful CPU than the Arduino family, the compiler or language interpreter can run directly on the Raspberry Pi and can generate machine code for itself, while on the Arduino, the compiler has to run on a host computer (PC) as a cross-compiler and the binary file has to be uploaded into the Arduino.

In the following, we will look at some essential tools for working with the Raspberry Pi and linking it to your PC:

- File exchange
- Remote command window or remote desktop
- Compiler and interpreter commands.

The file exchange tools allow you to keep your original files on your desktop or laptop PC and have a copy on the Raspberry Pi, while the remote access tools let you execute commands on the Raspberry Pi in a window from a Mac, Windows or Linux PC.

4.2.1 File Exchange for Raspberry Pi

The file system on the Raspberry Pi's SD card differs from the file system on Mac or Windows, so unfortunately, it is not possible to simply insert the Pi's SD card in the slot on PC to exchange data. Instead, we need a special tool for this task.

From the number of tools available, we recommend the free tool *FileZilla*, which has an intuitive user interface and comes in versions for Mac, Windows and Linux (Figure 4.5).

After establishing a connection with the Pi, you can simply drag and drop files or whole folders from your PC to the Pi and vice versa. If you prefer a simple command line interface, you can also use the *scp* command (*secure copy*), which is a standard on Mac and Linux and can be installed for Windows.

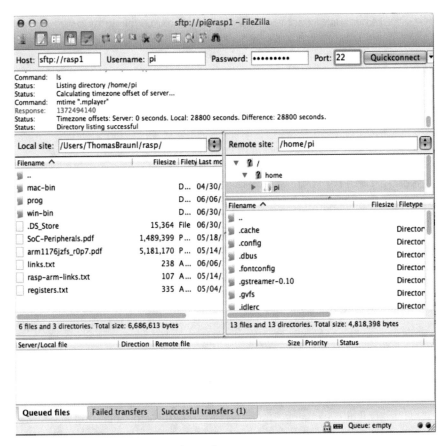

Figure 4.5: FileZilla file exchange

4.2.2 Remote Access for Raspberry Pi

The two levels of remote access are either a simple text window or a full graphics desktop. For remote access through a terminal window, open a standard terminal/command window and then issue the *ssh* or *slogin* command with the Pi's IP address.

When using a direct Ethernet link to the Pi, use our default Pi address 10.0.0.1, when connecting via Wi-Fi to the Pi's hotspot, use 10.1.1.1, and when connecting via a hardware router, use the Pi's DHCP address as displayed on the RoBIOS LCD home screen:

> ➢ `slogin -l pi 10.0.0.1`

when prompted for password enter: rasp

The commands *ssh* and *slogin* are a standard on Mac and Linux and can be installed for Windows. Alternatively for Windows, you can also use the tool *putty*.[3] Figure 4.6 shows the remote command login window. The prompt at

---

[3]Putty tool online: https://www.putty.org.

Figure 4.6: Remote command window

the end is the one from the Raspberry Pi and you can now enter any Linux command that will be executed on the Pi.

This method is limited to text transfer only. Whenever you want to display graphics from the Pi directly in a window, you need to run a remote desktop program. Two popular choices are the free tools RealVNC[4] (VNC stands for Virtual Network Connect) and Microsoft Remote Desktop.[5] You can start these tools just with the IP address of the Raspberry Pi as parameter. Then the full Pi screen with all tools will be displayed in a window (see Figure 4.7).

Figure 4.7: Remote desktop

[4]RealVNC, online: https://www.realvnc.com/en/connect/download/viewer/.

[5]Microsoft Remote Desktop, https://www.microsoft.com/en-us/p/microsoft-remote-desktop/9wzdncrfj3ps.

4.2.3 Compiling C and C++ Programs on Raspberry Pi

The programming languages we are concentrating on are C/C++ and Python for the Raspberry Pi. Compiling of C/C++ programs can be conducted directly on the Raspberry Pi by using the standard Linux command *gcc* or, if using any functions from the RoBIOS library, the script *gccarm*.

Program 4.1 shows the basic *"Hello, world!"* program. This does not use any RoBIOS functions and will print text directly to the console window. Hence, the standard compile command *gcc* can be used.

Program 4.1: Basic "Hello, World!"

```
#include <stdio.h>

int main()
{ printf("Hello, World!\n");
}
```

```
Compilation:
> gcc -o hello hello.c
```

```
Execution and Output:
> ./hello
Hello, world!
```

The *gcc* compile command specifies the output filename (binary executable) through option '*-o*' followed by the desired binary filename and then the name of the source file. Running the binary file is by simply typing its name. However, since this is happening in a user directory instead of a system directory, one has to use the path to the current directory ('*./*' in Linux) as a prefix, so together "*./hello*".

An extended *"Hello, World!"* program making use of some of the RoBIOS functions is shown in Program 4.2. The included *eyebot.h* file replaces the *<stdio.h>* include. Before the actual print statement, we use *LCDMenu* to label the four standard soft-key input buttons on our EyeBot LCD screen. Then we use *LCDPrintf* instead of plain *printf*, in order to print the argument onto the LCD screen instead of the text console. Finally, command *KEYWait(KEY4)* will pause execution until the user has pressed the rightmost of the four input buttons—the one we labeled *"END"* with the *LCDMenu* command. After that, the program terminates (see Figure 4.8).

Program 4.2 RoBIOS "Hello, World!"

```
#include "eyebot.h"

int main()
{ LCDMenu("", "", "", "END");
  LCDPrintf("Hello, World! \n");
  KEYWait(KEY4);
}
```

Compilation:
```
> gccarm -o hello_rob hello_rob.c
```

Execution (see output in Figure 4.8):
```
> ./hello_rob
```

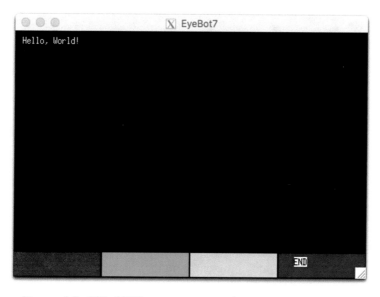

Figure 4.8: C/RoBIOS program execution

4.2.4 Interpreting Python Programs on Raspberry Pi

While C and C++ are compiled languages, Python is an interpreted programming language. This means a Python program is evaluated step by step by an interpreter program and does not have to be compiled into machine language first. The advantage of being able to run a Python program directly or try Python commands step by step comes at a cost. Interpreted programs run usually slower than the corresponding compiled program and they cannot do any error checking before program start. While a C/C++ compiler can find a large percentage of errors during the compile phase, the same bugs in a Python program could stay undetected until run-time—maybe even as late as in the deployment phase. Finally, the Python interpreter does require a significant memory overhead, which is why it cannot run on simple embedded controllers, such as the Arduino.

We use the *"Hello, World!"* example again, here for Python (see source file *hello.py* in Program 4.3).

Program 4.3 Python "Hello, World!"

```
from eye import *
LCDMenu ("","","","END")
LCDPrintf("Hello, World!")
KEYWait (KEY4)
```

There are different ways of running a Python program or executing Python commands:

1. Type *"python3"* to start the Python interpreter, then interactively enter individual commands:
   ```
   % python3
   Python 3.6.5 (v3.6.5:f59c0932b4, Mar 28 2018,
   05:52:31)
   [GCC 4.2.1 Compatible Apple LLVM 6.0 (clang-
   600.0.57)] on Darwin
   Type "help", "copyright", "credits" or "license" for
   more information.
    >>> print("Hello!")
   Hello!
   ```
2. Run a Python program from the command line:
   ```
   python3 hello.py
   ```
3. Embed the Python command in the Python source file as the first line:
   ```
   #!/usr/bin/env python3
   ```
   then make the source file executable (set file protections using command *"chmod a + x filename"*) and run the source file directly:
   ```
   ./hello.py
   ```

The program output is identical to the C version as shown in Figure 4.8.

4.2.5 Turnkey System

A turnkey system describes a system that starts up automatically without any user intervention required, as soon as power is turned on. Although this is the always the case for an application program on the Arduino, it is not the default for a program on the Raspberry Pi because of its complex Linux operating system.

Two steps have to be completed, in order to make a Pi to a turnkey system:

1. Auto-login without password has to be enabled.
   This can be set by running the command `raspi-config`.
   In our RoBIOS version of Raspbian (Pi OS), this has already been set as the default.

2. A binary application program has to activated for autostart.
   This can be achieved by editing the system file `/etc/xdg/autostart/myprog.desktop`
   and adding the following contents (for startup of binary file *MyProg*):

```
[Desktop Entry]
Type=Application
Name=MyProg
NoDisplay=false
Exec=/home/pi/MyProg
```

Please note that this procedure changes with subsequent Raspbian OS versions, so you need to check for the latest version.

In RoBIOS we have greatly simplified the procedure to implement a turnkey system. All that is required is to name this file 'startup' and place it in the user directory:

```
home/pi/usr/startup
```

4.3   Raspberry Pi Input/Output Lines

The pinouts for Raspberry Pi versions 2, 3 and 4 are identical and are shown in Figure 4.9. Some pins have dual functionality, e.g., the serial I/O lines can also be used as additional general purpose I/O lines (GPIO 14 for serial Tx and GPIO 15 for serial Rx). Please note that the pin labels used by chip manufacturer Broadcom (BCM) and the pin labels in the *WiringPi* library (discussed later) differ from each other. All of the Raspberry Pi's GPIO pins operate at 3.3 V levels and cannot sustain a higher voltage. The 5 V supply voltage on the two top right pins as shown in Figure 4.9 is for external devices only and must never be connected to an input pin.

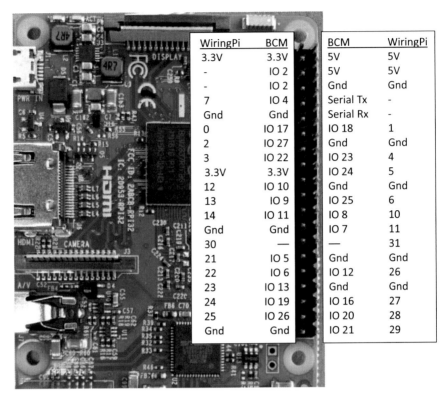

WiringPi	BCM	BCM	WiringPi
3.3V	3.3V	5V	5V
-	IO 2	5V	5V
-	IO 2	Gnd	Gnd
7	IO 4	Serial Tx	-
Gnd	Gnd	Serial Rx	-
0	IO 17	IO 18	1
2	IO 27	Gnd	Gnd
3	IO 22	IO 23	4
3.3V	3.3V	IO 24	5
12	IO 10	Gnd	Gnd
13	IO 9	IO 25	6
14	IO 11	IO 8	10
Gnd	Gnd	IO 7	11
30	—	—	31
21	IO 5	Gnd	Gnd
22	IO 6	IO 12	26
23	IO 13	Gnd	Gnd
24	IO 19	IO 16	27
25	IO 26	IO 20	28
Gnd	Gnd	IO 21	29

Figure 4.9: Raspberry Pi pinouts

We can set up similar experiments in linking I/O lines to switches and LEDs, like we have done for the Arduino in the previous chapter. As there are no simple commands to set individual I/O lines high or low in the Pi's Linux operating system, it will be a lot easier to install a library package that does this for us. From the several free libraries available, we chose the *WiringPi* library.[6] WiringPi is a small, compact library that provides the essential commands for setting I/O pins to high or low, as well as to generate pulse width modulation (PWM) signals—more on that topic later. Figure 4.9 shows the differing BCM and WiringPi pin labels side-by-side. The most important Wiring Pi commands are summarized in Table 4.1.

With these commands, we can now build a simple data read/write application, as we did for the Arduino on the previous chapter. Figure 4.10 shows the pin connections and Program 4.4 shows the code. We connected input pin 0 (bottom row) via a 10 kΩ resistor to +3.3 V and via a switch to ground. While the switch is open, input pin 0 will get a high potential (logic '1'), but when the switch is pressed, then pin 0 will get a low potential (logic '0').

On the output side (top row), we connect output pin 1 to the LED and then via a 330Ω resistor to ground. If we set pin 1 to high, the LED will be lit, otherwise it will be dark.

---

[6]Wiring Pi, GPIO Interface library for the Raspberry Pi, online at https://wiringpi.com, 2020.

4.3 Raspberry Pi Input/Output Lines

```
int wiringPiSetup (void);
    // Initialisation
void pinMode (int pin, int mode);
    // Set pin mode to either input or output
void pullUpDnControl (int pin, int pud);
    // Enable internal pull-up or pull-down resistors
void digitalWrite (int pin, int value);
    // Set output pin to a simple high or low value
void pwmWrite (int pin, int value);
    // Set output pin to a repeating pulse-width-modulation frequency
int digitalRead (int pin);
    // Read 0 or 1 from an input pin
analogRead (int pin);
    // Analog input
analogWrite (int pin, int value);
    // Analog output
void delay (unsigned int howMANYms);
    // Delay/sleep function
```

Table 4.1: Basic Wiring Pi commands

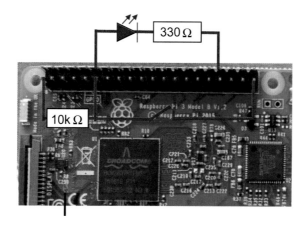

Figure 4.10: I/O pin connection example on Raspberry Pi

The code in Program 4.4 is rather simple as well. First, we initialize the WiringPi library by calling the *wiringPiSetup* function, then we set I/O pin 0 as input and I/O pin 1 as output. In a loop we then read the pin 0 input status and write this value via variable k to the output pin 1. If the button is not pressed, a logic '1' is being read and the LED is turned on. If the button is pressed, a logic '0' will be read and the LED turned off.

The *for*-loop in this program only runs for 1'000 iterations instead of infinitely long as in the Arduino example. The reason for this is that the Raspberry Pi does have an operating system and we do not want the program to run forever. Instead of the fixed number of iterations, we could add another input switch to stop program execution.

Program 4.4 Data Input/Output for Raspberry Pi

```
#include <wiringPi.h>
main()
{ int i,k;
  wiringPiSetup();
  pinMode (0, INPUT);
  pinMode (1, OUTPUT);
  for (i=0; i++; i<1000)
  { k = digitalRead(0); digitalWrite(1, k);
  }
}
```

4.4   Raspberry Pi Communication

The Raspberry Pi has a large variety of possible communication channels. Data can be sent via:

- Digital data pins (as '0' or '1' messages)
- I^2C communication
- Serial data link
- USB link
- Ethernet/LAN
- Wireless LAN
- Bluetooth

4.4.1 Pi Serial Communication

For serial communication, the *WiringSerial* library[7] provides a number of commands that simplify data exchange with another computer or controller, such as an Arduino. Table 4.2 lists the serial commands.

```
#include <wiringSerial.h>
// Pi serial port /dev/ttyACM0
int    serialOpen    (char *device, int baud);
void   serialPutchar (int fd, unsigned char c);
void   serialPuts    (int fd, char *s);
void   serialPrintf  (int fd, char *message, ...);
int    serialGetchar (int fd);
void   serialFlush   (int fd);
```

Table 4.2: WiringSerial commands

---

[7]WiringSerial Library, 2020, online: https://projects.drogon.net/raspberry-pi/wiringpi/serial-library/.

To link a Raspberry Pi with an Arduino (or another Pi or a PC) at a transmission rate of 9'600 Baud, we can then simply code on the Pi side as in the code snippet of Program 4.5

Program 4.5 Starting Serial Communication on Raspberry Pi

```
...
int port = serialOpen ("/dev/ttyACM0", 9600);
serialPrintf(port, "Hello from Pi!");
...
```

On the receiving side for an Arduino, we can use the serial communication commands *Serial.begin* and *Serial.read*, as described in the previous chapter.

4.4.2 Raspberry Pi LAN and WLAN Communication

As part of the RoBIOS library, we have implemented the *Radio* communication package, which lets a Pi communicate with other Pis (e.g., one robot talking to other robots) or with a base station, such as a PC or workstation. The library is based on IP addresses of the individual controllers/robots involved, so it can be used for wire-bound LAN communication as well as for wireless LAN. The Radio commands are listed in Table 4.3; they exist as a library for the Raspberry Pi, and also as libraries for Windows and MacOS PCs to enable a communication to a host computer.

```
int RADIOInit(void);                            // Start radio communicat
int RADIOGetID(void);                           // Get own radio ID
int RADIOSend(int id, char* buf);               // Send string  to ID
int RADIOReceive(int *id_no, char* buf, int size); // Wait for message
int RADIOCheck(void);                           // Check for waiting mes.
int RADIOStatus(int IDlist[]);                  // Return number of robot
int RADIORelease(void);                         // Terminate radio comm.
```

Table 4.3: RoBIOS Radio communication commands

The *Radio* library functions simplify the design of communication applications, as is explored in more detail in the chapter on communication.

4.5   Integration Development Environments

A number of powerful integrated development environments (IDEs) are available for the Raspberry Pi (some as well as for the Arduino). Using an IDE with built-in syntax editor and debugging tools greatly improves software development productivity and helps in reducing programming errors. We recommend using one of the following:

- PlatformIO,[8]
  a free open-source IDE, based on Microsoft's Visual Studio.
- CodeLite IDE,[9]
  a free, open-source, and "light-weight" IDE.
- CLion,[10]

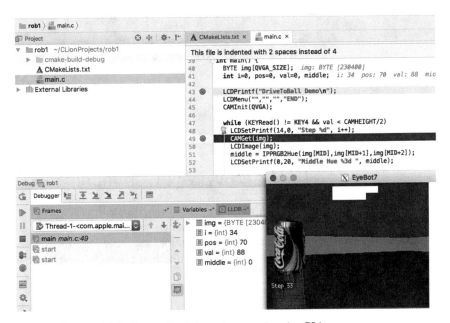

Figure 4.11: Example debugging session in CLion

When using an IDE together with the Raspberry Pi or the EyeSim simulator, make sure to set the include path and library path correctly. For example, when using CLion for EyeSim, the required settings are:

Mac and Linux

```
INCLUDE_DIRECTORIES(/usr/local/include /usr/X11/include)
LINK DIRECTORIES(/usr/local/lib /usr/X1/lib)
LINK LIBRARIES(eyesim X11)
```

Windows

```
INCLUDE_DIRECTORIES(.../cygwin/usr/local/include)
LINK DIRECTORIES(.../cygwin/lib)
LINK LIBRARIES(eyesim X11)
```

---

[8]PlatformIO IDE, *A new generation toolset for embedded C/C + + development*, online: https://platformio.org/platformio-ide.

[9]CodeLite IDE, online: https://codelite.org.

[10]Clion, *A cross-platform IDE for C and C + +* , online: https://www.jetbrains.com/clion/.a commercial IDE, free for students and educators (see Figure 4.11).

4.6   Tasks

1. Buy a Raspberry Pi controller, a Pi Camera (or a compatible USB camera), a fast 16GB micro-SD card, and optionally a pin-connected LCD (e.g. Waveshare) or a HDMI LCD.

2. Download the latest Raspbian-RoBIOS distribution from:
   https://robotics.ee.uwa.edu.au/rasp/
   e.g. image file:
   https://robotics.ee.uwa.edu.au/rasp/images-pi4_Buster/buster-2020-01-22-EyeBot-Buster.img.zip

3. Install Win32DiskImager (Windows), ApplePi-Baker (Mac) or Raspberry Pi Imager (Ubuntu/Linux). Use this tool to transfer the binary image file onto the micro-SD card.

4. Boot up the Raspberry Pi with the micro-SD card and familiarize yourself with its operating system. Use *FileZilla* to transfer files, use *putty* (Windows), or a terminal window with *ssh* (Mac, Linux) to connect to the Raspberry Pi. Use *VNC Viewer* or *Microsoft Remote Desktop* for a full desktop connection.

5. Use the script *gccarm* to compile programs on the Raspberry Pi. A number of example programs are under
   /home/pi/demo
   as part of the RoBIOS distribution.

6. Connect two I/O pins of the controller onto a breadboard and connect an input switch and an LED with the required resistors to it. Implement the program that lets the button turn the LED on/off using the *WiringPi* library.

7. Connect 8 LEDs to the controller outputs of a single port. Implement a binary counter that advances at a speed of 1 Hz.

8. Connect 8 LEDs to the controller outputs of a single port. Implement a single running light from right (bit 0) to left (bit 7), then from left to right, and so on. Set the speed to 3 Hz.

9. Connect a temperature sensor to the controller, then continuously display the sensor data as a text value and as a graph curve on the LCD.

Sensors and Interfaces

There are a vast number of different sensors being used in robotics, applying different measurement techniques for obtaining data and using different interfaces to connect to a controller. This, unfortunately, makes sensors a difficult subject to cover. Instead of looking at the sensors' internal characteristics, we will classify sensors by their interface and select a number of typical sensor systems to discuss their specifics in hardware and software. The scope of this chapter is therefore more toward interfacing sensors to controllers than on understanding the internal construction of sensors themselves.

It is always important to find the right sensor for a particular application. This involves the right measurement technique, the right size and weight, the right operating temperature range and power consumption, and of course the right price.

Data transfer from the sensor to the CPU can be either CPU-initiated (polling) or sensor-initiated (via interrupt). In case it is CPU-initiated, the CPU has to keep checking whether the sensor is ready by reading a status line in a loop. This is much more time-consuming than the alternative of a sensor-initiated data transfer, which requires the availability of an interrupt line. The sensor then signals that data is ready via an interrupt, and the CPU can react immediately to this request.

5.1 Sensor Categories

From an engineering point of view, it makes sense to classify sensors according to their output signals. This will be important for interfacing them to an embedded system. Table 5.1 shows a summary of sensor outputs together with their typical sensors. A different perspective is given by the classification based on application in Table 5.2.

T. Bräunl, *Embedded Robotics*,
https://doi.org/10.1007/978-981-16-0804-9_5

Sensor output	Typical sensor
Binary signal (0 or 1)	Tactile sensor
Analog signal (e.g., 0 … 5 V)	Inclinometer
Timing signal (e.g., PWM)	Gyroscope
Serial link (RS232 or USB)	GPS, IMU
Parallel link	Digital camera

Table 5.1: Sensor output

	Local		Global
Internal	Passive		Passive
	Battery sensor		–
	Chip-temperature sensor		
	Shaft encoders		
	Accelerometer		
	Gyroscope		
	Inclinometer		
	Compass		
	Active		Active
	–		–
	Local		**Global**
External	Passive		Passive
	On-board camera		Overhead camera
			Satellite GPS
	Active		Active
	Sonar sensor		Sonar (or other)
	Infrared distance sensor		Global Positioning
	Laser scanner		System

Table 5.2: Sensor classification based on application

From a robot's point of view, it is more important to distinguish:

- Local sensors
  sensors mounted on the robot
- Global sensors
  sensors mounted outside the robot in its environment and transmitting sensor data back to the robot.

For mobile robot systems, it is also important to distinguish:

- Internal (proprioceptive) sensors
  sensors monitoring the robot's internal state
- External sensors
  sensors monitoring the robot's environment.

A further distinction is between:

- Passive sensors
  sensors that monitor the environment without disturbing it,
  for example digital camera, gyroscope
- Active sensors
  sensors that stimulate the environment for their measurement,
  for example sonar sensor, Lidar laser scanner, infrared sensor.

Table 5.2 classifies a number of typical sensors for mobile robots according to these categories. A good source for information on sensors is the book Sensors for Mobile Robots by Everett.[1]

The output signal of digital sensors can have different forms. It can be a parallel interface (e.g., 8 or 16 digital output lines), a serial interface (e.g., following the RS232 or USB standard) or a synchronous serial interface (see below).

5.2 Synchronous Serial and I^{2}C Interfaces

The output signal of digital sensors can have different forms. It can be a parallel interface (e.g., 8 or 16 digital output lines), a serial interface (e.g., following the RS232 or USB standard), a *synchronous serial* interface or the more versatile I^{2}C (Inter-Integrated Circuit) interface.

5.2.1 Synchronous Serial Interface

The expression *synchronous serial* means that the converted data value is read bit by bit from the sensor. After setting the chip-enable line for the sensor, the CPU sends pulses via a digital output pin and at the same time reads one bit of information from the sensor (e.g., read one bit of data on each rising edge of the pulse pin). See Figure 5.1 for an example of a sensor with a 6-bit wide output word.

---

[1]H.R. Everett, *Sensors for Mobile Robots*, AK Peters, Wellesley MA, 1995.

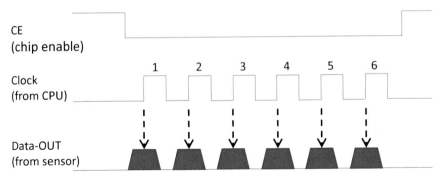

Figure 5.1: Signal timing for synchronous serial interface

5.2.2 I²C Interface

I²C (Inter-Integrated Circuit, also written as I2C or IIC) is a general-purpose extension of the synchronous serial interface. It allows to link one master module (microprocessor) with several slave modules (e.g., sensors) via a single port. Besides supply voltage V_{CC} and ground, the I²C interface only requires two signal lines:

- SDA   serial data line
- SCL   serial clock line.

Both signals have a pull-resistor to V_{CC} to indicate that the bus system is idle if none of the slave modules wants to transmit any data (see Figure 5.2).

The master module (in our case the CPU) generates the clock signal for either transmitting or receiving bitwise data via serial data line. In the standard I²C protocol, the master transmits in sequence:

- START signal
- 7-bit slave address
- read/write bit ('0' for write, '1' for read from the CPU's point of view).

After this follows the actual communication to or from the CPU. All data bits are transmitted via the SDA data line, while the CPU maintains timing via the SCL clock line. Communication is terminated by the CPU sending a STOP signal.

All slave modules need to have settable and unique address values for this mechanism to work. The roles between master and slave can even be swapped, e.g., when connecting several microprocessors via I²C.

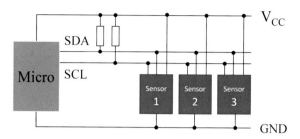

Figure 5.2: I²C connection schema

5.3   Binary Sensors

Binary sensors are the simplest type of sensors. They only return a single bit of information, either a 0 or a 1. A typical example is a tactile sensor on a robot, for example using a microswitch. Interfacing to a microcontroller can be achieved very easily by using a digital input on the controller. Figure 5.3 shows how to use a resistor to link to a digital input. In this case, a pull-up resistor will generate a high signal unless the switch is activated. This is called an *active low* setting.

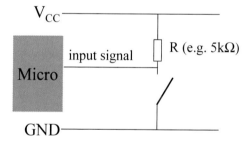

Figure 5.3: Interfacing a tactile sensor

A number of microcontrollers, such as the Arduino Nano, have built-in pull-up and pull-down resistors on their I/O lines. These can be activated via software and simplify the general system design as separate resistors are no longer required.

5.4   Shaft Encoders

Encoders are required as fundamental feedback sensors for motor control. There are several techniques for building an encoder. The most widely used ones are either magnetic encoders or optical encoders. Magnetic encoders use a Hall-effect sensor and a rotating disk on the motor shaft with a number of magnets (e.g., 16) mounted in a circle. Every revolution of the motor shaft drives the magnets past the Hall-effect sensor and therefore results in 16 pulses or *ticks* on the encoder line. The faster the encoder disk spins, the shorter and more frequent will the ticks be (see Figure 5.4).

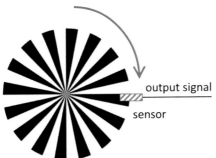

Signal for encoder disk spinning

· Slow:

· Fast:

· Not at all:

Figure 5.4: Encoder ticks

Standard optical encoders use a sector disk with black and white segments together with an LED and a photodiode (Figure 5.5, left). The photodiode detects reflected light during a white segment, but not during a black segment. So, if this disk has 16 white and 16 black segments, the sensor will generate 16 rectangle pulses during one full revolution.

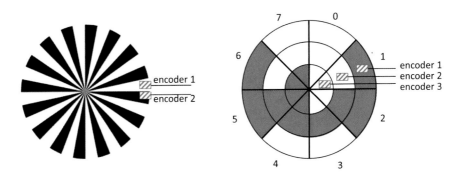

Figure 5.5: Incremental encoder (left) and absolute encoder using Gray code (right)

Encoders are usually mounted directly on the motor shaft (i.e., before the gearbox), so they have the full resolution compared to the much slower rotational speed at the geared-down wheel axle. For example, if we have an encoder which detects 16 ticks per revolution and a gearbox with a ratio of 100:1 between the motor and the vehicle's wheel, then this gives us an encoder resolution of 1,600 ticks per wheel revolution.

As has been mentioned above, an incremental encoder with only a single magnetic or optical sensor element can only count the number of segments passing by. But it cannot distinguish whether the motor shaft is moving clockwise or counterclockwise. This is especially important for applications such as robot vehicles which should be able to move forward and backward. For this reason, most incremental encoders are equipped with two sensors (magnetic or optical) that are positioned with a small phase shift to each other. With this arrangement, it is possible to determine the rotation direction of the motor shaft, since it is recorded which of the two sensors first receives the pulse for a new segment. If in Figure 5.5, left, *Enc1* receives the signal first, then the motion is clockwise; if *Enc2* receives the signal first, then the motion is counterclockwise.

Since each of the encoder sensors is just a binary digital sensor, we could interface them to a microcontroller by using two digital input lines. However, this would not be very efficient, since then the controller would have to constantly poll the sensor data lines in order to record any changes and update the sector count.

Luckily, this is not necessary, since many modern microcontrollers have special hardware built-in for cases like this. They are called *pulse counting registers* and can count incoming pulses up to a certain frequency completely independent of the CPU. This means the CPU is not being slowed down and is therefore free to work on higher-level application programs. Shaft encoders are standard sensors on mobile robots for determining their position, speed and orientation.

All encoders described above are called *incremental encoders*, because they can only count the number of segments passed from a certain starting point. They are not sufficient to locate the absolute position of the motor shaft. If this is required, a Gray-code disk (Figure 5.5, right) can be used in combination with a set of sensors. The number of sensors determines the maximum resolution of this encoder type (in the example, there are three sensors, giving a resolution of $2^3 = 8$ sectors). Note that for any transition between two neighboring sectors of the Gray-code disk, only a single bit changes (e.g., between $1 = 001$ and $2 = 011$ only the middle bit changes). This would not be the case for a standard binary encoding (e.g., $1 = 001$ and $2 = 010$, which differ by two bits). This is an essential feature of this encoder type, because it will still give a proper reading if the disk just passes between two segments. For binary encoding the result would be arbitrary when passing between 111 and 000, as it is uncertain which of the three bits would change first, so all bit patterns are possible intermediate outcomes.

5.5   A/D Converters

A number of sensors produce analog output signals rather than digital signals. This means an analog-to-digital converter (A/D, see further below) is required to connect such a sensor to a microcontroller. Typical examples of such sensors are:

- Microphone
- Infrared distance sensor
- Compass
- Barometer sensor.

An analog to digital (A/D) is required to convert an analog signal into a digital value. The characteristics of an A/D converter include:

- Accuracy
  expressed in the number of digits it produces per value, e.g., 10-bit A/D converter
- Speed
  expressed in maximum conversions per second, e.g., 500 conversions per second
- Measurement range
  expressed in volts, e.g., 0 V to 5 V.

A/D converters come in many variations, and their output format also varies. Typical are either a parallel interface (e.g., up to 8 bits of accuracy) or a synchronous serial interface. The latter has the advantage that it does not impose any limitations on the number of bits per measurement, for example 10 or 12 bits of accuracy. Figure 5.6 shows a typical arrangement of an A/D converter interfaced to a CPU.

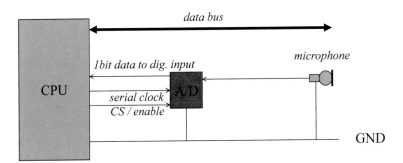

Figure 5.6: A/D converter interfacing

Many A/D converter modules include a multiplexer as well, which allows the connection of several sensors, whose data can be read and converted subsequently. In this case, the A/D converter module also has a 1-bit digital data input line, which allows the specification of a particular input line by using a synchronous serial transmission from the CPU to the A/D converter.

5.6   Position Sensitive Devices—Sonar, Infrared, Laser

Sensors for distance measurements, also called *position sensitive devices*, are among the most important components in robotics. For decades, mobile robots have been equipped with various sensor types for measuring distances to the nearest obstacle around the robot for navigation purposes.

5.6.1 Sonar Sensors

In the past, most robots had been equipped with sonar sensors. Because of the relatively narrow cone of these sensors, a typical configuration to cover the whole circumference of a round robot required 24 sensors, mapping about 15° each. Sonar sensors use the principle of a short acoustic signal of about 1 ms at an ultrasonic frequency of 50 kHz to 250 kHz, measuring the time between signal emission until echo reception to the sensor. The measured time-of-flight is proportional to twice the distance of the nearest obstacle in the sensor cone. If no signal is received within a certain time limit, then no obstacle is detected within the sensor's measuring range. Measurements are repeated about 20 times per second, which gives this sensor its typical clicking sound (see Figure 5.7).

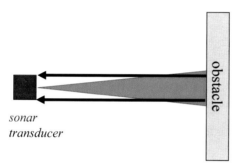

Figure 5.7: Sonar sensor

Sonar sensors have a number of disadvantages, but they are also a very powerful sensor system, as can be seen in the vast number of published articles dealing with them, see Barshan, Ayrulu, Utete[2] and Kuc.[3] The most significant problems of sonar sensors are inaccuracy, reflections and interference. When the acoustic signal is reflected, for example off a wall at a certain angle, then an obstacle seems to be further away than the actual wall that reflected the signal. Interference occurs when several sonar sensors are operated at once (among the 24 sensors of one robot or between several robots close to each other). Here, it can happen that the acoustic signal from one sensor is being picked up by another sensor, resulting in incorrectly assuming a closer than actual obstacle. Coded sonar signals can be used to prevent this, for example using pseudo-random codes as in Jörg, Berg.[4]

---

[2]B. Barshan, B. Ayrulu, S. Utete, *Neural network-based target differentiation using sonar for robotics applications*, IEEE Transactions on Robotics and Automation, vol. 16, no. 4, August 2000, pp. 435–442 (8).

[3]R. Kuc, *Pseudoamplitude scan sonar maps*, IEEE Transactions on Robotics and Automation, vol. 17, no. 5, 2001, pp. 767–770.

[4]K. Jörg, M. Berg, *Mobile Robot Sonar Sensing with Pseudo-Random Codes*, IEEE International Conference on Robotics and Automation 1998 (ICRA ë98), Leuven Belgium, 16–20 May 1998, pp. 2807–2812 (6).

5.6.2 Infrared Sensors

Infrared (IR) distance sensors do not follow the same principle as sonar sensors, since the time-of-flight for a photon would be much too short to measure with a simple and cheap sensor arrangement. Instead, these systems typically use a pulsed infrared LED at about 40 kHz together with a detection array (see Figure 5.8). The angle under which the reflected beam is received changes according to the distance to the object and therefore can be used as a measure of the distance. The wavelength used is typically 880 nm. Although this is invisible to the human eye, it can be transformed to visible light either by IR detector cards or by recording the light beam with an IR-sensitive camera.

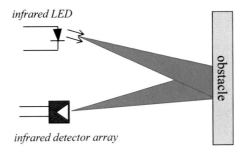

infrared LED

obstacle

infrared detector array

Figure 5.8: Infrared sensor

Figure 5.9 shows the Sharp sensor GP2D02[5] which is built in a similar way as described above. It exists in several variations, e.g.:

- Sharp GP2D12 with analog output
- Sharp GP2D02 with digital serial output.

The analog sensor simply returns a voltage level in relation to the measured distance (unfortunately not proportional, see Figure 5.9, right, and text below). The digital sensor has a serial digital interface, transmitting an 8-bit measurement bitwise over a single line, triggered by a clock signal from the CPU as shown in Figure 5.1.

In Figure 5.9, right, the relationship between digital sensor read-out (raw data) and actual distance information can be seen. From this diagram, it is clear that the sensor does not return a value linear or proportional to the actual distance, so some post-processing of the raw sensor value is necessary. The simplest way of solving this problem is to use a lookup table which can be calibrated for each individual sensor. Since only 8 bits of data are returned, the lookup table will have the reasonable size of 256 entries. Such a table is provided in the hardware description table (HDT) of our RoBIOS operating system. With this concept, calibration is only required once per sensor and is transparent to the application program.

---

[5]Sharp, Data Sheet GP2D02 - Compact, High Sensitive Distance Measuring Sensor, Sharp Co., data sheet, https://www.sharp.co.jp/ecg/, 2006.

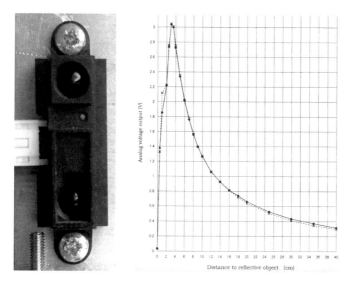

Figure 5.9: Sharp PSD sensor and sensor diagram (data source: Sharp 2006)

Another problem becomes evident when looking at the diagram for actual distances closer than about 6 cm. These distances are below the measurement range of this sensor and will result in an incorrect reading of a much higher distance. This is a serious problem, since it cannot be fixed in a simple way. One could, for example, continually monitor the distance of a sensor until it reaches a value in the vicinity of 6 cm. However, from then on it is impossible to know whether the obstacle is coming closer or going further away. The safest solution is to mechanically mount the sensor in such a way that an obstacle can never get closer than 6 cm or to use an additional (IR) proximity sensor to cover for any obstacles closer than this minimum distance.

IR proximity switches are of a much simpler nature than IR PSDs. They are the electronic equivalent of tactile binary sensors shown earlier. These sensors also return only a 0 or a 1, depending on whether there is sufficient free space (e.g., 1–2 cm) in front of the sensor or not. IR proximity switches can be used in lieu of tactile sensors for most applications that involve obstacles with reflective surfaces. They also have the advantage that no moving parts are involved compared to mechanical microswitches.

5.6.3 Laser Distance Sensors

Today, in many mobile robot systems, sonar and infrared sensors have been replaced by laser sensors. Inexpensive single-beam time-of-flight measurement laser sensors exist for Arduino and Raspberry Pi, such as the VL53L1X from ST (see Figure 5.10). They measure the distance between the sensor to the single nearest point along the direction of the sensor. The sensor is very small (less than $5 \times 3 \times 2$ mm^3) and requires surface-mount device (SMD) soldering technology, so it is best to buy them on a small carrier or breakout board.

Figure 5.10: Front and rear of single-beam laser time-of-flight sensors VL53L1X (left—note their tiny size from the attached ruler); same sensor on breakout board (right)

The sensor has a wide detection range of up to 4 m and uses a Class-1 (eye-safe) 940 nm invisible infrared laser with up to 50 distance measurements per second. The interface to a microcontroller is via I^{2}C interface.

5.7 Lidar Sensors

Used in advanced robotics applications as well as in autonomous vehicles, *Lidar* sensors (light detection and ranging) contain one or more laser diodes that point at a rotating mirror, so the sensor can deliver thousands of distance points per second (Figure 5.11). In the future, solid state Lidars without any moving parts will take over. For more details on Lidar sensors, see the book *Robot Adventures in Python and C*.[6]

---

[6]T. Bräunl, *Robot Adventures in Python and C*, Springer-Verlag Heidelberg, 2020.

5.7 Lidar Sensors

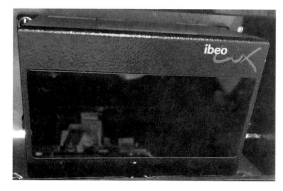

Figure 5.11: Lidar sensors with single beam (Sick, left) and multiple beams (ibeo, right)

Single-beam systems (typically used as industrial safety sensors) generate a single slice through the world. Figure 5.12 shows three example Lidar scans of 180° each from three different poses A, B and C in the same environment. Robot A faces a straight wall, which will translate into a parabola-like shape in the polar coordinates of the Lidar distance measurement. Robot B looks directly at a corner, while robot C looks at an inverse corner (see Figure 5.12). This assumes that the Lidar scans progresses clockwise, which is left-to-right from the robot's point of view.

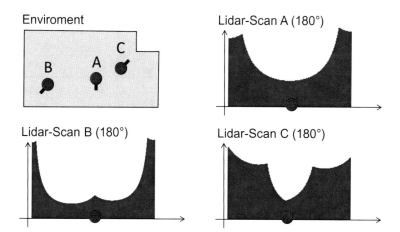

Figure 5.12: Sample Lidar scans from three different positions and orientations

Multi-beam systems (e.g., Sick Auto Ident[7]) can return an almost perfect local 2D field with distance data, so creating a 3D *depth map* or *point cloud*. Figure 5.13 shows the point cloud generated from multiple Lidar sensors on the REV/UWA autonomous shuttle bus *n*UWA*y*.

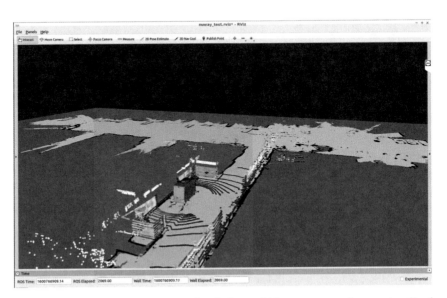

Figure 5.13: 3D point cloud from Lidar sensors, image by Yuchen Du, REV/UWA

5.8 Orientation Sensors

A compass sensor or magnetometer is a very useful tool for many mobile robot applications, especially for localization, when the robot can rely on its on-board sensors to keep track of its current position and orientation. The simplest method for achieving this in a driving robot is to use shaft encoders on each wheel and then apply a method called *dead reckoning*. This method starts with a known initial position and orientation (called a *pose*) and then adds all driving and turning actions to find the robot's current position and orientation. Unfortunately, due to wheel slippage and other factors, the *dead reckoning* error will grow larger and larger over time. Therefore, it is a good idea to have a compass sensor on-board, to be able to determine at least the robot's absolute orientation. In many applications, the correct robot orientation is even more important than the correct position.

---

[7]Sick, *Auto Ident Laser-supported sensor systems*, Sick AG, https://www.sick.de/de/products/categories/auto/en.html, 2006.

A number of magnetic compass sensor modules exist that can be used for work with Arduino or Raspberry Pi microprocessors, e.g., the HMC5883L[8] sensor chip. A project to build your own sensor board is listed at the Arduino page.[9] Since most of the modern sensor chips are surface mount, soldering them onto a PCB can be a problem. The alternative is to purchase a ready built compass module from one of the electronics suppliers. However, pure compass modules as in Figure 5.14 are rarely produced these days. Most modules combine a magnetic compass with three-axis accelerometers and/or gyros to build a more powerful—but also much more expensive—inertial measurement unit (IMU), as explained in the following section.

Figure 5.14: Vector 2X compass

5.9 Inertial Measurement Units

An inertial measurement unit (IMU) is an internal sensor system for measuring changes in a robot's spatial orientation. An IMU typically comprises three accelerometers and three gyroscopes (one for each axis), sometimes also magnetometers, plus an embedded processor for integrating the sampled data. IMU prices range from tens of dollars to tens of thousands of dollars, depending on their sensor and processing quality.

Figure 5.15 shows two inexpensive IMUs. On the left is the PhidgetSpatial[10] IMU, It comprises three accelerometer, three gyroscopes and a three-axis magnetic compass and communicates to a controller via its USB interface.

---

[8]Utsource, HMC5883L *Sensor MagMtr I2C 16ICC*, online: https://www.utsource.net/itm/p/8036866.html.

[9]Arduino Project Hub, *The 3 steps for you to build your compass with Arduino*, online: https://create.arduino.cc/projecthub/322974/the-3-steps-for-you-to-build-your-compass-with-arduino-56ebc6.

[10]Phidget Spatial Precision 3/3/3, online: https://www.phidgets.com/?&prodid=32.

The Bosch IMU BO055[11] is one of the cheapest systems in the market. In Figure 5.15, right, this sensor is shown on a DFRobot Intelligent 10 DoF[12] carrier board. It contains three accelerometers, three gyroscopes, three magnetometers and an integrated microprocess for fusing the generated data. From the carrier board, the Bosch IMU can be easily be interfaced to the I/O pins of a microprocessor either via its built-in I^2C interface or through its UART interface.

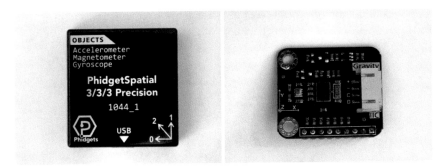

Figure 5.15: IMU sensors Phidget (left) and Bosch (right)

While individual accelerometer and gyros are a lot cheaper, their raw sensor data needs to be integrated over time to calculate the robot's continuously updated position and orientation (pose). Even a small sensor error will therefore generate a drift in the integrated pose over time and require significant processing effort and often additional sensors for compensation (see Figure 5.16).

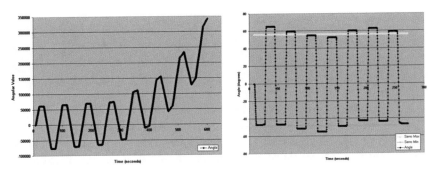

Figure 5.16: Drift of integrated gyroscope signal (left) and compensation (right) by Smith and Stamatiou, Robotics Lab UWA

---

[11]Bosch Smart Sensor:: BNO055, https://www.bosch-sensortec.com/products/smart-sensors/bno055.html.

[12]DFRobot SEN0253 Gravity, https://wiki.dfrobot.com/Gravity_BNO055_+_BMP280%20intelligent_10DOF_AHRS_SKU_SEN0253.

5.10   Global Navigation Satellite Systems

The *Global Positioning System* (GPS), started by the US Department of Defense in 1973, (civilian use since 1980) was the world's first Global Navigation Satellite System (GNSS). Since then, Russia's GLONASS, China's BeiDou and Europe's Galileo have been created as additional independent systems. Figure 5.17 shows a simple GPS receiver with USB connection (left) and a more sophisticated GPS/IMU combination from Xsens (right).

Figure 5.17: GPS receiver (left) and GPS/IMU combo (right)

A GNSS receiver module will automatically connect to several satellites and calculate its own position from their relative data. Orientation data is only available from triangulation when a GNSS receiver is in motion.

When a GPS sensor gets powered up, it continually sends position data, for example in National Marine Electronics Association (NMEA) format as shown in Table 5.3. NMEA presents the same data in different formats, but we only need to listen to the *$GPGGA* tokens at the beginning of a line. Their data record contains a quality indicator that can be either 0 or 1 (see red color), but only when it is 1, then the data record contains valid GPS coordinates. It takes some time until the GPS receiver locks onto a sufficient number of satellites. Then, the quality indicator changes from 0 to 1, and we can read out the GPS coordinates shown in blue color:

```
3152.6044,S,11554.2536,E
```

The location of our research center at the University of Western Australia in Perth.

The most serious and most frequent GNSS problem is losing satellite connection and thereby localization. A close second is the inaccuracy of localization. Even if the robot (or car) does not move at all, subsequent GNSS readings show different positions, often meters apart, so the robot or vehicle appears to be conducting a "random walk." Our tracked electric vehicle in Figure 5.18 appears to move more than 10 m although being stationary during this time. Methods to improve GNSS accuracy are presented in the following sections.

```
$TOW: 0
$WK: 1151
$POS: 6378137 0 0
$CLK: 96000
$CHNL:12
$Baud rate: 4800 System clock: 12.277MHz
$HW Type: S2AR
$GPGGA,235948.000,0000.0000,N,00000.0000,E,0,00,50.0,0.0,M,,,,0000*3A
$GPGSA,A,1,,,,,,,,,,,,,50.0,50.0,50.0*05
$GPRMC,235948.000,V,0000.0000,N,00000.0000,E,,,260102,,*12
$GPGGA,235948.999,0000.0000,N,00000.0000,E,0,00,50.0,0.0,M,,,,0000*33
$GPGSA,A,1,,,,,,,,,,,,,50.0,50.0,50.0*05
$GPRMC,235948.999,V,0000.0000,N,00000.0000,E,,,260102,,*1B
$GPGGA,235949.999,0000.0000,N,00000.0000,E,0,00,50.0,0.0,M,,,,0000*32
$GPGSA,A,1,,,,,,,,,,,,,50.0,50.0,50.0*05
...
$GPRMC,071540.282,A,3152.6047,S,11554.2536,E,0.49,201.69,171202,,*11
$GPGGA,071541.282,3152.6044,S,11554.2536,E,1,04,5.5,3.7,M,,,,0000*19
$GPGSA,A,2,20,01,25,13,,,,,,,,,6.0,5.5,2.5*34
$GPRMC,071541.282,A,3152.6044,S,11554.2536,E,0.53,196.76,171202,,*1B
$GPGGA,071542.282,3152.6046,S,11554.2535,E,1,04,5.5,3.2,M,,,,0000*1E
$GPGSA,A,2,20,01,25,13,,,,,,,,,6.0,5.5,2.5*34
$GPRMC,071542.282,A,3152.6046,S,11554.2535,E,0.37,197.32,171202,,*1A
$GPGGA,071543.282,3152.6050,S,11554.2534,E,1,04,5.5,3.3,M,,,,0000*18
$GPGSA,A,2,20,01,25,13,,,,,,,,,6.0,5.5,2.5*34
$GPGSV,3,1,10,01,67,190,42,20,62,128,42,13,45,270,41,04,38,228,*7B
$GPGSV,3,2,10,11,38,008,,29,34,135,,27,18,339,,25,13,138,37*7F
$GPGSV,3,3,10,22,10,095,,07,07,254,*76
```

Table 5.3: GPS data sequence

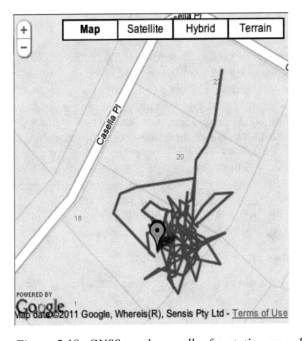

Figure 5.18: GNSS random walk of a stationary vehicle

5.10.1 Differential GNSS and RTK

Positioning accuracy has improved, especially for the GPS system, when the U.S. Department of Defense lifted its *selective availability* in the year 2000. However, cheap GPS receivers still have an error of several meters, especially near buildings. Differential GPS (DGPS) is one method to improve accuracy by adding a stationary reference station with a second GPS receiver that is linked to the GPS receiver on the mobile robot through a wireless connection. As both receivers of a DGPS suffer from the same positioning errors (*pseudoranges*), the stationary receiver can send a correction signal to the mobile GPS receiver to improve its positioning to several centimeters.

Real-time kinematic (RTK) is another technique to improve positioning accuracy of a GNSS receiver. A stationary GNSS receiver in the vicinity of the mobile GNSS receiver is using a phase measurement of the GNSS's carrier signal, in order to calculate a carrier-phase enhancement signal, which is then transmitted via a wireless link (or even through the Internet plus a 4G/5G receiver) to the mobile GNSS receiver. With the help of the RTK correction signal, the mobile GNSS receiver can then improve its localization down to several centimeters.

5.10.2 GNSS/IMU Combinations

Although a great invention in principle, GNSS modules often cause problems on autonomous robots and vehicles, because of connectivity problems. GNSS receivers only work outdoors and in unobstructed areas, so even when driving close to a building or a tree, the GNSS connection can get lost and the robot will lose one of its most valuable senses. This is why combined GNSS/IMU systems (Figure 5.17, right) are available as advanced localization tools. They can compensate a short-term loss of the GNSS signal by dead reckoning of the robot's pose through the IMU. On the other hand, any sensor drift of the IMU can be corrected from the GNSS signal.

Combined GNSS/IMU systems are highly accurate solutions for industrial problems, which unfortunately makes them very expensive. Most systems are around the $5,000–$10,000 price bracket, especially when also enabled for RTK. Prominent manufacturers and models are:

- Xsens,[13] Netherlands (MTi-7, MTi-G-710)
- SBG,[14] France (Ellipse, Ekinox, Apogee)
- Advanced Navigation,[15] Australia (Spatial, Certus)
- NovAtel,[16] Canada (PwrPak7D, MarinePak7, SPAN CPT7).

---

[13]Xsens, online: https://www.xsens.com/gnss.

[14]SBG Ellipse, online: https://www.sbg-systems.com/products/ellipse-series/.

[15]Advanced Navigation, online: https://www.advancednavigation.com/products/applications/motorsport.

[16]Hexagon NovAtel, SPAN Combined Systems, online: https://novatel.com/products/span-gnss-inertial-navigation-systems/span-combined-systems.

5.11 Digital Image Sensors

Digital cameras are the most complex sensors used in robotics. The central idea behind the EyeBot development in 1995 was to create a small, compact embedded vision system, and it became the first of its kind. Today, mobile phones with cameras are commonplace and digital cameras have pushed conventional film-based cameras out of the market.

For mobile robot applications, we are interested in a high frame rate, because our robot is moving, and we want updated sensor data as fast as possible. Since there is always a trade-off between high frame rate and high resolution, we are not so much concerned with camera resolution. For many applications for small mobile robots, a QVGA or even a QQGA resolution is sufficient (a quarter of a quarter of the 480 × 640 VGA[17] resolution from early IBM-PC days, so 120 × 160 pixels). Even from such a small resolution, we can detect, for example, colored objects or obstacles in the way of a robot (see even smaller 60 × 80 sample images from robot soccer in Figure 5.19). At QQVGA resolution, processed frame rates of up to 30 fps (frames per second) are achievable on a Raspberry Pi. The frame rate will drop, however, depending on the image processing algorithms applied.

Figure 5.19: Sample images with 60 × 80 resolution

The image resolution must be high enough to detect a desired object from a specified distance. When the object in the distance is reduced to a mere few pixels, then this is not sufficient for a detection algorithm. Many image processing routines are nonlinear in time requirement, but even simple linear filters, for example Sobel edge detectors, require a loop over all pixels, which creates a considerable CPU load (see *Parallel Image Processing*[18]). Even at the low resolution of 120 × 160 pixels in color (3 bytes per pixel), this amounts to

---

[17]Virtual Graphics Adapter; the name is also used for the 15-pin monitor adapter on the original IBM PS/2.

[18]T. Bräunl, *Parallel Image Processing*, Springer-Verlag, Berlin Heidelberg, 2001.

57,600 bytes per image frame. When processing 20 frames per second (fps), the required data throughput is over 1 MB/s. The Pi camera module is shown in Figure 5.20.

Figure 5.20: Raspberry Pi camera module

5.11.1 Camera Sensor Data

We distinguish between grayscale and color cameras, although, as we will see, there is only a minor difference between the two. The simplest available sensor chips provide a grayscale image of 120 lines by 160 columns with 1 byte per pixel. This is sufficient for many applications for small mobile robots, as a higher resolution also requires a much higher processing speed. However, the Raspberry Pi camera has the quite high maximum resolution of 8 megapixels.

In a grayscale camera, a value of zero represents a black pixel, while a value of 255 is a white pixel; everything in between is a shade of gray. Figure 5.21 illustrates such an image. The camera transmits the image data in row-major order, usually after a certain frame-start sequence.

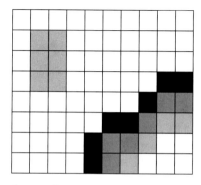

Figure 5.21: Grayscale image

Creating a color camera sensor chip from a grayscale camera sensor chip is quite simple. All it needs is a layer of paint over the pixel mask. The standard technique for pixels arranged in a grid is the Bayer pattern (Figure 5.22). Pixels in odd rows (1, 3, 5, etc.) are colored alternating in green and red, while pixels in even rows (2, 4, 6, etc.) are colored alternating in blue and green. With this colored filter over the pixel array, each pixel only records the intensity of its painted-on color component. For example, a pixel with a red filter will only record the red intensity at its position. At first glance, this requires 4 bytes per color pixel: green and red from one line and blue and green (again) from the line below. This would result effectively in a color image of one quarter of the original grayscale resolution, with an additional redundant green byte per pixel.

However, there is one thing that is easily overlooked. The four components red, green1, blue, and green2 are not sampled at the same position. For example, the blue sensor pixel is below and to the left of the red pixel. So by treating the four components as one pixel, we have already applied some sort of filtering and lost information.

Figure 5.22: Color image (Bayer pattern)

A technique called *demosaicing*[19] can be used to restore the image in its full original resolution and in full color. This technique basically recalculates the three color component values (R, G and B) at each pixel position. A simple example would be averaging the four closest component neighbors of the same color. Figure 5.23 shows the 13 pixels used for demosaicing the red, green and blue components of the pixel at position [3, 2] (assuming the image starts in the top left corner with [0, 0]).

Averaging, however, does not produce the best results. A number of articles have presented better algorithms for demosaicing, such as by Kimmel[20] and by Muresan and Parks.[21]

[19]C. Nix, *BryceCicada / demosaic*, online: https://github.com/BryceCicada/demosaic.

[20]R. Kimmel, *Demosaicing: Image Reconstruction from Color CCD Samples*, IEEE Transactions on Image Processing, vol. 8, no. 9, Sept. 1999, pp. 1221–1228 (8).

[21]D. Muresan, T. Parks, *Optimal Recovery Demosaicing*, IASTED International Conference on Signal and Image Processing, SIP 2002, Kauai Hawaii, https://dsplab.ece.cornell.edu/papers/conference/sip_02_6.pdf, 2002, pp. (6).

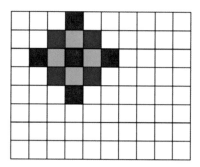

Figure 5.23: Demosaic of single pixel position

5.11.2 Camera RoBIOS Interface

The program listed in Program 5.1 shows how to initialize the camera module and continuously read a frame and display it on the connected LCD (or on a remote desktop). This code applies to all versions of the Raspberry camera, but can also be applied to external USB-connected cameras.

Program 5.1: Color Camera Access and Image Display in C

```
#include "eyebot.h"

int main()
{ BYTE img[QVGA_SIZE];

  LCDMenu("", "", "", "END");
  CAMInit(QVGA);
  do { CAMGet(img);
       LCDImage(img);
  } while (KEYRead() != KEY4);

  return 0;
}
```

In RoBIOS, each image is stored as a one-dimensional array. The standard image sizes are:

- QQVGA 120 × 160 × 3 bytes (RGB color)
  or 120 × 160 × 1 bytes (grayscale)
- QVGA 240 × 320 × 3 bytes (RGB color)
  or 240 × 320 × 1 bytes (grayscale)
- VGA 480 × 640 × 3 bytes (RGB color)
  or 480 × 640 × 1 bytes (grayscale)
- HDTV 1080 × 1920 × 3 bytes (RGB color)
  or 1080 × 1920 × 1 (grayscale)

For grayscale, each pixel uses 1 byte, with values from 0 (black) over 128 (medium-gray) to 255 (white). For color, each pixel comprises 3 bytes in the order: red, green, blue. For example, medium green is represented by (0, 128, 0), full red is (255, 0, 0), bright yellow is (200, 200, 0), black is (0, 0, 0), white is (255, 255, 255).

Before reading the first image, the camera needs to be initialized and the resolution set via a parameter. For example, for QVGA as in Program 3.1:

```
CAMInit(QVGA);
```

Valid parameters are QQVA, QVGA, VGA, CAM1MP (1 megapixel), CAMHD (HD) and CAM5MP (5 megapixels).

After this, a color or grayscale image can be read by using the appropriate RoBIOS command.

```
CAMGet(img); // for reading a color image or
CAMGetGray(img); // for reading a grayscale image
```

The camera image can immediately be displayed on the controller's LCD with this command:

```
LCDImage(img); // for displaying a color image or
LCDImageGray(img); // for displaying a grayscale image
```

The program continues to run until *KEY4* on the touch screen is being pressed (associated with menu text *"END"*). The equivalent program in Python to read and display a color image is shown in Program 5.2.

Program 5.2: Color Camera Read and Image Display in Python

```
from eye import *

LCDMenu("", "", "", "END")
CAMInit(QVGA)
while (KEYRead() != KEY4):
    img = CAMGet()
    LCDImage(img)
```

5.12 Tasks

1. Implement a pulse counter on your controller, ideally using hardware support to increase the maximum frequency. Connect a simple sensor, e.g., binary switch or IR detector to your door, and let the controller count how many people enter every day.
   If you have a camera connected, take a photograph every time somebody enters. If your controller is Wi-Fi connected, you can send each photograph to your email address.

2. Use the same pulse counter to determine the propeller speed of a drone or boat.
   This requires an IR LED on one side of the robot and the IR detector on the other.
   Make sure to divide the measurement by the number of blades on the rotor to get the correct rotation speed.

3. Get an inexpensive GPS receiver and connect it to your controller via USB. Continuously read out the NMEA tokens and plot the controller's current position as an overlay over OpenStreetMap:
   https://www.openstreetmap.org/
   https://wiki.openstreetmap.org/wiki/Overlay_API

4. Interface your controller to an IR distance sensor. Continuously perform a distance measurement and display the distance as a numeric value on the LCD. You can also generate a graph by continuously plotting distance values on the LCD.

5. Connect your controller to an IMU and plot movements for all three axes on the LCD. Mount controller and IMU together on a timber board that is hinged to a base plate, so it can only rotate about one axis. Measure the sensor output while repeatedly moving the sensor–controller pair along the hinge joint. Assess the quality of the sensor output.

6. Connect a camera to the Raspberry Pi and then write a program to find the brightest spot in a grayscale image and continuously mark it with a dot on the LCD.

ACTUATORS

6

There are many different ways that robotic actuators can be built. Most prominently, these are electric motors or pneumatic actuators with electric valves. In this chapter, we will concentrate on electric actuators using direct current (DC) instead of alternating current (AC) power.

6.1 DC Motors

Electric motors can be:

- AC or DC motors
- Brushed or brushless
- Single-phase or three-phase
- Special motors, such as stepper motors or servos.

Single-phase DC electric motors are arguably the most commonly used method for locomotion in mobile robots. DC motors are clean, quiet and can produce sufficient power for a variety of tasks. They are much easier to control than pneumatic actuators, which are frequently used when high torques are required and an umbilical cord for an external compressor pump is possible—so usually not an option for mobile robots, unless they carry their own compressor and reservoir.

- Standard DC motors revolve continuously and therefore require a feedback mechanism using shaft encoders (see Figure 6.1).
- Stepper motors also move continuously, but only advance a fixed step angle (e.g., 2°) forward or backward at a time.
- Servos do not have a continuous movement, instead they can only rotate back and forth about a limited angular range (e.g., ±90°). In addition to power, they have a PWM signal input that tells them the desired angular position.

© The Author(s), under exclusive license to Springer Nature Singapore Pte Ltd. 2022
T. Bräunl, *Embedded Robotics*,
https://doi.org/10.1007/978-981-16-0804-9_6

Figure 6.1: Motor–encoder combination

The first step when building robot hardware is to select the appropriate motor system. The best choice is an encapsulated motor combination comprising:

- DC motor
- Gearbox
- Optical or magnetic encoder
  (dual phase-shifted encoders for detection of speed and direction).

Using encapsulated motor systems has the advantage that the solution is much smaller than that using separate modules, plus the system is dust-proof and shielded against stray light, which is required for optical encoders. The disadvantage of using a fixed assembly like this is that the gear ratio may only be changed with difficulty, or not at all.

A magnetic encoder comprises a disk equipped with a number of magnets and a Hall-effect sensor. An optical encoder has a disk with black and white sectors (reflective encoder) or a disk with slots (transmissive decoder) in combination with an LED and a light sensor. If two sensors are positioned with a phase shift, it is possible to detect which one is triggered first. This information can be used to determine whether the motor shaft is being turned clockwise or counterclockwise. The output of either encoder type is a rectangle signal with a frequency proportional to the motor speed.

A number of companies offer small, powerful precision motors with encapsulated gearboxes and encoders. They all have a variety of motor and gearbox combinations available, so it is important to do some power-requirement calculations first, in order to select the right motor and gearbox for a new robotics project. For example, in the Faulhaber[1] motor series with power range from 2 to 4 W, there are gear ratios available from around 3-to-1 to 1 million-to-1.

6.1.1 Motor Model

Figure 6.2 illustrates an effective linear model for the DC motor, and Table 6.1 contains a list of all relevant variables and constant values. A voltage V_a is applied to the terminals of the motor, which generates a current i in the motor armature. The torque τ_m produced by the motor is proportional to the current, and K_m is the motor's torque constant:

---

[1]Faulhaber DC minimotors, series 2230, online: https://www.faulhaber.com/de/produkte/serie/2230s/.

$$\tau_m = K_m \cdot i$$

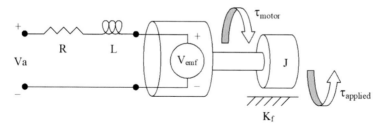

Figure 6.2: Motor model

θ	Angular position of shaft [rad]	R	Nominal terminal resistance [Ω]
ω	Angular shaft velocity [rad/s]	L	Rotor inductance [H]
α	Angular shaft accel. [rad/s^2]	J	Rotor inertia [kg·m^2]
i	Current through armature [A]	K_f	Frictional const. [N·m·s/rad]
V_a	Applied terminal voltage [V]	K_m	Torque constant [N·m/A]
V_e	Back *emf* voltage [V]	K_e	Back *emf* constant [V·s/rad]
τ_m	Motor torque [N·m]	K_s	Speed constant [rad/(V·s)]
τ_a	Applied torque (load) [N·m]	K_r	Regulation constant [(V·s)/rad]

Table 6.1: DC motor variables and constant values

It is important to select a motor with the right output power for a desired task. The output power P_o is defined as the rate of work, which for a rotational DC motor equates to the angular velocity of the shaft ω multiplied by the applied torque τ_a (the torque of the load):

$$P_o = \tau_a \cdot \omega$$

The input power P_i, supplied to the motor is equal to the applied voltage multiplied by the current through the motor:

$$P_i = V_a \cdot i$$

The motor also generates heat as an effect of the current flowing through the armature. The power lost to thermal effects P_t is equivalent to:

$$P_t = R \cdot i^2$$

The efficiency η of the motor is a measure of how well electrical energy is converted to mechanical energy. This can be defined as the output power produced by the motor divided by the input power required by the motor:

$$\eta = P_o/P_i = (\tau_a \cdot \omega)/(V_a \cdot i)$$

The efficiency is not constant for all speeds, which needs to be kept in mind if the application requires operation at different speed ranges. The electric system of the motor can be modeled by a resistor–inductor pair in series with a voltage V_{emf}, which corresponds to the back electromotive force (see Figure 6.2). This voltage is produced because the coils of the motor are moving through a magnetic field, which is the same principle that allows an electric generator to function. The voltage produced can be approximated as a linear function of the shaft velocity; K_e is referred to as the *back emf* constant:

$$V_e = K_e \cdot \omega$$

6.1.2 Simplified Motor Model

In the simplified DC motor model, motor inductance and motor friction are negligible and set to zero, and the rotor inertia is denoted by J. The formulas for current and angular acceleration can therefore be approximated by:

$$i = -\omega \cdot K_e/R + V_a/R$$

$$d\omega/dt = (K_m \cdot i - \tau_a)/J$$

Figure 6.3 shows the ideal DC motor performance curves. With increasing torque, the motor velocity is reduced linearly, while the current increases linearly. Maximum output power is achieved at a medium torque level, while the highest efficiency is reached for relatively low torque values. For further reading, see Bolton[2] and El-Sharkawi.[3]

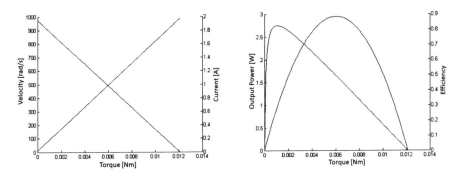

Figure 6.3: Ideal DC motor performance curve

[2]Bolton, W. *Mechatronics – Electronic Control Systems in Mechanical Engineering*, Addison Wesley Longman, Harlow UK, 1995.

[3]El-Sharkawi, M. Fundamentals of Electric Drives, Brooks/Cole Thomson Learning, Pacific Grove CA, 2000.

6.2 H-Bridge

For most applications, we want to be able to do two things with a motor:

1. Run it in forward and backward directions.
2. Modify its speed.

In order to change polarity of the supply voltage on the motor, we need a so-called *H-bridge*. This will enable a motor to spin forward or backward at a fixed speed. In the next section, we will discuss a method called *pulse width modulation* to change the motor speed. Figure 6.4 demonstrates the H-bridge setup, which received its name from its resemblance to the letter *H*. We have a motor with two terminals a and b and a power supply with '+' and '−'. Closing switches 1 and 2 will connect *a* with '+' and *b* with '−': The motor runs forward. In the same way (after opening 1 and 2 again), closing switches 3 and 4 instead will connect *a* with '−' and *b* with '+': The motor runs backward.

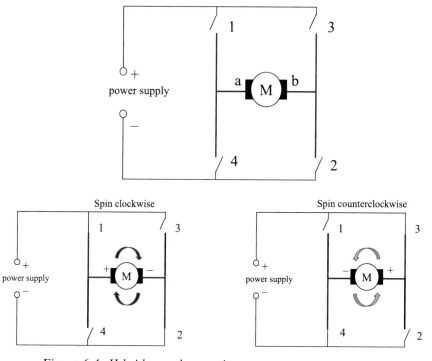

Figure 6.4: H-bridge and operation

The way to implement an H-bridge when using a microcontroller is to use a power amplifier chip in combination with the digital output pins of the controller or an additional latch. This is required because the digital outputs of a microcontroller have very severe output power restrictions. They can only be used to drive other logic chips, but never a motor directly. Since a motor can draw a lot of power (e.g., 1A or more), connecting digital outputs directly to a motor can destroy the microcontroller.

A typical power amplifier chip containing two separate amplifiers is L293D from ST SGS-Thomson. Figure 6.5 demonstrates the schematics. The two inputs x and y are needed to switch the input voltage, so one of them has to be '+', the other has to be '−'. Since they are electrically decoupled from the motor, x and y can be directly linked to digital outputs of the microcontroller. So the direction of the motor can then be specified by software, for example setting output x to logic 1 and output y to logic 0. Since x and y are always the opposite of each other, they can also be substituted by a single output port and a negator. The rotation speed can be specified by the 'speed' input (see the next section on pulse width modulation).

There are two principal ways of stopping the motor:

- set both x and y to logic 0 (or both to logic 1) *or*
- set speed to 0.

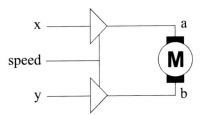

Figure 6.5: Power amplifier as motor driver

6.3 Pulse Width Modulation

Pulse width modulation or PWM for short is a smart method for avoiding analog power circuitry by utilizing the fact that mechanical systems have a relatively high latency. Instead of generating an analog output signal with a voltage proportional to the desired motor speed, it is sufficient to generate digital pulses at the full system voltage level (e.g., 5 V). These pulses are generated at a fixed frequency, for example 20 kHz, so they are beyond the human hearing range.

By varying the pulse width in software (see Figure 6.6, top versus bottom), we also change the equivalent or effective analog motor signal and therefore control the motor speed. One could say that the motor system behaves like an integrator of the digital input impulses over the cycle time. The quotient t_{on}/t_{period} is called the *pulse width ratio* or *duty cycle*.

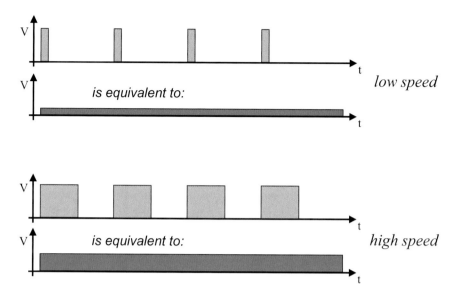

Figure 6.6: PWM

The PWM signal can be generated by software, and many microcontrollers have already special modes and output ports built in to support this operation. The digital output port for the PWM signal is then connected to the *speed* pin of the power amplifier in Figure 6.5.

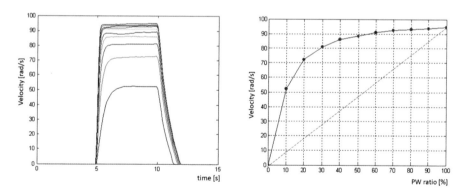

Figure 6.7: Measured motor step response and speed versus pulse width ratio

Figure 6.7, left, shows the motor speed over time for PWM settings of 10%, 20%, ..., 100%. In each case, the PWM signal is issued at time $t = 5$ s. The motor then reaches the desired velocity with only a small delay and stays constant until the signal is shut down at $t = 10$ s. After that, the motor speed drops off somewhat slower due to its internal inertia. This type of behavior is called a *step response*, since the motor input signal jumps in a step function from zero to the desired PWM value.

Unfortunately, the generated motor speed is normally not a linear function of the PWM signal ratio, as can be seen when comparing the measurement of a Faulhaber 2230 motor in Figure 6.7, right, to the dashed diagonal line. Note that these measurements have been done for an idling motor without any load. Depending on load, these curves will change significantly.

In our RoBIOS application programming interface (API), the function *MOTORDrive* can be used to specify PWM ratios for a motor. For example, *MOTORDrive(1, 50)* will apply a 50% PWM signal to motor 1.

Motor calibration can be done by measuring the motor speed at various settings between 0 and 100, and then entering the PW ratio required to achieve the desired actual speed in a motor calibration table of the HDT (hardware description table). The Faulhaber motor's maximum speed is about 1'300 rad/s at a PW ratio of 100%. It reaches 75% of its maximum speed (975 rad/s) at a PW ratio of 20%, so the entry for value 75 in the motor calibration HDT should be 20. Values between the 10 measured points can be interpolated.

Motor calibration is especially important for robots with differential drive, because in these configurations, normally one motor runs forward and one backward, in order to drive the overall robot forward. Many DC motors exhibit some differences in speed versus PW ratio between forward and backward direction. This can be eliminated by using motor calibration.

We are now able to achieve the two goals we set earlier: We can drive a motor forward or backward, and we can change its speed. However, without sensors we have no way of telling at what speed the motor is actually running. Note that the actual motor speed does depend not only on the PWM signal supplied, but also on external factors such as the load applied (e.g., the weight of a vehicle or the steepness of its driving area). What we have achieved so far is called open-loop control. With the help of feedback sensors, we will achieve closed-loop control (often simply called *control*; see next chapter), which is essential to run a motor at a desired speed under varying load.

6.4 Stepper Motors

There are two motor designs which are significantly different from standard DC motors. These are stepper motors discussed in this section and servos, introduced in the following section.

Stepper motors differ from standard DC motors in such a way that they have two independent coils which can be independently controlled. As a result, stepper motors can be moved by pulses to proceed exactly a single step forward or backward, instead of a smooth continuous motion in a standard DC motor. A typical number of steps per revolution is 200, resulting in a step size of 1.8°. Some stepper motors allow half steps, resulting in an even finer step size. There is also a maximum number of steps per second, depending on load, which limits a stepper motor's speed.

Figure 6.8 demonstrates the stepper motor schematics. Two coils are independently controlled by two H-bridges (here marked A, A' and B, B'). Each four-step cycle advances the motor's rotor by a single step if executed in order 1–4. Executing the sequence in reverse order will move the rotor one step back. Note that the switching sequence pattern resembles a gray code, only a single bit changes at each transition. For details on stepper motors and interfacing, see Harman.[4]

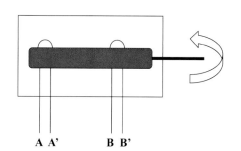

Switching Sequence:

Step	A	B
1	ON	ON
2	ON	OFF
3	OFF	OFF
4	OFF	ON

A A'   B B'

Figure 6.8: Stepper motor schematics

Stepper motors seem to be a simple choice for building mobile robots, considering the effort required for velocity control and position control of standard DC motors. However, stepper motors are rarely used for driving mobile robots, since they lack any feedback on load and actual speed (e.g., a missed step execution). In addition to requiring double the power electronics, stepper motors also have a worse weight/performance ratio than DC motors.

6.5 Servos

DC motors are sometimes also referred to as *servo motors*. This is not what we mean by the term *servo*. A servo motor is a high-quality DC motor that qualifies to be used in a *servoing application*, i.e., in a closed-control loop. Such a servo motor must be able to handle fast changes in position, speed and acceleration and must be rated for high intermittent torque.

---

[4]Harman, T. The Motorola MC68332 Microcontroller - Product Design, Assembly Language Programming, and Interfacing, Prentice Hall, Englewood Cliffs NJ, 1991.

Figure 6.9: Servo

A *servo*, on the contrary, is a DC motor with encapsulated electronics for PWM control and is mainly used for hobbyist purposes, as in model airplanes, cars, or boats (see Figure 6.9).

A servo has three wires: V_{CC}, ground and the PWM input control signal. Unlike PWM for DC motors, the input pulse signal for servos is not transformed into a velocity. Instead, it is an analog timing control input to specify the desired position of the servo's rotating disk head. A servo's disk cannot perform a continuous rotation like a DC motor. It only has a range, depending on the model, of about $\pm 120°$ from its middle position. Internally, a servo combines a DC motor with a simple feedback circuit, often using a potentiometer sensing the servo head's current position.

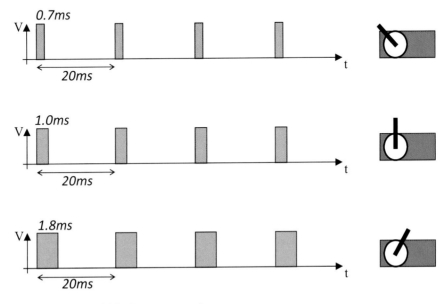

Figure 6.10: Servo control

The PWM signal used for servos always has a frequency of 50 Hz, so pulses are generated every 20 ms. The width of each pulse specifies the desired position of the servo's disk (Figure 6.10). For example, a width of 0.7 ms will rotate the disk to the leftmost position (+120°), and a width of 1.8 ms will rotate the disk to the rightmost position (−120°). Exact values of pulse duration and angle depend on the servo brand and model.

Like stepper motors, servos seem to be a good and simple solution for robotics tasks. However, servos have the same drawback as stepper motors: They do not provide any feedback to the outside. When applying a certain PWM signal to a servo, we do not know when the servo will reach the desired position or whether it will reach it at all, e.g., because of a too high load or because of an obstruction.

6.6   Tasks

1. Write a program to generate a PWM output. Connect the output line to an oscilloscope and verify the timing.

2. Connect an electric motor via a motor controller to your controller. Write a program to continuously ramp the speed up and down via a PWM signal.

    Change the PWM ratio from 0% to 100% and then back to 0%. Determine the output percentage (or voltage) required to get the motor moving when stationary. Determine the lowest and the maximum possible idle speed.

3. Connect a load to the motor by letting it lift a weight via a pulley. Determine the output percentage (or voltage) required to get the motor moving when stationary. Determine the lowest and the maximum possible speed under load.

4. Connect a motor-encoder combination to your controller. This may require an additional I/O board, such as the EyeBot IO7 board. While letting the motor spin at various PWM settings, calculate the motor speed by subtracting subsequent encoder readings and display the result on the LCD as a numeric value as well as a graph.

5. Connect a servo directly to a controller output line, and connect a potentiometer to the controller's analog input (or ADC, if required). Write a procedure to generate a 50 Hz PWM signal with a variable uptime.

    Write a procedure that reads the analog value and translates it into the range [0, 100]. Combine the two procedures into a program where the analog input determines the signal uptime and therefore the servo angle in the range [0.7 ms, 1.8 ms]. Record the required uptimes for the leftmost and rightmost servo positions for your particular servo type.

CONTROL

7

losed-loop control is an essential topic for embedded systems, bringing together actuators and sensors through the software control algorithm. The central point of this chapter is to use motor feedback via encoders for velocity control of a motor. We will exemplify this by a stepwise introduction of proportional, integral, derivative (PID) control. We will then also look at position control of motors and the problem of synchronizing two independent motors, which is required for driving a robot vehicle in a straight line.

Previously, we showed how to drive a motor forward or backward and how to change its speed. However, if there is no sensor feedback available, the actual motor speed cannot be verified. This is important because supplying the same analog voltage (or equivalent: the same PWM signal) to a motor does not guarantee that the motor will run at the same speed under all circumstances. For example, a motor under the same PWM signal will run faster when spinning freely than under load (for example, when driving a vehicle). In order to control the motor speed, we do need feedback from the motor shaft encoders. Feedback control is also called *closed-loop control* (or simply called *control* in the following), as opposed to *open-loop control*, which was discussed in the chapter on actuators.

7.1 On-Off Control

Feedback is everything! As we established before, we require feedback on a motor's current speed in order to control it. Setting a certain PWM level alone will not help, since the motor's speed also depends on its load.

The idea behind feedback control is very simple. We have a desired speed, specified by the user or the application program, and we have the current actual speed, measured by the shaft encoders. Measurements and subsequent actions can be taken very frequently, for example, from 100 times per second to 20,000 times per second. The action taken depends on the control model, several of which are introduced in the following sections, however, in principle it looks like the following:

- **If** actual speed is **lower** than desired speed:
  → *Increase motor power.*
- **If** actual speed is **higher** than desired speed:
  → *Decrease motor power.*

In the simplest case, power to the motor is either switched on (when the actual speed is too low) or switched off (when the actual speed is too high). This control law is represented by the formula below, with:

R(t) motor output function over time t

$v_{act}(t)$ actual measured motor speed at time t

$v_{des}(t)$ desired motor speed at time t

$$R(t) = \begin{cases} 1 & \text{if } v_{act}(t) < v_{des}(t) \\ 0 & \text{else} \end{cases}$$

What has been defined here is the concept of an *on-off controller*, also known as *piecewise constant controller* or *bang-bang controller*. The motor output is set to one if the measured velocity is too low, otherwise, it is set to zero. Note that this controller only works for a positive value of v_{des}. The schematics diagram is shown in Figure 7.1 and the motor function is shown in Figure 7.2, left.

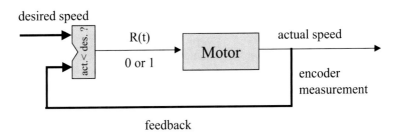

desired speed

R(t)

0 or 1

Motor

actual speed

encoder measurement

feedback

Figure 7.1: On-off controller

The behavior over time of an on-off controller is shown in Figure 7.2, right. Assuming the motor is at rest in the beginning and the desired speed is non-zero, then the control signal for the motor will be 1. This will stay until at some stage the actual speed becomes larger than the desired speed. Then, the motor control signal is changed to 0. Over some time, the actual speed will slow down and once it falls below the desired speed, the control signal will again be set to 1. This algorithm continues indefinitely and can accommodate any changes in the desired speed.

Note that the motor control signal is not instantaneously updated. Instead, it takes some time to read and process the sensor signal, so output changes occur at fixed time intervals (e.g., every 10 ms in Figure 7.2). This delay creates a small overshooting or undershooting of the actual motor speed and thereby introduces some hysteresis.

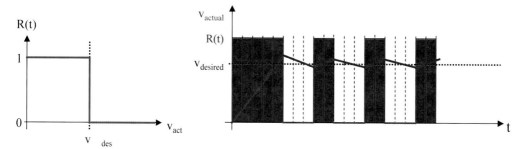

Figure 7.2: On-off control signal: control signal (left); motor output over time (right)

The on-off controller is the simplest possible method of control. Many technical systems use it, not limited to controlling a motor. Examples are a refrigerator, heater, thermostat, light dimmer, etc. Some of these technical systems use a hysteresis band, which consists of two desired values, one for switching on and one for switching off. This prevents a high switching frequency near the desired value, in order to avoid excessive wear. The formula for an on-off controller with hysteresis is:

$$R(t + \Delta t) = \begin{cases} 1 & \text{if } v_{act}(t) < v_{on}(t) \\ 0 & \text{if } v_{act}(t) > v_{off}(t) \\ R(t) & \text{else} \end{cases}$$

Note that this definition is not a function in a mathematical sense, because the new motor output for an actual speed between the two band values is equal to the previous motor value. This means it can be equal to 1 or 0, depending on its history. Figure 7.3 shows the hysteresis curve and the corresponding control signal.

All technical systems have some delay and therefore exhibit some inherent hysteresis, even if it is not explicitly built in.

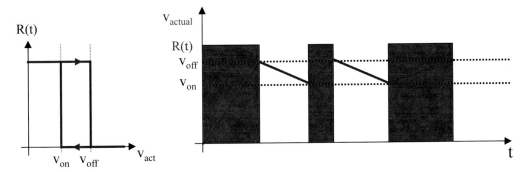

Figure 7.3: On-off control signal with hysteresis band (left); motor output over time (right)

Once we understand the theory, we would like to put this knowledge into practice and implement an on-off controller in software. We will proceed step by step:

1. We need a subroutine for calculating the motor signal as defined by the formula.
   This subroutine has to:

   a. Read encoder data (*input*)
   b. Compute new output value R(t)
   c. Set motor speed (*output*)

2. The subroutine has to be called periodically (for example, every 1/100 s).
   Since motor control is a *low-level* task, we would like this to run in the background and not interfere with any user program.

Let us take a look at how to implement step 1. For implementing on an Arduino/Atmel controller, we can read the current encoder value as a one-byte value from input port B (PINB) and we can write the motor output function to the lowest bit of port D using PORTD.

The program code for this subroutine then looks like Program 7.1. Variable r_mot denotes the control parameter R(t). The dotted line has to be replaced by the control function for R(t).

Program 7.1: Control Subroutine Framework on Arduino

```
void controller()
{ int enc_new, r_mot;
  enc_new = PINB;   // read encoder (1 byte)

  r_mot = ...

  PORTD = r_mot;   // set motor pin (1 bit)
}
```

Program 7.2 shows the completed control program, assuming this routine is called every 10 ms (100 Hz).

Program 7.2: On-Off Controller on Arduino

```
int v_des;    // user input in ticks/s

void onoff_controller()
{ int enc_new, v_act, r_mot;
  static int enc_old;

  enc_new = PINB;
  v_act = (enc_new-enc_old) * 100; // called every 1/100s
  if (v_act < v_des) r_mot = 1;
             else  r_mot = 0;
  PORTD = r_mot;
  enc_old = enc_new;
}
```

The three lines of the motor output part can be shortened to a single line:
PORTD = (v_act < c_des);

For the Raspberry Pi, we use the *WiringPi*[1] library to simplify digital I/O operations. We chose digital I/O lines 26 + 27 for positive and negative pins the of motor amplifier, denoted as A and B in the program. As there is no parallel data input on the Pi, we use the RoBIOS routine *ENCODERRead* instead. Program 7.3 shows the details.

Program 7.3: On-Off Controller on Raspberry Pi

```
#define A 26   // positive pin of motor controller
#define B 27   // negative pin of motor controller
int v_des;    // user input in ticks/s
...
wiringPiSetup();
pinMode (A, OUTPUT);  // both are output pins
pinMode (B, OUTPUT);
digitalWrite(B,0);   // only drive forward in this example
...

void onoff_controller()
{ int enc_new, v_act;
  static int enc_old;

  enc_new = ENCODERRead(1);  // Read Encoder no. 1
  v_act = (enc_new-enc_old) * 100; // called every 1/100s
```

---

[1]Wiring Pi, *GPIO Interface library for the Raspberry Pi*, online: http://wiringpi.com.

```
    if (v_act < v_des) digitalWrite(A,1); //ON
                else  digitalWrite(A,0); //OFF
    enc_old = enc_new;
}
```

Again, we can shorten the motor output function to a single line:
`digitalWrite(A, (v_act<c_des));`

The RoBIOS operating system contains more complex functions for reading encoder input and setting motor output via an additional I/O board (see below). Function *MOTORDrive* already has a control loop implemented, so the motor output part of Program 7.3 is a possible implementation for it. Program 7.4 shows the equivalent control framework from Program 7.1, however, using the RoBIOS motor and encoder functions for Raspberry Pi.

```
int MOTORDrive(int motor, int speed);  // Set motor [1..4] speed in percent
int ENCODERRead(int quad);             // Read quadrature encoder [1..4]
```

Program 7.4: Control Subroutine Framework on Raspberry Pi with RoBIOS Functions

```
void controller()
{ int enc_new, r_mot;
  static int enc_old;

  enc_new = ENCODERRead(1);       // read encoder no.1 (4 bytes)
  v_act   = (enc_new-enc_old)*100; // speed calculation 1/100s
  enc_old = enc_new;               // copy for next call

  r_mot = ...                      // insert controller formula

  MOTORDrive(1, r_mot);            // drive motor no.1 (8 bits)
}
```

So far, we have not considered any potential problems with counter overflow or underflow. However, as the following examples will show, it turns out that *roll-over* and *roll-under* transitions across zero still result in the correct difference values when using signed integers.

Roll-over example from negative to positive values:
Assume the following are two consecutive encoder values:

$$FF\ FF\ FF\ FD_{16} \quad = -3_{10}$$
$$00\ 00\ 00\ 04_{16} \quad = +4_{10}$$

The difference of second value minus first value is: $4 - (-3) = +7$, which means driving forward and is the correct number of encoder ticks.

Roll-under example from positive to negative values:

$$00\ 00\ 00\ 12_{16} \qquad = +18_{10}$$
$$FF\ FF\ FF\ FA_{16} \qquad = -6_{10}$$

The difference of second value minus first value results in -24, which means driving backward and is the correct number of ticks elapsed.
What must be avoided, however, is an overflow/underflow error, i.e., the transition from the highest positive number to the lowest negative number or vice versa:

$$7F\ FF\ FF\ FF_{16} \qquad = +2147483647_{10}$$
$$+1 = 80\ 00\ 00\ 00_{16} \qquad = -2147483648_{10} \quad \rightarrow \text{overflow error}$$

Let us now look at how to implement step 2 of the on-off controller. Program 7.5 shows a straightforward implementation using the RoBIOS routines below.

```
int    OSWait(int n);                              // Wait for n/1000 sec
TIMER OSAttachTimer(int scale, void (*fct)(void)); // Add fct to timer
int    OSDetachTimer(TIMER t);                     // Remove fct from timer
```

In the otherwise idle *while*-loop, any top-level user programs can be executed. The loop continues until the specified end-button has been pressed.

Program 7.5: Timer Start

```
int main()
{ TimerHandle t1;

  t1 = OSAttachTimer(10, onoff_controller);
  while (KEYRead() != KEY4)    // until button press
  { /* other tasks or idle */ }
  OSDetachTimer(t1);           // free timer
  return 0;
}
```

Figure 7.4 shows a typical measurement of the step response of an on-off controller. The saw-tooth shape of the velocity curve is clearly visible.

7.2   PID Control

The simplest method of control is not always the best. A more advanced controller and an industry standard is the PID controller. It comprises a proportional, an integral and a derivative control part (PID = P + I + D). The controller parts are introduced in the following sections individually and in combined operation.

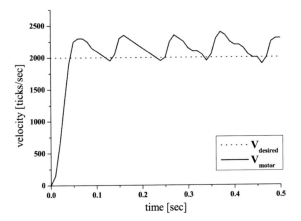

Figure 7.4: Measured step response of an on-off controller

7.2.1 Proportional Controller

For many control applications, the abrupt motor control change between fully on and off does not result in a smooth control behavior. We can improve this by using a linear or proportional term instead. The formula for the proportional controller (P controller) is:

$$R(t) = K_P \cdot (v_{des}(t) - v_{act}(t))$$

The difference between the desired and actual speed is called the *error function*. Figure 7.5 shows the schematics for the P controller, which differs only slightly from the on-off controller. Figure 7.6 shows measurements of characteristic motor speeds over time. Varying the *controller gain* K_P will change the controller behavior. The higher a value for K_P is chosen, the faster the controller responds; however, a too high value will lead to an undesirable oscillating system. Therefore, it is important to choose a value for K_P that guarantees a fast response, but does not lead the control system to overshoot too much or even oscillate.

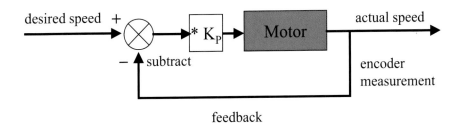

Figure 7.5: Proportional controller

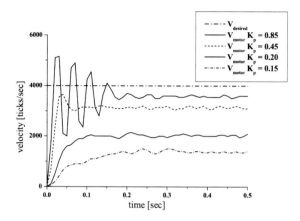

Figure 7.6: Step response for proportional controller

Note that the P controller's equilibrium state is not at the desired velocity. If the desired speed is reached exactly, the motor output is reduced to zero, as defined in the P controller formula shown above. Therefore, each P controller will keep a certain *steady-state error* from the desired velocity, depending on the controller gain K_P. As can be seen in Figure 7.6, the higher the gain K_P, the lower the steady-state error. However, the system starts to oscillate if the selected gain is too high.

Program 7.6 shows the brief P controller code that can be inserted into the control frame of Program 7.4, in order to form a complete program.

Program 7.6: P Controller Code

```
e_func = v_des - v_act;  /* error function */
r_mot  = Kp * e_func;    /* motor output */
```

7.2.2 Integral Controller

Unlike the P controller, the I controller (integral controller) is rarely used alone, but mostly in combination with the P or PD controller. The idea for the I controller is to reduce the steady-state error of the P controller. With an additional integral term, this steady-state error can be reduced to zero, as seen in the measurements in Figure 7.7. The equilibrium state is reached somewhat later than with a pure P controller, but the steady-state error has been eliminated.

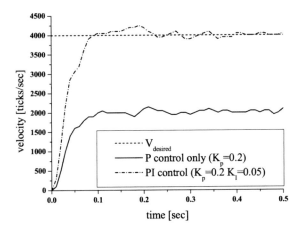

Figure 7.7: Step response for integral controller

When using e(t) as the error function, the formula for the PI controller is:

$$R(t) = K_P \cdot [e(t) + 1/T_I \cdot \int_0^t e(t)dt]$$

We rewrite this formula by substituting $Q_I = K_P/T_I$, so we receive independent additive terms for P and I:

$$R(t) = K_P \cdot e(t) + Q_I \cdot \int_0^t e(t)dt$$

The naïve way of implementing the I controller part is to transform the integration into a sum of a fixed number (for example, 10) of previous error values. These 10 values would then have to be stored in an array and added for every iteration.

The proper way of implementing a PI controller starts with discretization, replacing the integral with a sum, using the trapezoidal rule:

$$R_n = K_P \cdot e_n + Q_I \cdot t_{delta} \cdot \sum_{i=1}^{n} \frac{e_i + e_{i-1}}{2}$$

Now we can get rid of the sum by using the preceding output value R_{n-1}:

$$R_n - R_{n-1} = K_P \cdot (e_n - e_{n-1}) + Q_I \cdot t_{delta} \cdot (e_n + e_{n-1})/2$$

Therefore (substituting K_I for $Q_I \cdot t_{delta}$):

$$R_n = R_{n-1} + K_P \cdot (e_n - e_{n-1}) + K_I \cdot (e_n + e_{n-1})/2$$

Program 7.7: PI Controller Code

```
static int r_old=0, e_old=0;
...
e_func = v_des - v_act;
r_mot  = r_old + Kp*(e_func-e_old) + Ki*(e_func+e_old)/2;
r_mot = min(r_mot, +100);  // limit motor output
r_mot = max(r_mot, -100);  // limit motor output
r_old = r_mot;
e_old = e_func;
```

So, we only need to store the previous control value and the previous error value to calculate the PI output in a much simpler formula. Here it is important to limit the controller output to the correct value range (for example, $-100 \ldots +100$ in RoBIOS) and also store the limited value for the subsequent iteration. Otherwise, if a desired speed value cannot be reached, both controller output and error values can become arbitrarily large and invalidate the whole control process, see Kasper.[2]

Program 7.7 shows the program fragment for the PI controller, to be inserted into the framework of Program 7.4.

7.3 Derivative Controller

Similar to the I controller, the D controller (derivative controller) is rarely used by itself, but mostly in combination with the P or PI controller. The idea for adding a derivative term is to speed up the P controller's response to a change of input. Figure 7.8 shows the measured differences of a step response between the P and PD controller (top), and the PD and PID controller (bottom). The PD controller reaches equilibrium faster than the P controller, but still has a steady-state error. The full PID controller combines the advantages of PI and PD. It has a fast response and suffers no steady-state error.

---

[2]M. Kasper, *Rug Warrior Lab Notes*, Internal report, Univ. Kaiserslautern, Fachbereich Informatik, 2001.

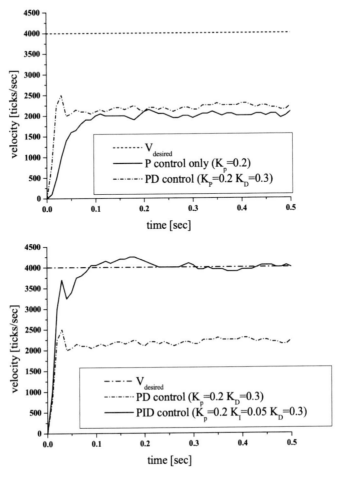

Figure 7.8: Step response for derivative controller and full PID controller

When using e(t) as the error function, the formula for a combined PD controller is:

$$R(t) = K_P \cdot [e(t) + T_D \cdot de(t)/dt]$$

The formula for the full PID controller is:

$$R(t) = K_P \cdot [e(t) + 1/T_I \cdot \int_0^t e(t)dt + T_D \cdot de(t)/dt]$$

Again, we rewrite this by substituting T_D and T_I, so we receive independent additive terms for P, I and D. This is important, in order to experimentally adjust the relative gains of the three terms.

$$R(t) = K_P \cdot e(t) + Q_I \cdot \int_0^t e(t)dt + Q_D \cdot de(t)/dt$$

Using the same discretization as for the PI controller, we will get:

$$R_n = K_P \cdot e_n + Q_I \cdot t_{delta} \cdot \sum_{i=1}^{n} \frac{e_i + e_{i-1}}{2} + Q_D/t_{delta} \cdot (e_n - e_{n-1})$$

Again, using the difference between subsequent controller outputs, this results in:

$$R_n - R_{n-1} = K_P \cdot (e_n - e_{n-1}) + Q_I \cdot t_{delta} \cdot (e_n + e_{n-1})/2 + Q_D/t_{delta} \cdot (e_n - 2 \cdot e_{n-1} + e_{n-2})$$

Finally (substituting K_I for $Q_I \cdot t_{delta}$ and K_D for Q_D/t_{delta}):

$$R_n = R_{n-1} + K_P \cdot (e_n - e_{n-1}) + K_I \cdot (e_n + e_{n-1})/2 + K_D \cdot (e_n - 2 \cdot e_{n-1} + e_{n-2})$$

Program 7.8: PD Controller Code

```
static int e_old=0;
...
e_func = v_des - v_act;            /* error function */
deriv = e_old - e_func;            /* diff. of error fct. */
e_old = e_func;                    /* store error function */
r_mot = Kp*e_func + Kd*deriv;      /* motor output */
r_mot = min(r_mot, +100);          /* limit output */
r_mot = max(r_mot, -100);          /* limit output */
```

Program 7.9: PID Controller Code

```
static int r_old=0, e_old=0, e_old2=0;
...
e_func = v_des - v_act;
r_mot = r_old + Kp*(e_func-e_old) + Ki*(e_func+e_old)/2
               + Kd*(e_func - 2* e_old + e_old2);
```

```
r_mot = min(r_mot, +100);  /* limit output */
r_mot = max(r_mot, -100);  /* limit output */
r_old = r_mot;
e_old2 = e_old;
e_old  = e_func;
```

Program 7.8 shows the program fragment for the PD controller, while Program 7.9 shows the full PID controller. Both are to be inserted into the framework of Program 7.4.

7.3.1 PID Parameter Tuning

The tuning of the three PID parameters K_P, K_I, and K_D is an important issue. The following guidelines can be used for experimentally finding suitable values (adapted after Williams[3]):

1. Select a typical operating setting for the desired speed, turn off integral and derivative parts, then increase K_P to the maximum or until oscillation occurs.
2. If system oscillates, divide K_P by 2.
3. Increase K_D and observe the behavior when increasing/decreasing the desired speed by about 5%. Choose a value of K_D which gives a damped response.
4. Slowly increase K_I until oscillation starts. Then divide K_I by 2 or 3.
5. Check whether overall controller performance is satisfactorily under typical system conditions.

Further details on digital control can be found in articles by Aström, Hägglund[4] and Bolton.[5]

7.4 Velocity Control and Position Control

So far, we are able to control a single motor at a certain speed, but we are not yet able to drive a motor at a given speed for a number of revolutions and then come to a stop at exactly the right motor position. The former, maintaining a certain speed, is called velocity control, while the latter, reaching a specified position, is called position control.

---

[3]C. Williams, *Tuning a PID Temperature Controller*, web: http://newton.ex.ac.uk/teaching/CDHW/Feedback/Setup-PID.html, 2006.

[4]K. Aström, T. Hägglund, *PID Controllers: Theory, Design, and Tuning*, 2nd Ed., Instrument Society of America, Research Triangle Park NC, 1995.

[5]W. Bolton, *Mechatronics – Electronic Control Systems in Mechanical Engineering*, Addison Wesley Longman, Harlow UK, 1995.

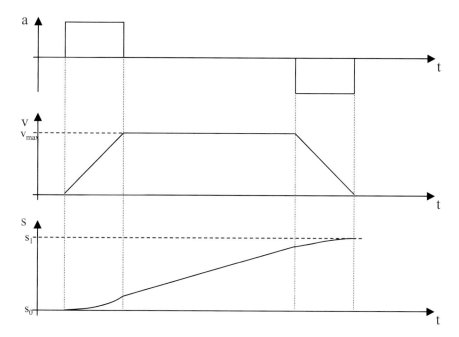

Figure 7.9: Position control

Position control requires an additional controller on top of the previously discussed velocity controller. The position controller sets the desired velocities in all driving phases, especially during the acceleration and deceleration phases (starting and stopping).

Let us assume a single motor is driving a robot vehicle that is initially at rest and which we would like to stop at a specified position. Figure 7.9 demonstrates the *speed ramp* for the starting phase, constant speed phase and stopping phase. When ignoring friction, we only need to apply a certain force (here constant) during the starting phase, which will translate into an acceleration of the vehicle. The constant acceleration will linearly increase the vehicle's speed v (integral of acceleration a) from 0 to the desired value v_{max}, while the vehicle's position s (integral of v) will increase quadratically.

When the force (acceleration) stops, the vehicle's velocity will remain constant, assuming there is no friction, and its position will increase linearly.

During the stopping phase (deceleration, braking), a negative force (negative acceleration) is applied to the vehicle. Its speed will be linearly reduced to zero (or may even become negative if applied too long—the vehicle would then be driving backward). The vehicle's position will increase slowly, following the square root function.

The tricky bit now is to control the amount of acceleration in such a way that the vehicle:

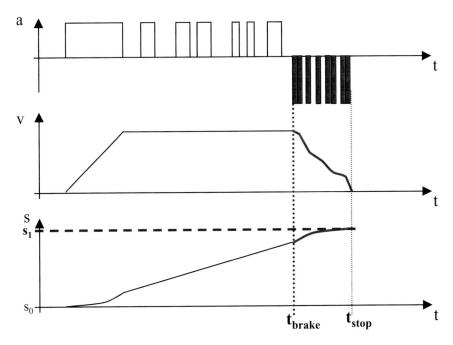

Figure 7.10: Braking adaptation

    a.   Comes to rest
        (not moving slowly forward to backward)
    b.   Comes to rest at exactly the specified position
        (for example, we want the vehicle to drive exactly 1 m and stop within
        ±1 cm).

Figure 7.10 shows a way of achieving this by controlling (continuously updating) the braking acceleration applied. This control procedure has to take into account not only the current speed as a feedback value, but also the current position, since previous speed changes or inaccuracies may have already had an effect on the vehicle's position.

In a real application scenario, there are of course roll resistance and air resistance, so the control algorithm also has to apply acceleration forces during the constant speed phase to maintain the desired speed (also shown in Figure 7.10). Braking forces using friction brakes can only reduce a vehicle's speed to zero, not below.

As each vehicle has a certain inertia, the control forces have to be calibrated to be able to move the vehicle accurately at low speeds before it stalls. If our vehicle comes to a rest, e.g., 1 cm too short or too far from the desired position, then it may not be possible to shift it by only 1 cm. You can sometimes see this problem when an elevator cabin does not line up properly with the floor (see Figure 7.11).

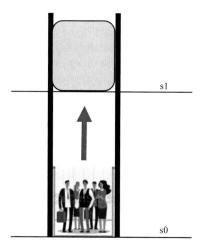

Figure 7.11: Lift positioning example

7.5   Multiple Motors—Driving Straight

There is another motor control problem we need to discuss. So far, we have only looked at a single isolated motor with velocity control and briefly at position control. The way that a robot vehicle is constructed, however, shows us that a single motor is not enough (see Figure 7.12).

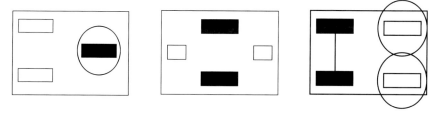

Figure 7.12: Wheeled robots

All these robot constructions require two motors, with the functions of driving and steering either separated or dependent on each other. In the design on the left and the design on the right, the driving and steering functions are separated. It is therefore quite easy to drive in a straight line (simply keep the steering fixed at the angle representing *straight*) or drive in a circle (keep the steering fixed at the appropriate angle). The situation is quite different in the *differential steering* design shown in the middle of Figure 7.12, which is a very popular design for small mobile robots. Here, one has to constantly monitor and update both motor speeds in order to drive straight. Driving in a circle can be achieved by adding a constant offset to one of the motors. Therefore, a synchronization of the two motor speeds is required.

There are a number of different approaches for driving straight. The idea presented in the following is from Jones, Flynn, Seiger.[6] Figure 7.13 shows the first try for a differential drive vehicle to drive in a straight line. There are two separate control loops for the left and the right motor, each involving feedback control via a P controller. The desired forward speed is supplied to both controllers concurrently. Unfortunately, this design will not produce a nice straight line of driving. Although each motor is controlled in itself, there is no control link for any slight speed discrepancies between the two motors. Such a setup will most likely result in the robot driving in a wiggly line, but not straight.

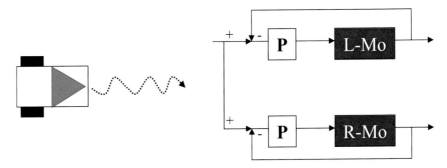

Figure 7.13: Driving straight—first try

An improvement of this control structure is shown in Figure 7.14 as a second try. We now also calculate the difference in motor movements (position, not speed) and feed this back to both P controllers via an additional I controller. The I controller integrates (or rather sums up) the differences in position, which will subsequently be eliminated by the two P controllers. Note the different signs for the input of this additional value, which is the opposite sign of the corresponding I controller input. Also, this principle relies on the fact that the whole control circuit is very fast compared to the actual motor or vehicle speed. Otherwise, the vehicle might end up in a trajectory parallel to the desired straight line.

---

[6]Jones, J., Flynn, A., Seiger, B. *Mobile Robots - From Inspiration to Implementation*, 2nd Ed., AK Peters, Wellesley MA, 1999.

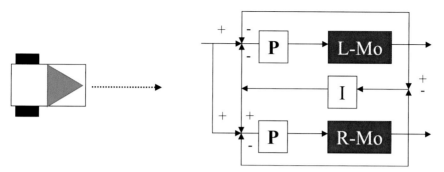

Figure 7.14: Driving straight—second try

For the final version in Figure 7.15 (adapted from Jones, Flynn, Seiger 1999), we added another user input with the curve offset. With a curve offset of zero, the system behaves exactly like the previous one for driving straight. A positive curve offset, however, will let the robot drive in a counter-clockwise circle (a negative offset will let it drive in a clockwise circle). The radius can be calculated from the curve offset value.

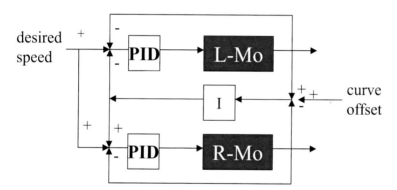

Figure 7.15: Driving straight or in curves

The controller used in the RoBIOS vω-library is a bit more complex than the one shown in Figure 7.15. It uses a PI controller for handling the rotational velocity ω in addition to the two PI controllers for left and right motor velocity. More elaborate control models for robots with differential drive wheel arrangements can be found in papers by Kim, Tsiotras[7] (conventional controller models) and by Seraji, Howard[8] (fuzzy controllers).

---

[7]Kim, B., Tsiotras, P. *Controllers for Unicycle-Type Wheeled Robots: Theoretical Results and Experimental Validation*, IEEE Transactions on Robotics and Automation, vol. 18, no. 3, June 2002, pp. 294-307 (14).

[8]Seraji, H., Howard, A. *Behavior-Based Robot Navigation on Challenging Terrain: A Fuzzy Logic Approach*, IEEE Transactions on Robotics and Automation, vol. 18, no. 3, June 2002, pp. 308-321 (14).

7.6 V-Omega Interface

When programming a robot vehicle, we want to concentrate on the higher levels and therefore need an abstraction level for low-level motor control as shown in this chapter. The preferred choice is a user-friendly driving function interface that automatically takes care of all the feedback control issues discussed.

We have implemented such a driving interface in the RoBIOS operating system, called the *v-omega interface*, because it allows us to specify a linear and rotational speed of a robot vehicle. There are low-level functions for direct control of the motor speed, as well as high-level functions for driving a complete trajectory in a straight line or in a curve (Table 7.1).

```
int VWSetSpeed(int linSpeed, int angSpeed);    // Continuous drive
int VWGetSpeed(int *linSspeed, int *angSpeed); // Read curr. speed
int VWSetPosition(int x, int y, int phi);      // Set curr. pose
int VWGetPosition(int *x, int *y, int *phi);   // Read curr. pose
int VWStraight(int dist, int lin_speed);    // Drive fixed distance
int VWTurn(int angle, int ang_speed);          // Turn fixed angle
int VWCurve(int dist, int angle, int lin_speed); // Drive curve
int VWDrive(int dx, int dy, int lin_speed);      // Drive coord.
int VWRemain(void);     // Check remaining distance in mm
int VWDone(void);       // Check if drive operation has finished
int VWWait(void);       // Wait until drive operation finished
int VWStalled(void);    // Check if wheels have stalled
```

Table 7.1: vω interface functions

Program 7.10 shows a simple application program using the vω driving interface for completing a square in C (Program 7.11 shows the identical program in Python). In a loop for the four sides, the robot executes in each iteration a *VWStraight* command (400 mm distance at 300 mm/s speed) followed by a *VWTurn* command (90° at angular speed of 90°/s). As all driving commands start the desired action and then immediately return control to the calling program, we have to add a call to *VWWait* for both of them. Otherwise, the first *VWStraight* would immediately be overwritten by the subsequent *VWTurn* before the robot could have even started to drive—and the same for all following driving commands.

Program 7.10: Driving a Square in C

```
#include "eyebot.h"

int main()
{ for (int i=0; i<4; i++)      // run 4 sides
  { VWStraight(400, 300); VWWait(); // straight - then wait
    VWTurn(90, 90);        VWWait(); // turn - then wait
  }
}
```

Program 7.11: Driving a Square in Python

```
#!/usr/bin/env python3
from eye import *

for x in range (0,4): # start at 0, stop at 4-1 = 3 (run 4 times)
    VWStraight(300,500)
    VWWait()
    VWTurn(90, 100)
    VWWait()
```

Figure 7.16 shows the results of either program in the EyeSim simulator.

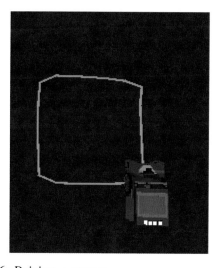

Figure 7.16: Driving a square

VWSetSpeed controls the robot at a more basic level, setting fixed wheel speeds for left and right, which need to be altered or stopped by a subsequent command depending on a condition, such as reaching a desired destination or encountering an obstacle. Program 7.12 implements driving straight until an obstacle is encountered and Figure 7.17 shows the result, visualizing the robot's track. For checking the distance in front of the robot, we call the *PSDGet* function for sensor *PSD_FRONT*, which gives us the free distance in mm. If there is sufficient free space (more than 200 mm), then we wait 100 ms in every iteration. Once the loop has terminated, which means there is an obstacle at 200 mm or nearer, we stop the robot by calling *VWSetSpeed* with parameters (0,0).

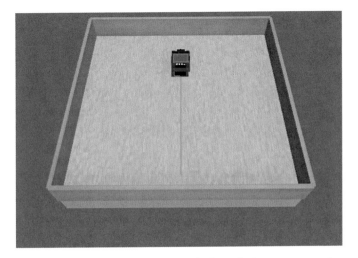

Figure 7.17: Driving straight until obstacle is encountered

Program 7.13 once again shows the same program in Python. Note the importance (*and susceptibility to error*) of correct indentation, especially for the final *VWSetSpeed* command.

Program 7.12: Driving Straight Until Obstacle in C

```c
#include "eyebot.h"

int main()
{VWSetSpeed(100,0);                                    /* drive      */
  while (PSDGet(PSD_FRONT) > 200) OSWait(100); /* wait 100ms */
  VWSetSpeed(0,0);                                     /* stop       */
}
```

Program 7.13: Driving Straight Until Obstacle in Python

```python
#!/usr/bin/env python3
from eye import *

VWSetSpeed(100,0)
while PSDGet(PSD_FRONT) > 200:
   OSWait(100)
VWSetSpeed(0, 0)
```

Function *VWGetPosition* can be used to read a robot's pose (position and orientation), which the robot automatically and continually updates by evaluating its two shaft encoders. This process is called *dead reckoning* and gives somewhat accurate position and orientation values, but only over a short driving distance, after which the overall error accumulates. When starting, a robot's position will always be $(x,y) = (0,0)$ and orientation $a = 0°$, irrespective of the world coordinates and orientation where it starts. Function *VWSetPosition* can be used to initialize the robot's pose to any desired value. (Note that this will not actually drive the robot to a desired location).

Function *VWStalled* compares desired and actual motor speeds in order to check whether the robot's wheels have stalled, possibly because it has collided with an obstacle. Calling this function in the main control loop can be used to avoid an infinite loop, in case the robot never reaches its destination because of motors stalled at an obstacle that has not been detected otherwise.

7.7 Tasks

1. Connect a motor-encoder combination to your controller.
   This will require some motor driving hardware plus some encoder reading hardware.

2. Write a program for an on-off (bang-bang) controller.
   Plot the controller's idle speed response from zero to a desired speed. Repeat the same experiment under load (e.g. lifting a weight).

3. Write a program for a PID controller and its P, I, D components.
   Plot the controller's idle speed response from zero to a desired speed. Repeat the same experiment under load (e.g. lifting a weight).

4. Tune the coefficients for the P, I and D terms as lined out in this chapter.

5. On a robot vehicle with two motors in differential drive, implement two independent PID control loops, one for each wheel. Then set both motors to the same desired speed and measure the path the robot is driving. Is it driving in a straight line?

6. Link the two PID control loops with an I-controller as described in this chapter.
   Repeat the driving straight experiment. Did the outcome improve? Compare the driving performance of your program with the built-in RoBIOS *vw* routines.

7. Implement a program for position control. The application can either be a load that is to be lifted to a certain height (e.g. elevator) or a robot vehicle that should drive at a constant speed, but stop at a specified distance.
   Implement speed ramps for accelerating and decelerating as described in this and the previous chapter. Note the maximum as well as the minimum possible speeds and acceleration values of your scenario and make sure the control algorithm stays clear of all these values.
   What is the best position accuracy you can achieve for your motor/load/sensor setting?

MULTITASKING

8

Concurrency is an essential part of every robot program. A number of more or less independent threads have to be taken care of, which requires some form of multitasking, even if only a single processor is available on the robot's controller.

Imagine a robot program that should do some image processing and at the same time monitor the robot's infrared sensors in order to avoid an obstacle collision. Without the ability for multitasking, this program would comprise one large loop for processing one image, then reading infrared data. But if processing an image takes much longer than the time interval required for updating the infrared sensor signals, we have a problem. The solution is to use separate threads for each activity and let the operating system switch between them.

The Raspberry Pi runs a version of Linux as its operating system, which supports multitasking, but in its standard version does not allow real-time processing. While multitasking is essential for robotics, real-time processing—for most cases—is not. If we are doing sensor processing for a driving robot and there are some slight timing variations, we might miss a sensor value, but as long as we are doing this fast enough (i.e., 10–30 frames per second; fps), this will not be problem.

When generating control signals, such as for a servo, however, real-time ability will be an advantage. This is why in our configuration we handle all these tasks through a low-level Atmel board or Arduino processor, which can generate these signals in hardware in real-time.

While the Raspberry Pi can run several processes concurrently (multitasking), we are more interested in running multiple threads of control within a single robot process. This is called multithreading and causes less runtime overhead than multitasking, as no memory operation to swap out code is required. On the Raspberry Pi, we use the standard *Posix Threads* or

T. Bräunl, *Embedded Robotics*,
https://doi.org/10.1007/978-981-16-0804-9_8

abbreviated *Pthreads* package for multithreading; see LLNL[1] for details. In this chapter, we will look at preemptive multithreading, mutex synchronization and timer interrupts.

8.1 Preemptive Multithreading

Our first multithreading example is shown in Program 8.1. The *slave* procedure prints a line five times (including its character parameter and the loop counter) within the words *START* and *STOP*, however, split over two print commands.

Program 8.1: Basic Multithreading

```
#include <pthread.h>
#include "eyebot.h"
void *slave(void *arg)
{ for (int i=1; i<=5; i++)
  { LCDPrintf("START-%c -%d- ", (char)((int) arg), i);
    LCDPrintf("%c-STOP\n", (char)((int) arg));
  }
 return NULL;
}
```

```
int main()
{ pthread_t t1, t2;
  XInitThreads();
  LCDMenu(" ", " ", " ", "END");

  pthread_create(&t1, NULL, slave, (void *) 'A');
  pthread_create(&t2, NULL, slave, (void *) 'B');

  KEYWait(KEY4);
}
```

The main program first initializes the multithreading system by calling *XInitThreads* and then calls *pthread_create* twice for the same function, which creates two versions of *slave*, running in parallel. The parameters for *pthread_create* are:

1. Thread variable (return parameter)
    Needed for outside access to a thread, e.g., for terminating it.

2. Attributes (pointer to structure)
    Needed only for setting certain thread features; we just use *NULL*.

---

[1]LLNL, *POSIX Threads Programming*, online: https://computing.llnl.gov/tutorials/pthreads/.

3. Function name to be started as a thread (must be of type "*void* *")
   In our case *slave*.

4. Any arguments (must be of type "*void* *")
   We give each thread its single-character parameter, e.g., (*void* *) 'A.'

The main program now uses *KEYWait* to pause until the corresponding button has been pressed on the touch screen, leaving the slave threads run in the meantime. After this, the whole program terminates.

On the screen, we now get a wild mix of start and stop messages, not even a single print statement (*LCDPrintf*) is being completed—these functions are not what we call *thread-safe*! In the sample output in Figure 8.1, we see in the first line:

```
STASR ...
```

The first three characters, "*STA*" are probably from the first thread trying to print *START*, but after three characters, the second thread takes over and writes its first "*S*," then the first thread comes back and writes its "*R*" and so on. Some characters may be even lost completely, as the function is not thread-safe.

Uncontrolled thread activation can work, but not if the threads have to share a medium, such as the screen in our case. For this, we need some synchronization and we will sort this out in the next section.

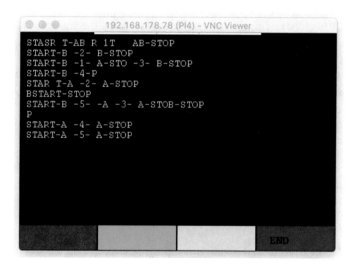

Figure 8.1: Unsynchronized multithreading output

8.2   Synchronization

In almost every application of preemptive multithreading, some synchronization scheme is required. Whenever two or more threads exchange information blocks or share resources, e.g., the LCD for printing messages, reading data from sensors or setting actuator values, synchronization is essential. The standard synchronization methods[2] are:

- Mutex
- Semaphore
- Monitor
- Message passing.

Here, we will concentrate on synchronization using a *mutex* (*mutual exclusion*), as it is called in *Pthreads*. A mutex is a low-level synchronization tool and therefore especially useful for embedded controllers.

8.2.1 Mutex Example

The concepts of mutex and semaphore have been around for a long time and were formalized by Dijkstra[3] as a model resembling railroad signals. For further historic notes, see also Hoare,[4] Brinch Hansen[5] and the more recent collection Brinch Hansen.[6]

A mutex is a synchronization object that can be in either of two states: *free* or *occupied*. Each thread can perform two different operations on a mutex: *lock* or *unlock*. When a thread locks a previously *free* mutex, it will change the mutex's state to *occupied*. The first thread locking the mutex can continue processing, but any subsequent thread trying to lock the now occupied mutex will be blocked until the first thread releases the mutex. This will then only momentarily change the mutex's state to free—the next waiting thread will be unblocked, which then re-locks the mutex.

In Pthreads, the mutex type is called *pthread_mutex_t* and declaration and initialization of a mutex *m* uses the function *pthread_mutex_init*:

```
pthread_mutex_t m;
pthread_mutex_init(&m, NULL);
```

For locking and unlocking a mutex, Pthreads provides the functions *lock* and *unlock*:

```
pthread_mutex_lock (&m);
pthread_mutex_unlock(&m);
```

[2]Bräunl, T. *Parallel Programming - An Introduction*, Prentice Hall, Englewood Cliffs NJ, 1993.

[3]Dijkstra, E. *Communicating Sequential Processes*, Technical Report EWD-123, Technical University Eindhoven, 1965.

[4]Hoare, C.A.R. *Communicating sequential processes*, Communications of the ACM, vol. 17, no. 10, Oct. 1974, pp. 549–557 (9).

[5]Brinch Hansen, P. *The Architecture of Concurrent Programs*, Prentice Hall, Englewood Cliffs NJ, 1977.

[6]Brinch Hansen, P. (Ed.) *Classic Operating Systems*, Springer-Verlag, Berlin, 2001.

In C Program 8.2, we can now use the lock and unlock functions as a bracket around the shared *critical section* of the two print statements, where both slave processes want to print. The mutex is declared as a global variable, so the slaves and the main program can access it. After its declaration, the mutex is automatically unlocked and will let only a single thread through at a time. If all threads would wait for an unavailable resource after calling *lock*, then we would have encountered a *deadlock* situation from which the program cannot free itself.

Program 8.2: Synchronized Multithreading

```
#include <pthread.h>
#include "eyebot.h"

pthread_mutex_t print;

void *slave(void *arg)
{ for (int i=1; i<=5; i++)
  { pthread_mutex_lock(&print);
      LCDPrintf("START-%c -%d- ", (char) ((int)arg), i);
      LCDPrintf("%c-STOP\n", (char) ((int)arg));
    pthread_mutex_unlock(&print);
    OSWait(1);
  }
  return NULL;
}

int main()
{ pthread_t t1, t2;

  XInitThreads();
  LCDMenu(" ", " ", " ", "END");
  pthread_mutex_init(&print, NULL);
  pthread_create(&t1, NULL, slave, (void *) 'A');
  pthread_create(&t2, NULL, slave, (void *) 'B');
  KEYWait(KEY4);
}
```

Figure 8.2 now shows the neat sequence of print statements:

```
START-A -1- A-STOP
START-B -1- B-STOP
```

Note, however, that there is no strict alternation between threads A and B. For example, the 4th and 5th print line are both from thread B, so there is no forced sequencing.

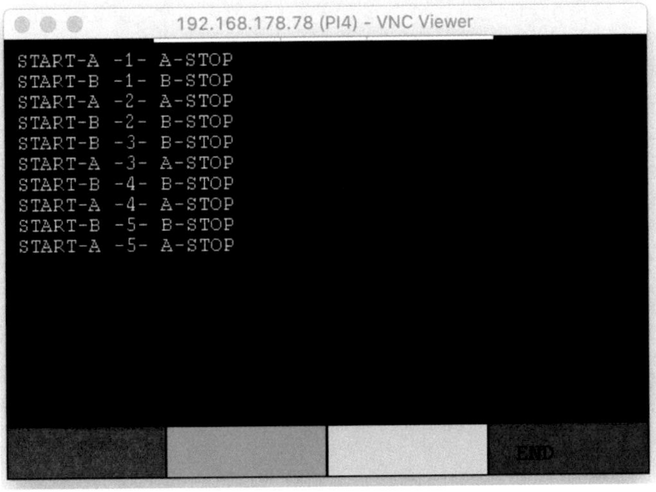

Figure 8.2: Synchronized multithreading output

8.2.2 Master with Multiple Slaves

In the following, we introduce a more complex example, running four threads in total, with three slaves and one master. Each slave is controlled by a mutex *s* *[i]* that is set by the master, and all print instructions are protected by the shared mutex *print*. The full C code is shown in Program 8.3.

Each slave now has to go past two mutexes, its own lock *s[i]* and—after being successful—through the *print* mutex. Note that the nested order of the lock operations is important. If a slave would successfully acquire the *print* mutex and then have to wait for its *s[i]*, then it would block any printouts until it gets freed by the master.

As a general rule, mutexes should be released in *reverse order* of acquiring (so in our case release *print* first, then *s[i]*), as otherwise under some circumstances a deadlock can occur.

Program 8.3: Master and Slave Threads

```c
#include <pthread.h>
#include "eyebot.h"

pthread_mutex_t s[3], print;
int running = 1;

void *slave(void *arg)
{ int i = (int) arg;
  while(running)
  { pthread_mutex_lock(&s[i]);
    pthread_mutex_lock(&print);
```

```
    LCDPrintf("%d", i);
    pthread_mutex_unlock(&print);
    pthread_mutex_unlock(&s[i]);
  }
  return NULL;
}
```

```
void *master(void *arg)
{ while(running)
  { switch (KEYRead())
    { case KEY1: pthread_mutex_unlock(&s[0]); break;
      case KEY2: pthread_mutex_unlock(&s[1]); break;
      case KEY3: pthread_mutex_unlock(&s[2]); break;
      case KEY4: running = 0; break;
    }
  }
  for (int i=0; i<3; i++)
    pthread_mutex_unlock(&s[i]); // free all waiting slaves
  return NULL;
}
```

```
int main()
{ pthread_t t[4];

  XInitThreads();
  LCDPrintf("Multithreading\n");
  LCDMenu("THREAD 0", "THREAD 1", "THREAD 2", "END");

  pthread_mutex_init(&print, NULL);
  for (int i=0; i<3; i++)
  { pthread_mutex_init(&s[i], NULL);
    pthread_mutex_lock(&s[i]);
  }
  pthread_create(&t[0], NULL, slave, (void *) 0);
  pthread_create(&t[1], NULL, slave, (void *) 1);
  pthread_create(&t[2], NULL, slave, (void *) 2);
  pthread_create(&t[3], NULL, master,(void *) 3);

  pthread_exit(0); // terminate program
}
```

The master thread listens to any key presses on the touch screen. The first three buttons free the corresponding slave thread, which will then print its ID number on the screen. Pressing *KEY4* will set the global variable *running* to 0, so each thread can terminate its loop. However, this can only happen, if a thread is not stuck in a locked mutex. This is why the master thread executes as its last instruction *pthread_mutex_unlock* in a loop, to free up all slaves.

At the end of the main program, *pthread_exit* waits for all threads of the program to terminate, before it terminates the whole program.

Figure 8.3 (left) shows the program running with only a slave 0 being unlocked. On the right, all three slave mutexes have been unlocked and the threads are sharing the screen for writing their characters.

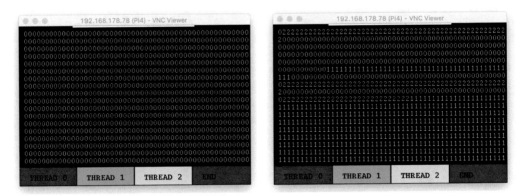

Figure 8.3: Multiple slaves, single unlocked (left) and all unlocked (right)

8.3 Scheduling

A scheduler is an operating system component that handles thread switching. Thread switching occurs in preemptive multithreading when a thread's time slice has run out or when the thread is being blocked (e.g., through a *mutex_lock* operation).

Each thread can be in exactly one of three states (Figure 8.4):

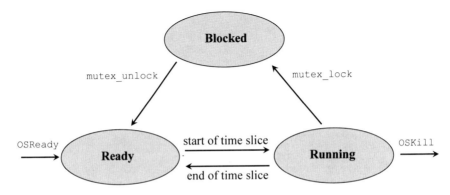

Figure 8.4: Thread model

- **Ready**
  A thread is ready to be executed and waiting for its turn.
- **Running**
  A thread is currently being executed.
- **Blocked**
  A thread is not ready for execution, e.g., because it is waiting for a mutex.

Each thread is identified by a thread control block that contains all required control data for a thread. This includes the thread's start address, a thread number, a stack start address and stack size, a text string associated with the thread and an integer priority.

Without the use of priorities, i.e., with all threads being assigned the same priority as in our examples, thread scheduling is performed in a *round-robin* fashion. For example, if a system has three ready threads, t_1, t_2 and t_3, the execution order would be:

t_1, t_2, t_3, t_1, t_2, t_3, t_1, t_2, t_3, ...

Indicating the *running* thread in bold face and the *ready* waiting list in square brackets, this sequence looks like the following:

t_1 [t_2, t_3]
t_2 [t_3, t_1]
t_3 [t_1, t_2]
...

Each thread gets the same time slice duration, so it gets its fair share of the available processor time. If a thread is blocked during its running phase and is later unblocked, it will re-enter the list as the last *ready* thread, so the execution order may be changed.

For example:
t_1 *(block t_1)* t_2, t_3, t_2, t_3 *(unblock t_1)* t_2, t_1, t_3, t_2, t_1, t_3, ...

Using square brackets to denote the list of ready threads and curly brackets for the list of all blocked processes, this example of scheduled threads look as follows:

t_1	[t_2, t_3]	{}	$\rightarrow$	*t1 is being blocked*
t_2	[t_3]	{t_1}		
t_3	[t_2]	{t_1}		
t_2	[t_3]	{t_1}		
t_3	[t_2, t_1]	{}	$\rightarrow$	*t3 unblocks t1*
t_2	[t_1, t_3]	{}		
t_1	[t_3, t_2]	{}		
t_3	[t_2, t_1]	{}		
...				

Whenever a thread is put back into the *ready* list (either from running or from blocked), it will be put at the end of the list of all waiting threads with the same priority. So, if all threads have the same priority, the new *ready* thread will go to the end of the complete list.

8.3.1 Static Priorities

The situation gets more complex if different priorities are involved. Assume that threads can be started with static priorities 1 (lowest) to 8 (highest). Now, a new *ready* thread will follow after the last thread with the same priority and before all threads with a lower priority. Scheduling remains simple since only a single waiting list has to be maintained. However, *starvation* of threads can occur, as shown in the following example.

Assuming threads t_A and t_B have the higher priority 2 and threads t_a and t_b have the lower priority 1, then in the following sequence threads t_a and t_b are being kept from executing (*starvation*), unless t_A and t_B are both blocked by some events.

t_A	$[t_B, t_a, t_b]$	$\{\}$		
t_B	$[t_A, t_a, t_b]$	$\{\}$		
t_A	$[t_B, t_a, t_b]$	$\{\}$	$\rightarrow$	t_A blocked
t_B	$[t_A, t_a, t_b]$	$\{t_A\}$	$\rightarrow$	t_B blocked
t_a	$[t_b]$	$\{t_A, t_B\}$		
...				

8.3.2 Dynamic Priorities

For the reasons shown in the previous section, many operating systems use the more complex dynamic priority model. For n different priorities, the scheduler maintains n distinct *ready* waiting lists, one for each priority (eight in our example). After spawning, threads are entered in the *ready* list matching their priority and each queue for itself implements the *round-robin* principle shown before. So the scheduler only has to determine which queue to select next.

Each queue (not thread) is assigned a static priority (1...8) and a dynamic priority, which is initialized with twice the static priority times the number of *ready* threads in this queue. This factor is required to guarantee fair scheduling for different queue lengths (see below). Whenever a thread from a *ready* list is executed, the dynamic priority of this queue is decremented by 1. Only after the dynamic priorities of all queues have been reduced to zero are the dynamic queue priorities reset to their original values.

The scheduler now simply selects the next thread to be executed from the (non-empty) *ready* queue with the highest dynamic priority. If there are no eligible threads left in any *ready* queue, the multithreading system terminates and continues with the calling main program (or the operating system). This model prevents starvation and still gives threads with higher priorities more frequent time slices for execution. See the example below with three priorities, with static priorities shown on the right, dynamic priorities on the left. The highest dynamic priority after decrementing and the thread to be selected for the next time slice are printed in bold type:

Run	DPrio	SPrio	Thread list	
–	6	3	$[t_A]$	$DPrio = (2 \bullet SPrio \bullet$
				$number\_of\_threads = 2 \bullet 3 \bullet 1 = 6)$
	8	**2**	$[\mathbf{t_a,t_b}]$	$(2 \bullet 2 \bullet 2 = 8) \leftarrow$ highest dynamic priority
	4	1	$[t_x,t_y]$	$(2 \bullet 1 \bullet 2 = 4)$
t_a	6	3	$[t_A]$	
	7	**2**	$[\mathbf{t_b}]$	*decrement dynamic priority*
	4	1	$[t_x,t_y]$	
t_b	**6**	**3**	$[\mathbf{t_A}]$	
	6	2	$[t_a]$	*decrement dynamic priority*
	4	1	$[t_x,t_y]$	
t_A	5	3	$[]$	*decrement dynamic priority*
	6	**2**	$[\mathbf{t_a,t_b}]$	
	4	1	$[t_x,t_y]$	
t_a	**5**	**3**	$[\mathbf{t_A}]$	
	5	2	$[t_b]$	*decrement dynamic priority*
	4	1	$[t_x,t_y]$	
...				
t_a	3	3	$[t_A]$	
	3	2	$[t_b]$	
	4	**1**	$[\mathbf{t_x,t_y}]$	
t_x	**3**	**3**	$[\mathbf{t_A}]$	
	3	2	$[t_a,t_b]$	
	3	1	$[t_y]$	*decrement dynamic priority*
...				

8.4 Interrupts and Timer-Activated Tasks

A different way of designing a concurrent application is to use interrupts, which can be triggered by external devices or by a built-in timer. Both are very important techniques; external interrupts can be used for reacting to external sensors, such as counting ticks from a shaft encoder, while timer interrupts can be used for implementing periodically repeating tasks with fixed time frame, such as motor control routines.

The event of an external interrupt signal will stop the currently executing task and instead execute a so-called *interrupt service routine* (ISR). As a general rule, ISRs should have a short duration and are required to clean up any stack changes they performed, in order not to interfere with the foreground task. Initialization of ISRs often requires assembly commands, since interrupt lines are directly linked to the CPU and are therefore machine-dependent (Figure 8.5).

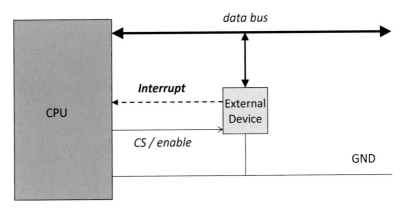

Figure 8.5: Interrupt generation from external device

Somewhat more general are interrupts activated by software instead of external hardware. Most important among software interrupts are timer interrupts, which occur at regular time intervals.

In the RoBIOS operating system, we have implemented a general-purpose 1000 Hz (1 kHz) timing interrupt. User programs can attach or detach timer interrupt ISRs at a rate between up to 1000 Hz, by specifying an integer frequency divider. This is achieved with the following RoBIOS operations:

```
TIMER OSAttachTimer(int scale, void (*fct)(void));
int   OSDetachTimer(TIMER t);
```

The integer timing scale parameter is used to divide the 1 kHz timer and thereby specifies the number of timer calls per second (e.g., 1 for 1000 Hz, 10 for 100 Hz, etc.). Parameter *fct* is simply the name of a function without parameters or return value (*void*).

An application program can now implement a background task, for example a PID motor controller (see chapter on control), which will be executed several times per second. Although activation of a timer function is handled in the same way as preemptive multitasking, a timer function itself will not be preempted, but will keep processor control until it terminates. This is one of the reasons why any ISR should be rather short in time. It is also obvious that the execution time of an ISR (or the sum of all ISR execution times in the case of multiple ISRs) must not exceed the time interval between two timer interrupts, or regular activations will not be possible.

Program 8.4: Timer Interrupt Example

```
#include "eyebot.h"
int count = 0;

void cleanup()
{ count = 0; // reset through timer interrupt
}
```

```
int main()
{ TIMER t;
  LCDMenu(" "," "," ","END");

  // start background timer
  t = OSAttachTimer(1000, cleanup); // once a second

  // foreground task
  while (!KEYRead())
  { LCDPrintf("%d\n", count);
    count++;
    OSWait(100); // slow things down
  }
  OSDetachTimer(t);
}
```

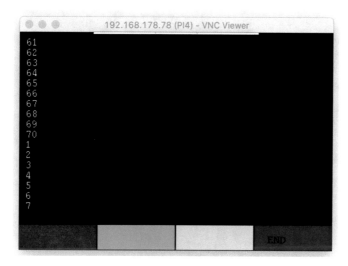

Figure 8.6: Timer interrupt during print sequence

The example in Program 8.4 shows the timer routine and the corresponding main program. The main program sets the timer interrupt to once every second. While the foreground task prints consecutive numbers to the screen, the background task generates strikes once a second to reset the global counter. Figure 8.6 shows the result on the screen.

8.5   Tasks

1. Write a robot program as a collection of parallel threads. Each sensor and each actuator will get its own thread. Communication between threads and the main program will be via shared global data structures and requires a *mutex* for synchronization.
    Let each thread run unrestrictedly at its maximum speed.
    Determine update times in Hz for the various sensors, e.g. Infrared PSD versus camera.

2. Write a program with 5 parallel threads and a single *token* (data item). Only the thread with the token is allowed to run and print text onto the LCD.
    Passing on the token is being controlled from the main program on pressing one of the *KEY buttons.*

3. Write a program with two parallel threads, one producer and one consumer.
    The task of the producer is to transmit the following text byte by byte to the consumer:
    the quick brown fox jumps over the lazy dog[7]
    Transmit each character via a single global variable.
    The consumer's task is to print this text byte by byte onto the LCD. Did any characters get lost?

4. Improve the producer-consumer implementation by introducing mutexes for *handshakes*. You will need one handshake for the producer to say "new byte is prepared" and one handshake for the consumer to say "ready to accept a new byte".

---

[7]Check the significance of this sentence on the Internet.

COMMUNICATION

The Raspberry Pi has built-in Ethernet, Wi-Fi and Bluetooth communication channels, all of which are very useful for robotics applications. In the following sections, we will look at each of these channels in detail and outline the communication parameter defaults we have chosen for our Raspberry–EyeBot distribution.

9.1 Communication Channels

9.1.1 Ethernet

Wired Ethernet or LAN is the most basic of the three communication channels. It requires an Ethernet patch cable to link the Raspberry Pi with a router, Ethernet switch or directly with a PC. The Raspberry's Ethernet port is smart enough to detect whether it is linked to a router or a PC, so no special crossover cable is required for a direct PC connection.

- Wired LAN to an Ethernet router
  This is a typical setup during software development and the most reliable form of communication. In our configuration, when the Pi is connected to a router, it uses DHCP to accept an IP address. The home screen will then display the IP address, which we will need to remotely connect to the Pi.

- Wired LAN directly to a laptop PC
  This setup does not require a router, but it does require setting a manual IP address on your PC within the same range as the Pi. In the RoBIOS system, the Raspberry Pi's default, IP address is simple and easy to remember:
  ```
  10.0.0.1
  ```

© The Author(s), under exclusive license to Springer Nature Singapore Pte Ltd. 2022
T. Bräunl, *Embedded Robotics*,
https://doi.org/10.1007/978-981-16-0804-9_9

9.1.2 Wi-Fi

More practical for driving robots is a wireless connection. We implemented two standard setups. One where each robot creates its own hotspot to which users can connect their PC, tablet or phone as a communication device, and one where multiple robots as well as other user devices are linked to the same router.

- WLAN Master (Hotspot)
  Under RoBIOS, the Raspberry Pi creates a hotspot with the network name (SSID) *Pi_xyz*, where *xyz* is derived from the Pi's hardware MAC address. The generated SSID will be displayed on the Pi's home screen (see Figure 9.1), which is important to distinguish multiple robots in the same room. In order to remote control one or several robots, the user can then connect his/her laptop to the Pi's hotspot and connect to the Pi with this equally simple and easy to remember IP address:
  `10.1.1.1`
  The laptop will then automatically get an IP address assigned in this range, e.g., `10.1.1.2`

- WLAN Slave
  When using multiple robots together, a wireless router has to be used. In this case, the router gives each Raspberry Pi an IP address within its address range. This is equivalent to a wired LAN connection to a router. The Pi's home screen again displays the assigned IP address, so the user can connect his/her laptop PC to right Pi.

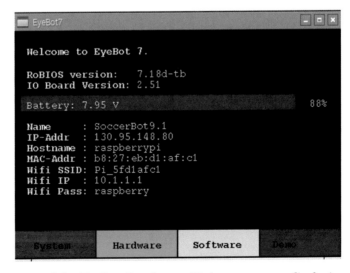

Figure 9.1: EyeBot–Raspberry Pi home screen displaying SSID and IP addresses

9.1.3 Bluetooth

Bluetooth is another versatile communication link for the Raspberry Pi. It can be used for connecting keyboards or mice to the Pi, or for connecting a game controller, which is most useful during running a robot application (see Figure 9.2).

As our standard user interface has four on-screen buttons, it is very helpful to be able to use the buttons on a remote game controller instead of having to chase the robot and push the buttons on its touch screen. Also, the game controller's joysticks can be used for steering a robot directly and its additional buttons can be used for directly accessing additional features instead of having to navigate through several menu screens. Finally, if the game controller has some level of force feedback (*rumble effect*), this can be used to transfer event information from the robot back to the operator, e.g., about any robot collisions.

Figure 9.2: PS3 game controller connecting via Bluetooth

9.2 File Transfer and Remote Access

During software development, we need essential tools for transferring files from a host to the robot and for remotely accessing the robot, e.g., for editing, compiling and running programs, exchanging data, or for monitoring the controller's display. The tools we use for these are:

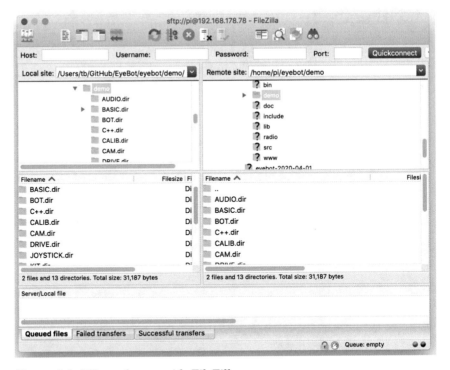

Figure 9.3: File exchange with FileZilla

- File exchange: FileZilla (for Windows, MacOS, Linux)
  A free tool to move files between a PC and the Raspberry Pi (see Figure 9.3).

- Command window (shell): Putty (Windows), Terminal (MacOS/Linux)
  A remote shell command window is a simple tool for entering text commands at the robot's command console (see Figure 9.4). Any graphics elements, however, require a remote desktop connection.

- Remote Desktop: VNC-Viewer (Windows/MacOS/Linux) or
  Microsoft Remote Desktop (Windows/MacOS)
  Remote desktop either just provides a 1:1 copy of the robot's touch-screen LCD (VNC; see Figure 9.5) or opens a larger full Raspberry Pi desktop (Microsoft Remote Desktop). This is a valuable tool for program development and debugging.

Each of these tools is started with the robot's (Raspberry Pi's) IP address. If the Pi is equipped with an LCD touch screen, our software will display this address on the home screen. If not, you can connect to the Pi as a host either via an Ethernet cable (address 10.0.0.1) or as a WLAN hotspot (address 10.1.1.1). Alternatively, when using a DHCP router, you can look up the Pi's IP address in the router's online directory.

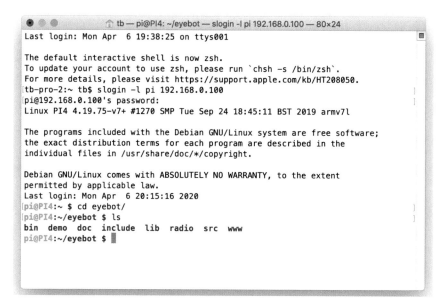

Figure 9.4: Remote shell with text terminal

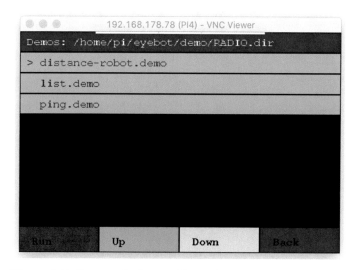

Figure 9.5: Remote desktop with VNC

9.3 Radio Library

The RoBIOS library contains a set of functions that form the *Radio* library (see Table 9.1). These simplify communication between a group of robots or between a robot and a PC base station.

```
int RADIOInit(void);                              // Start communication
int RADIOGetID(void);                             // Get own radio ID
int RADIOSend(int id, char* buf);                 // Send string to "id"
int RADIOReceive(int *id_no, char* buf, int size); // blocking
int RADIOCheck(void);                             // Check for message
int RADIOStatus(int IDlist[]);                    // List all network IDs
int RADIORelease(void);                           // Terminate communication
```

Table 9.1: RoBIOS radio functions

The Radio library uses the Raspberry Pi's WLAN connection and assumes that all robots and any communicating PCs are in the same subnet. This could be set up by a DHCP router or as a direct link between a PC and a robot. Functions *RADIOInit* and *RADIORelease* start and terminate the radio link for a particular node. Function *RADIOGetID* returns a robot's own ID number, which is required as an address for sending and receiving messages. In our IPv4 implementation, each robot's ID number is the last byte of its IPv4 address. So, for example, the robot with the IP address

10.1.1.5 → has ID number 5.

This also limits the total number of robots communicating in the same network to 254, as addresses 0 and 255 are not available. Functions *RADIO-Send* send a string to a destination ID, while *RADIOReceive* receives a string and also returns the sender's ID. *RADIOReceive* is blocking, so program control will halt until the next message has been received. As this can easily result in a deadlock, it is advisable to either run the receiving function in a parallel thread (see Chap. 8) or to use function *RADIOCheck* first, which returns 1 if a message is waiting and 0 otherwise. Finally, function *RADIOStatus* returns a list of all robots and PCs registered in the same network, which is very helpful to let each robot know the total number of participants and their respective ID numbers.

9.4 Robot to Robot Communication

When multiple robots are connected to the same WLAN network, they can be programmed to exchange data via the RoBIOS radio functions in order to cooperate. Program 9.1 shows an example where two robots exchange messages back and forth, similar to a ping-pong match.

Both robots run the same program and the user can select which one plays the role of master and which one the role of slave, by pressing the corresponding buttons on its touch screen. If *master* (*KEY1*) was selected, the robot has to first find the other robot's address. For this it calls *RADIOStatus* and uses the first entry *id[0]*. It then initiates the data exchange by sending the first message (string "00000") to its destination: *RADIOSend(partnerid, buf)*.

The following loop is then executed by both master and slave robot. Each first waits for an incoming message via *RADIOReceive*, prints the message contents to the screen, increments the received number and sends it back. The resulting data exchange will be:

Master		Slave
00000	$\rightarrow$	00000
		$\downarrow$ inc.
00001	$\leftarrow$	**00001**
$\downarrow$ inc.		
00002	$\rightarrow$	00002
		and so on

Program 9.1: Robot to Robot Communication

```
#include "eyebot.h"
#define MAX 10

int main ()
{ int k, i, num, ret;
  int id[256];
  int myid, partnerid;
  char buf[MAX] = "00000";

  LCDMenu("MASTER", "SLAVE", " ", "END");
  RADIOInit();
  myid = RADIOGetID();
  LCDPrintf("my id is: %d\n", myid);

  k = KEYGet();
  if (k==KEY4) return 0; // exit
  if (k==KEY1) // master only
    { LCDPrintf("scanning ... ");
      ret = RADIOStatus(id);
      if (ret<0) LCDPrintf("error RADIOStatus\n");
      partnerid = id[0];
      LCDPrintf("partner is %d\n", partnerid);
      RADIOSend(partnerid, buf);
    }
  LCDPrintf("I am waiting for partner\n");
```

```
  for (i=0; i<10; i++)
  { RADIOReceive(&partnerid, buf, MAX);
    LCDPrintf("received from %d text %s\n", partnerid, buf);

    sscanf ((char*)buf, "%d", &num);
    num++;
    sprintf((char*)buf, "%05d", num);
    RADIOSend(partnerid, buf);
  }
  KEYWait(KEY4);
}
```

Figure 9.6 shows the displays of the two robots side by side, executing the *ping* program. For function *RADIOStatus* to work, the underlying Linux function requires that the Raspberry Pi has talked to each of the other robot controllers at least once before. In a new environment, this can be achieved by *pinging* all other robots from the master; e.g., if the master has address 192.168.0.100, this can be achieved by calling:

```
ping 192.168.0.101
ping 192.168.0.102
ping 192.168.0.103
```

… and so on.

9.5 Robot to PC Communication

Instead of a group of robots exchanging data, one or more of the communication nodes can also be a PC. The PC Program 9.2 sends distance measurement requests to a robot, which will then enter it into an array data structure. The robot returns data from its front PSD sensor, as shown in Raspberry Pi Program 9.3.

Program 9.2: PC to Robot Communication (Runs on PC)

```
#include "radio.h"
#include <stdio.h>
#define MAX   8
#define ITER 10

int main ()
{ int  k, i, num, ret ,myid, robot, who, id[256];
  float sum, avg;
  char rbuf[MAX], sbuf[MAX] = "PSD";
```

```
   printf("PC program for robot communication\n");
   RADIOInit();
   myid = RADIOGetID();
   printf("my id is: %d\n", myid);

   // get robot ID
   { printf("scanning IDs\n");
     ret = RADIOStatus(id);
     if (ret<0) printf("error RADIOStatus\n");
       else printf("number of robots: %d\n", ret);
     if (ret<1)
     { printf("no robots found\n"); return 0; //exit
     }
     for (i=0; i<ret; i++) printf(" %d", id[i]); // print IDs
     printf("\n");                    // new line
     robot = id[ret-1];
     printf("robot ID is %d\n", robot);
   }

   sum=0.0;
   for (i=0; i<ITER; i++)
   { RADIOSend(robot, sbuf);
     RADIOReceive(&who, rbuf, MAX);
     printf("%d: received from robot %d value %s\n", i, who, rbuf);
     sscanf ((char*)rbuf, "%d", &num);
     sum +=num;
   }
   avg = sum/ITER;
   printf("Average distance: %5.1f mm\n", avg);
}
```

Program 9.3: Robot to PC Communication (Runs on Raspberry Pi Robot)

```
#include "eyebot.h"
#define MAX  8
#define ITER 10

int main ()
{ int  i, num, myid, PC;
  char sbuf[MAX], rbuf[MAX];

  LCDPrintf("Robot-PC Communication\n");
  LCDMenu(" ", " ", " ", "END");
  RADIOInit();
```

```
myid = RADIOGetID();
LCDPrintf("my id is: %d\n", myid);
LCDPrintf("Waiting for PC request\n");

for (i=0; i<ITER; i++)
 { RADIOReceive(&PC, rbuf, MAX);
   LCDPrintf("received from %d text %s\n", PC, rbuf);
   num = PSDGet(PSD_FRONT);  // read front distance
   sprintf((char*)sbuf, "%05d", num);
   RADIOSend(PC, sbuf);
 }
 KEYWait(KEY4);
}
```

Figure 9.6: Robot-to-robot communication with ping program

Figure 9.7, left, shows the output on the PC (here a Mac), while Figure 9.7, right, shows the corresponding output on the robot's screen.

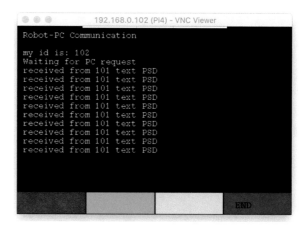

```
tb-pro:RADIO.dir tb$ ./distance-PC
PC program for robot communication
IP 192 168 0 101
my id is: 101
scanning IDs
number of robots: 2
 100 102
robot ID is 102
0: received from robot 102 value 00270
1: received from robot 102 value 00280
2: received from robot 102 value 00270
3: received from robot 102 value 00270
4: received from robot 102 value 00280
5: received from robot 102 value 00270
6: received from robot 102 value 00270
7: received from robot 102 value 00270
8: received from robot 102 value 00270
9: received from robot 102 value 00270
Average distance: 272.0 mm
```

Figure 9.7: PC–robot wireless communication

9.6   Tasks

1. Write a robot program that transmits a robot's pose continuously to a base station.
   Write a base station program that graphically displays the robot's position and orientation in its rectangular driving area.

2. Write a robot program that transmits a robot's camera image continuously to a base station.
   Write a base station program that continuously displays the robot's camera data.

3. Write a robot program to find a colored object and run it on a group of robots.
   The first robot to find it sends the global object coordinates to all other robots, which should then all drive toward this object.

4. The robot master is being defined as the robot with the lowest ID number. Let robots exchange ID number in regular intervals of 30 s. Each robot keeps a list of active robots (list of ID numbers), as new robots can be switched on and other robots may get switched off or drive outside the communication range. If a new lowest number is detected, then this robot will become the new master.

5. Construct a relay chain of four robots to extend the communication range between robots. All messages from robot 1 will be relayed via robot 2 and robot 3 to reach robot 4 and vice versa.

Part II
Robot Hardware

DRIVING ROBOTS

10

U sing two DC motors and two wheels is the easiest way to build a mobile robot. In this chapter, we will discuss several robot designs such as differential drive, synchro-drive, Ackermann steering and omni-directional drives. A collection of related research papers can be found in Rückert, Sitte, Witkowski[1] and Cho, Lee.[2] Introductory textbooks are by Borenstein, Everett, Feng,[3] Arkin,[4] Jones, Flynn, Seiger,[5] and McKerrow.[6]

10.1 Single Wheel Drive

Having a single wheel that is both driven and steered is the simplest conceptual design for a mobile robot. This design also requires two passive caster wheels in the back, since three contact points are always required for a stable base.

Linear velocity and angular velocity of the robot are completely decoupled. So for driving straight, the front wheel is positioned in the middle position and driven at the desired speed. For driving in a curve, the wheel is positioned at an angle matching the desired curve.

Figure 10.1 shows the driving action for different steering settings. Curve driving is following the arc of a circle; however, this robot design cannot turn on the spot. With the front wheel set to 90°, the robot will rotate about the

---

[1]U. Rückert, J. Sitte, U. Witkowski (Eds.) *Autonomous Minirobots for Research and Edutainment – AMiRE2001*, Proceedings of the 5th International Heinz Nixdorf Symposium, HNI-Verlagsschriftenreihe, no. 97, Univ. Paderborn, Oct. 2001.

[2]H. Cho, J.-J. Lee (Eds.), *Proceedings of the 2002 FIRA World Congress*, Seoul, Korea, May 2002.

[3]J. Borenstein, H. Everett, L. Feng, *Navigating Mobile Robots: Sensors and Techniques*, AK Peters, Wellesley MA, 1998.

[4]R. Arkin, *Behavior-Based Robotics*, MIT Press, Cambridge MA, 1998.

[5]J. Jones, A. Flynn, B. Seiger, *Mobile Robots—From Inspiration to Implementation*, 2nd Ed., AK Peters, Wellesley MA, 1999.

[6]P. McKerrow, *Introduction to Robotics*, Addison-Wesley, Reading MA, 1991.

© The Author(s), under exclusive license to Springer Nature Singapore Pte Ltd. 2022
T. Bräunl, *Embedded Robotics*,
https://doi.org/10.1007/978-981-16-0804-9_10

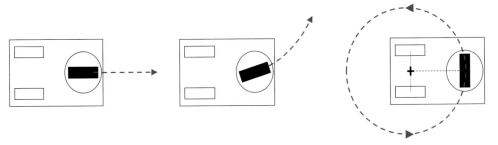

Figure 10.1: Driving and rotation of single wheel drive

midpoint between the two caster wheels (see Figure 10.1, right). So the minimum turning radius is the distance between the front wheel and the midpoint of the back wheels.

10.2 Differential Drive

The differential drive design has two motors mounted in fixed positions on the left and right sides of the robot, independently driving one wheel each. Since at least three ground contact points are necessary, this design requires one or two additional passive caster wheels or sliders, depending on the location of the driven wheels. Differential drive is mechanically simpler than the single wheel drive, because it does not require rotation of a driving motor. However, directional control for differential drive is more complex than for single wheel drive, because it requires the coordination of two driven wheels, as explained in the chapter on control.

The minimal differential drive design with only a single passive wheel cannot have the driving wheels in the middle of the robot for stability reasons. So when turning on the spot, the robot will rotate about the off-center midpoint between the two driven wheels. The design with two passive wheels or sliders, one at the front and one at the back of the robot, allows rotation about the center of the robot. However, this design can introduce surface contact problems, because it is using four contact points instead of three.

Figure 10.2 demonstrates the driving actions of a differential drive robot. If both motors run at the same speed, the robot drives straight forward or backward, if one motor is running faster than the other, the robot drives in a curve along the arc of a circle, and if both motors run at the same speed in opposite directions, the robot turns on the spot.

- Driving straight forward: $v_L = v_R$, $v_L > 0$
- Driving in a right curve: $v_L > v_R$
- Turning clockwise on the spot:$v_L = -v_R$, $v_L > 0$

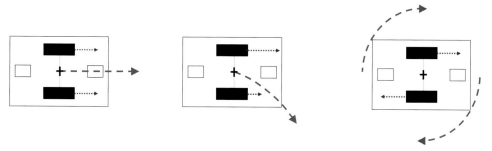

Figure 10.2: Driving and rotation of differential drive

10.2.1 Mini Robot Platform FT-DC-002

The *Mini Smart Robot Mobile Platform*[7] (FT-DC-002), which we dubbed *Little Red Riding Hood*, can be bought for around only $15, including all mechanical parts and two motors, but no encoders. The kit is very easy to assemble—it only takes 10 min and only a screwdriver is required. Figure 10.3, left, shows the assembled kit with the two yellow motors; Figure 10.3, right, shows the robot carrying a Raspberry Pi piggybacked on a USB power bank and fixed in place with Velcro dots. Other battery options that usually allow a higher current are a rechargeable LiPo battery pack or a battery holder and voltage regulator for two 18,650 Lithium-ion battery cells (3.7 V each). Using an Arduino or an ESP32 would also be possible for an inexpensive, although computationally less powerful setup.

Figure 10.3: Little Red Riding Hood platform and carrying battery and Raspberry Pi

---

[7]Elecrow, *2WD Mini Smart Robot Mobile Platform Kit*, FT-DC-002, online: https://www.elecrow.com/feetech-2wd-mini-smart-robot-mobile-platform-kit-ft-dc-002.html.

As already mentioned in the chapter on actuators, we cannot directly connect a motor to a Raspberry Pi or Arduino digital output. The high current would immediately destroy the output port. Instead, we have to use a motor controller, such as the L298N Dual H-Bridge[8] motor controller as shown in Figure 10.4. This board can be used for DC motors as well as for stepper motors and the large heat sink protects it from overheating. Such generic controllers are quite inexpensive (about $5 for a dual controller), so using one of them is a much cheaper solution than developing a specific circuit board.

Completing a full driving robot will now be very easy, as shown in the circuit diagram in Figure 10.5. The power bank supplies current to the Raspberry Pi (or Arduino), as well as to the motor controller logic. The motors can be driven from a separate power source through pin V_{CC} (up to 35 V for this particular controller), or simply also from 5 V (as shown in the figure), by linking up the 5 V cable with the V_{CC} pin. The two DC motors are directly connected to the two output ports or the motor controller. Polarity (+/− lines) of each motor determines its forward/backward direction, but this can also be inverted later in software, if required.

Figure 10.4: Dual motor controller (left); DC-DC voltage adapter board (right)

This only leaves the four motor signal lines to be connected (output from the microcontroller, input into the motor controller inputs: In1, ..., In4). We chose digital I/O lines 26, ..., 29 on the Pi, but any other free lines would work as well. Only the port numbers selected for the hardware connection need to match the numbers in software (see Program 10.1).

Since both the USB power bank and the Raspberry Pi or Arduino have USB sockets, connecting them with a USB cable is easy. Since the motor controller in Figure 10.4, left, gets its input power from a screw-in terminal, we can cut a second spare USB cable and screw in the positive and negative voltage wires are required. However, these cables are quite thin and difficult to isolate. So

---

[8]Banggood, *L298N Dual H Bridge Stepper Motor Driver Board*, online: https://au. banggood.com/Wholesale-L298N-Dual-H-Bridge-Stepper-Motor-Driver-Board-p-42826.html.

alternatively, or when different voltages for microcontroller and drive motors are required, a DC-DC voltage adapter and breakout board[9] as shown in Figure 10.4, right, is a good solution. This board for only $2 has a standard USB socket as input and provides single-pin connections as output for the motor controller power. Using pin cables from the Raspberry Pi's 5 V and ground pins would only be possible if the motors do not draw too much power, so this is not the best option.

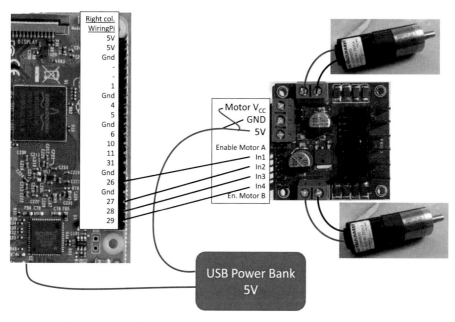

Figure 10.5: Circuit diagram for motor connection

We can now implement the basic driving routines as shown in Program 10.1. Note that this simple setup does not include motor encoders, so the driving and positioning functions will not be as accurate as when using encoders.

Subroutine *motors* takes the four digital output values and sets them as the motor controller input by writing their values to the Raspberry pins 26...29, after the *main* program has defined them all as output pins using the *pinMode* function. For driving each of the two motors forward, we have to set its two pins to values (1, 0), e.g., pins 26 + 27 for the left motor. To reverse a motor, we need the setting (0, 1) and to stop it the setting (0, 0). The same holds for the right motor and pins 28 + 29.

So, e.g., for driving the left motor forward and the right motor backward, we can call the function *motors* as:

```
motors(1,0, 0,1);
```

---

[9]DC-DC Adapter and Breakout Board, online: https://www.ebay.com.au/i/112518801378.

Program 10.1: Motor Driving Functions Without Encoders

```
#include "eyebot.h"
void motors(int a, int b, int c, int d)
{ digitalWrite(26, a);
  digitalWrite(27, b);
  digitalWrite(28, c);
  digitalWrite(29, d);
}

int main (void)
{ LCDPrintf("init pins\n");
  wiringPiSetup();
  pinMode(26, OUTPUT);
  pinMode(27, OUTPUT);
  pinMode(28, OUTPUT);
  pinMode(29, OUTPUT);

  LCDPrintf("Left Motor\n");
  motors(1, 0, 0,0); delay(1000); // forward
  motors(0, 1, 0,0); delay(1000); // reverse
  motors(0,0, 0,0); delay(1000); // stop

  LCDPrintf("Right Motor\n");
  motors(0,0, 1,0); delay(1000); // forward
  motors(0,0, 0,1); delay(1000); // reverse
  motors(0,0, 0,0); delay(1000); // stop

  return 0 ;
}
```

Although Program 10.1 lets us drive by turning motors on and off at full speed, it is not at all a smooth motor control software. For setting variable motor speeds, we need PWM outputs as discussed in the chapter on actuators. This has been implemented in Program 10.2 with the help of the WiringPi[10] function *softPwmCreate*.

Subroutine *pwm_init* initializes each of the output pins as a software PWM pin for the range [0, 100]. Subroutine *motors* then just sets left and right motors to the desired percentage. Only driving forward is considered in this example program. In the main program, we can then set individual motor speeds by calling *motors* (*left, right*) with two different values. In the *main* function of Program 10.2, we demonstrate this by implementing a speed ramp that incrementally increases the speed in both motors simultaneously.

---

[10]Wiring Pi GPIO Interface library for the Raspberry Pi, online: http://wiringpi.com.

Program 10.2: Motor Driving Functions with PWM

```
#include "eyebot.h"
int Left1, Left2, Right1, Right2;

void pwm_init(int a, int b, int c, int d)
{ wiringPiSetup();
  Left1 = a; Left2 = b; Right1 = c; Right2 = d; // global variables
  softPwmCreate(a, 0, 100); // setup pin for PWM [0..100]
  softPwmCreate(b, 0, 100); // setup pin for PWM [0..100]
  softPwmCreate(c, 0, 100); // setup pin for PWM [0..100]
  softPwmCreate(d, 0, 100); // setup pin for PWM [0..100]
}
```

```
void motors(int a, int b)
{ softPwmWrite(Left1, a);
  softPwmWrite(Left2, 0); // only needed for reversing
  softPwmWrite(Right1, b);
  softPwmWrite(Right2, 0); // only needed for reversing
}
```

```
int main (void)
{ LCDPrintf("init pins\n");
  pwm_init(26, 27, 28, 29);

  LCDPrintf("Motor Left and Right ramp up\n");
  for (int i=0; i<100; i++)
  { motors(i, i); delay(50);}

  LCDPrintf("Motor Left and Right ramp down\n");
  for (int i=100; i>0; i-)
  { motors(i, i); delay(50);}

  return 0 ;
}
```

In order to drive forward as well as backward, we only need to make a small modification to function *motors* and introduce a new function *drive* as shown in Program 10.3. For positive speeds, we set pins 1 and 2 of the motor as before (motor pin 2 is zero), but for negative speeds we reverse the two pin settings (motor pin 1 is zero) and use the negated (now positive) value for motor pin 2.

Program 10.3: Motor Driving Functions with Reversing

```
void drive(int Mot1, int Mot2, int speed)
{ if (speed > +100) speed = 100;   // limit range
  if (speed < -100) speed = -100; // limit range

  if(speed>0)
  { softPwmWrite(Mot1, speed); // Drive motor forward
    softPwmWrite(Mot2, 0);
  }
  else
  { softPwmWrite(Mot1, 0); // Drive motor backward
    softPwmWrite(Mot2, -speed);
  }
}
```

```
void motors(int a, int b)
{ drive(Left1, Left2, a);
  drive(Right1,Right2, b);
}
```

Finally, in Program 10.4, we are adding a simple $v\omega$ function for driving straight as well as curves. The first parameter v specifies the linear speed for the vehicle, and the second parameter ω specifies the angular rotation speed. We achieve this by subtracting ω from the left motor's linear speed and adding it to the right motor's, as a positive rotation should be counterclockwise.

The *main* program then demonstrates several straight and curved driving maneuvers. A stop command $vw(0,0)$ is inserted after every drive command.

Program 10.4: Simple $v\omega$ Driving Interface

```
void vw(int speed, int angular)
{ motors(speed-angular, speed+angular);
}
```

```
int main (void)
{ LCDPrintf("init pins\n");
  pwm_init(26, 27, 28, 29);

  LCDPrintf("Drive straight\n");
  vw(50, 0); delay(2000);
  vw( 0, 0); delay(1000);
```

```
LCDPrintf("Drive curve left\n");
vw(70, 30); delay(2000);
vw( 0, 0); delay(1000);

LCDPrintf("Drive curve right\n");
vw(70, -30); delay(2000);
vw( 0, 0); delay(1000);

LCDPrintf("Turn left\n");
vw(0, 25); delay(2000);
vw(0, 0); delay(1000);

LCDPrintf("Turn right\n");
vw(0, -25); delay(2000);
vw(0, 0); delay(1000);

return 0 ;
}
```

10.2.2 Four-Wheel-Drive Robot HC-4

Although using four wheels with four motors, the HC-4 robot is still a differential drive system, as there is no steering involved (compare with Ackermann steering later in this chapter). This robot kit is available from various sources[11] for around $20 including all mechanical parts and four motors.

Figure 10.6 shows all parts in the robot kit and the assembled vehicle. The two motors on the left side and the two motors on the right side are electrically linked together, so they get the same signal and always drive in the same direction at the same speed. This design creates a larger robot base capable of a higher payload for controller, sensors and battery, and eliminates the need for passive caster wheels.

The assembly of this kit is a bit more complex and requires soldering of the motor cables. Overall it will take about 1–2 hours. All software procedures are identical to the mini platform in the previous section.

---

[11]Joom, *4-wheel Smart Robot Car Chassis Kits HC-SR04*, online: https://www.joom.com/en/products/5d91a36c28fc7101013ab9d2.

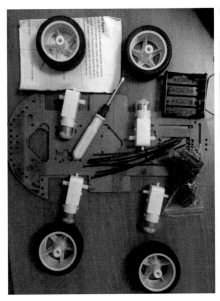

Figure 10.6: Four-wheel-drive robot HC-4 parts (left) and assembled (right)

10.2.3 SoccerBot S4

The *SoccerBot S4* robot (see Figure 10.8) was developed at the Robotics and Automation Lab at UWA and was the result of several generations of mobile robots over the years (see EyeBot robot generations in Figure 10.7). SoccerBot has a narrow wheelbase, which was accomplished by using gears and placing the motors side by side. Two servos are used as additional actuators, one for panning the camera and one for activating the ball kicking mechanism. Three PSDs are used for distance measuring to the left, front and right. Encoder feedback function *VWStalled* can also be used to detect collisions. The Raspberry Pi camera provides image input, and robots can send messages to each other or a base station via the Wi-Fi library *Radio* (see chapter on communication).

All low-level sensors, motors and encoders connect to the Atmel-based EyeBot7[12] I/O-board, which then connects to the Raspberry Pi (or any other embedded controller) via USB. Communication between the high-level controller and the IO-board is via simple ASCII messages,[13] such as

```
m 1 100
```

for driving motor no. 1 at full speed forward (+100%). See the Appendix for details.

---

[12]T. Bräunl, *EyeBot 7*, online: https://robotics.ee.uwa.edu.au/eyebot/.

[13]T. Bräunl, *EyeBot IO7 Interface Description*, online: https://robotics.ee.uwa.edu.au/eyebot/IO7.html.

Figure 10.7: EyeBot robot generations

Figure 10.8: SoccerBot S4

UWA Engineering's *CIIPS Glory* team of SoccerBot robots participated in both the RoboCup[14] small size league and FIRA RoboSot[15] competition. However, only RoboSot is a competition for actual autonomous mobile robots. The RoboCup small size league does allow the use of an overhead camera as a global sensor, remote computing on a central host system and remote controlling of all robots in a team. Therefore, its focus is more on centralized real-time image processing than on autonomous robot control.

Another typical laboratory task is to have one or two robots drive in an enclosed environment and search and collect cans (Figure 10.9). Each robot has to avoid obstacles (walls and other robots) and uses image processing to collect a can. Once a can has been detected, the robot should push it to the collection area.

---

[14]Asada, M., RoboCup-98: Robot Soccer World Cup II, Proceedings of the Second RoboCup Workshop, Paris, 1998.

[15]H. Cho, J.-J. Lee (Eds.), *Proceedings of the 2002 FIRA World Congress*, Seoul, Korea, May 2002.

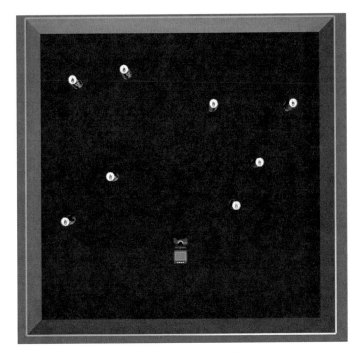

Figure 10.9: Can collection task in EyeSim simulator environment

As already discussed previously, driving a differential drive robot with the RoBIOS system and the Atmel-based EyeBot IO7 interface board can be accomplished at three different levels:

1. Motor level
   Setting individual speeds for each motor directly.
   Each subsequent MOTOR command will override the previous one.

   Commands:

   ```
   int MOTORDrive(int motor, int speed); // motor speed in percent
   int MOTORSpeed(int motor, int ticks); // motor speed in ticks
   ```

2. Vehicle speed control level
   Setting a linear speed and an independent angular speed (turning rate).
   Each subsequent *VW* command will override the previous one.

   Commands:

   ```
   int VWSetSpeed(int  linSpeed, int  angSpeed); // in mm/s and °/s
   int VWGetSpeed(int *linSpeed, int *angSpeed); // read speeds
   ```

3. Vehicle position control level
   These commands now specify a speed as well as a distance to be driven (and/or angle to be turned). Although control returns immediately back to

the calling program, the robot will continue on the specified path and stop automatically once the desired pose has been reached.

Any subsequent *VW* command will override the previous one immediately; i.e., the previous driving operation will not be completed in this case.

Commands:

```
int VWStraight(int dist, int lin_speed);      // Drive straight
int VWTurn(int angle, int ang_speed);         // Turn on the spot
int VWCurve(int dist, int angle, int lin_speed);// Drive curve
int VWDrive(int dx, int dy, int lin_speed);   // Drive xy distance
```

As an example for driving straight, either of the following program codes will work:

(a) Drive straight using *MOTOR* commands:

```
MOTORDrive(1, 50);
MOTORDrive(2, 50);
```

In this example, both motors will run at 50% of the maximum speed. Note that our *hardware description table* (HDT) links motor numbers (1, 2, 3, 4) to the actual motor output provides a lookup table to linearize motor speeds, and specifies the motor rotation direction. Setting the direction is especially important, as for a differential drive configuration, one motor has to run clockwise and the other one counterclockwise in order for the vehicle to drive forward.

To stop the driving motion, the program has to issue the commands:

```
MOTORDrive(1, 0);
MOTORDrive(2, 0);
```

(b) Drive straight using *VW* velocity control:

```
VWSetSpeed(1000,0);
```

This command will drive the vehicle forward at 1'000 mm/s or 1 m/s straight with zero angular speed. As before, HDT entries on wheel radius and the number of encoder ticks per revolution are required to calculate the required motor speed to achieve this overall vehicle speed. In our system, these calculations happen on the low-level Atmel controller, so a fast real-time feedback loop can be generated. The corresponding sensor commands for this implicit feedback are:

```
e1 = ENCODERRead(1);
e2 = ENCODERRead(2);
```

To stop the motion, the program has to issue the command:

```
VWSetSpeed(0,0);
```

(c)   Drive straight using *VW* position control:

```
VWStraight(500, 200);
```

This command will execute a complete controlled driving command over a distance of 500 mm at a speed of 200 mm/s. The vehicle will automatically come to a stop when it has reached the destination. Similar commands for turning on the spot or driving a curve are *VWTurn* and *VWCurve*.

No explicit command is required to stop the robot, but any subsequent *VW* command will let the vehicle continue with the new driving trajectory. Command *VWDone* can be used to check whether the driving operations has finished and *VWWait* can be used to suspend foreground processing until the robot has finished its current drive command. Therefore, a useful terminating command for this driving operation is:

```
VWWait();
```

Program 10.5 shows how to use low-level MOTOR commands to achieve a vω velocity control.

Program 10.5: Motor Control with Encoders

```
#include "eyebot.h"

void Mdrive(char* txt, int left, int right)
{ LCDPrintf("Distance %4d Direction %s\n", PSDGet(1), txt);
  MOTORDrive(1,left);
  MOTORDrive(2,right);
  OSWait(1500); // Drive for 1.5 sec
  LCDPrintf("Encoder %5d %5d\n", ENCODERRead(1),
  ENCODERRead(2));
}

int main ()
{ Mdrive("Forward", 60, 60);
  Mdrive("Backward", -60,-60);
  Mdrive("Left Curve", 20, 60);
```

```
    Mdrive("Right Curve", 60,  20);
    Mdrive("Turn Spot L",-20,  20);
    Mdrive("Turn Spot R", 20,-20);
    Mdrive("Stop",         0,   0);
    return 0;
}
```

10.3 Tracked Robots

A tracked mobile robot can be seen as a special case of a wheeled robot with differential drive. In fact, the only difference is the robot's better maneuverability in rough terrain and its higher friction in turns, due to its tracks and multiple points of contact with the surface.

Figure 10.10 shows EyeTrack, a model snow truck that was modified into a mobile robot. As discussed earlier, a model car can be simply connected to an embedded controller by driving its speed controller and steering servo from a microprocessor instead of a remote-controlled receiver. Normally, a tracked vehicle would have two driving motors, one for each track. In this particular model, however, there is only a single driving motor driving both tracks. The steering servo can selectively brake the left or right track, which allows to turn the vehicle.

Figure 10.10: EyeTrack robot and bottom view with sensors attached

EyeTrack is equipped with a number of sensors required for navigating rough terrain. Most of the sensors are mounted on the bottom of the robot. In Figure 10.10, right, the following are visible—top: PSD sensor; middle (left to right): digital compass, braking servo, electronic speed controller; and bottom: gyroscope. The sensors are in detail:

- Digital color camera

  The camera is mounted in the "driver cabin" and can be steered in all three axes by using three servos. This allows the camera to be kept stable when combined with the robot's orientation sensors shown below. The camera will stay actively locked on to a desired target, while the robot chassis is driving over the terrain.

- Digital compass

  The compass allows the determination of the robot's orientation at all times. This is especially important because this robot does not have two shaft encoders like a differential drive robot.

- Infrared PSDs

  The PSDs on this robot are not just applied to the front and sides in order to avoid obstacles. PSDs are also applied to the front and back at an angle of about $45°$, to detect steep slopes that the robot can only descend/ascend at a very slow speed or not at all.

- Piezo-gyroscopes

  Two gyroscopes are used to determine the robot's roll and pitch orientation, while yaw is covered by the digital compass. Since the gyroscopes' output is proportional to the rate of change, the data has to be integrated in order to determine the current orientation.

- Digital inclinometers

  Two inclinometers are used to support the two gyroscopes. The inclinometers used are fluid-based and return a value proportional to the robot's orientation. Although the inclinometer data does not require integration, they exhibit a time lag and are prone to oscillation. A sensor fusion of gyroscopes and inclinometers is used to overcome these problems.

There are numerous application scenarios for tracked robots with local intelligence. An important application is the use as a *rescue robot* in disaster areas. For example, the robot could still be remote-controlled and transmits a video image and sensor data; however, it might automatically adapt its speed according to its on-board orientation sensors, or even refuse to execute a driving command when its local sensors detect a potentially dangerous situation like a steep decline, which could lead to the loss of the robot.

10.4 Synchro-Drive

Synchro-drive is an extension to the robot design with a single driven and steered wheel. Here, however, we have three wheels that are all driven and all being steered. The three wheels are rotated together so they always point in the same driving direction (see Figure 10.11). This can be accomplished, for example, by using a single motor and a chain for steering and a single motor for driving all three wheels. Therefore, overall a synchro-drive robot still has only two degrees of freedom.

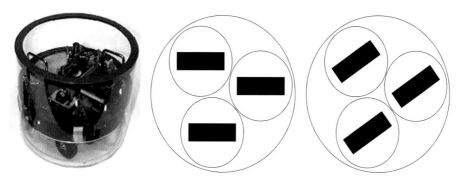

Figure 10.11: Xenia with schematic diagrams, TU Kaiserslautern

A synchro-drive robot is almost an omni-directional (holonomic) vehicle, in the sense that it can drive in any desired direction (for this reason, it usually has a cylindrical body shape). However, there are some limitations. The robot has to stop and realign its wheels when going from driving forward to driving sideways. Also, the robot cannot drive and rotate at the same time. Truly holonomic vehicles are introduced later in this chapter.

An example task that demonstrates the advantages of a synchro-drive is "complete area coverage" of a robot in a given environment. The real-world equivalent of this task is cleaning or vacuuming floors.

A behavior-based approach has been developed to perform a goal-oriented complete area coverage task. The algorithm was tested in simulation first and thereafter ported to the synchro-drive robot Xenia for validation in a real environment. It uses an external laser positioning system to provide absolute position information for the robot. Figure 10.12 depicts the result of a typical run using an occupancy grid in an area of 3.3 m × 2.3 m. The photograph was taken with a wide-angle overhead camera, which explains the cushion distortion. For details see Kamon, Rivlin;[16] Kasper, Fricke, von Puttkamer;[17] Peters et al.;[18] and von Puttkamer.[19]

---

[16]I. Kamon, E. Rivlin, *Sensory-Based Motion Planning with Global Proofs*, IEEE Transactions on Robotics and Automation, vol. 13, no. 6, Dec. 1997, pp. 814–822 (9).

[17]M. Kasper, G. Fricke, E. von Puttkamer, *A Behavior-Based Architecture for Teaching More than Reactive Behaviors to Mobile Robots*, 3rd European Workshop on Advanced Mobile Robots, EUROBOT '99, Zürich, Switzerland, September 1999, IEEE Press, pp. 203–210 (8).

[18]F. Peters, M. Kasper, M. Essling, E. von Puttkamer, *Flächendeckendes Explorieren und Navigieren in a priori unbekannter Umgebung mit low-cost Robotern*, 16. Fachgespräch Autonome Mobile Systeme AMS 2000, Karlsruhe, Germany, Nov. 2000.

[19]E. von Puttkamer, *Autonome Mobile Roboter*, Lecture notes, TU Kaiserslautern, Fachbereich Informatik, 2000.

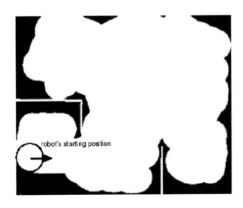 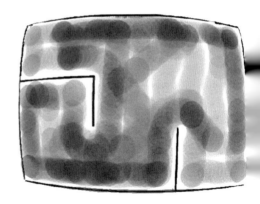

Figure 10.12: Result of a cleaning run, map and photograph, TU Kaiserslautern

10.5   Ackermann Steering

The standard drive and steering system of a rear-wheel-drive automobile consists of a motor that drives both rear wheels (linked though a differential) and two jointly steered passive front wheels. This is known as Ackermann steering, which has advantages and disadvantages when compared to differential drive:

+   Driving straight is not a problem, since the rear wheels are driven via a common axis.
−   Vehicle cannot turn on the spot, but requires a certain minimum radius.

Obviously, a different driving interface is required for Ackermann steering. Linear velocity and angular velocity are completely decoupled since they are generated by independent motors. This makes control a lot easier, especially the problem of driving straight. The driving library contains two independent velocity/position controllers, one for the rear driving wheels and one for the front steering wheels. The steering system requires a position controller, since the front wheels need to be set to a particular steering angle as opposed to the velocity controller for maintaining a constant speed on the back wheels. Additional sensors are required to indicate the zero position (and possibly the maximum left and right positions) for steering the front wheels.

Figure 10.13 shows the *Four Stooges* robot soccer team from The University of Auckland, which competed in the RoboCup Robot Soccer World Cup. Each robot has a model car base and is equipped with an EyeBot controller and a digital camera as its only sensor.

Figure 10.13: The Four Stooges, J. Baltes, University of Auckland

Arguably, the cheapest way of building a mobile robot is to use a model car. We retain the chassis, motors and servos, add a number of sensors and replace the remote-controlled receiver with an embedded controller. This gives us a ready-to-drive mobile robot in about an hour, as for the example in Figure 10.13.

The driving motor and steering servo of the model car are now directly connected to the controller and not to the receiver. However, we could retain the receiver and connect it to additional digital or analog inputs of the controller. This would allow us to transmit *high-level commands* to the controller from the car's remote control.

Model car with servo and speed controller.
Connecting a model car to an embedded controller is easy. Higher-quality model cars usually have proper servos for steering and either a servo or an electronic power controller for speed. Such a speed controller has the same connector and can be accessed similar to a servo. On the EyeBot IO7 interface board, we can simply plug the steering servo and speed controller into two of its servo outputs. That is all—the new autonomous vehicle is ready to go.

Driving control for steering and speed is achieved by using the command *SERVOSet*. One servo channel is used for setting the driving speed, and one servo channel is used for setting the steering angle. The servo value range [0, 255] is the mapped to the application, i.e.,

- Driving speed: 0 = full speed back, 128 = stop,   255 = full speed forward
- Steering angle: 0 = max. angle left; 128 = straight; 255 = max. angle right

Model car with integrated electronics.
The situation is a bit more complex for small, cheap model cars. These sometimes do not have proper servos, but for cost reasons contain a single

electronics box that comprises receiver and motor controller in a single unit. This is still not a problem, since the EyeBot IO7 board has two motor drivers already built in. For other embedded controllers, a separate motor driver has to be used, as outlined in the sections above. The model car's steering motor is being connected to the motor controller, and the steering sensor (usually a potentiometer) is connected to an analog input. We can then program the software equivalent to a servo by having the embedded controller in the control loop for the steering motor.

Figure 10.14 shows the wiring details. The driving motor has two wires, which need to be connected to the pins Motor + and Motor– of the first motor driver output. The steering motor has five wires, two for the motor and three for the position feedback. The two motor wires need to be connected to Motor + and Motor– of the second motor driver output. The connectors of the feedback potentiometer need to be connected to V_{CC} (3.3 V) and Ground, while the slider of the potentiometer is connected to a free analog input pin. Note that standard servos are only rated for 4.8 V, while others can take a higher voltage. Using a too high voltage can lead to severe motor damage.

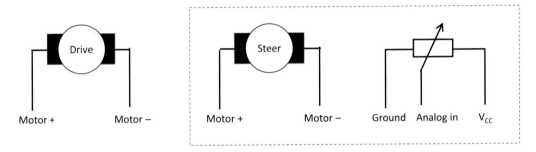

Figure 10.14: Model car connection diagram with pin numbers

Driving such a model car is a bit more complex than in the servo case. We can use the library routine *MOTORDrive* for setting the linear speed of the driving motors. However, we need to implement a simple PID or on/off controller for the steering motor, using the analog input from the potentiometer as feedback. As many cheap model cars cannot position their steering accurately, a reduced steering setting with only five or just three values (left, straight, right) can be sufficient.

Things get a lot easier, if a proper servo is available for steering. Program 10.6 demonstrates the control of an Ackermann vehicle, using the *MOTOR* command for controlling the drive motor and the *SERVO* command for setting the steering servo.

Program 10.6: Ackermann Drive with Steering Servo

```
#include "eyebot.h"

void Mdrive(char* txt, int drive, int steer)
/* Print txt and drive motors and encoders for 1.5s */
{ LCDPrintf("%s\n", txt);
  MOTORDrive(1, drive);
  SERVOSet (1, steer);
}

int main ()
{ LCDPrintf("Ackermann Steering\n");
  Mdrive("Forward",      20, 128); OSWait(1000);
  Mdrive("Backward",    -20, 128); OSWait(1000);
  Mdrive("Left Curve",   10,   0); OSWait(2000);
  Mdrive("Right Curve",  10, 255); OSWait(2000);
  Mdrive("Stop",          0,   0);
  return 0;
}
```

Figure 10.15 (left) shows a converted model car with Raspberry Pi, camera and Lidar sensor on top. The alternative to modifying a model car is to purchase an Ackermann robot kit, such as the SunFounder[20] Smart Video Car Kit for Raspberry Pi (at around $100) as shown in Figure 10.15 (right).

Figure 10.15: Converted Ackermann model car (left) and Ackermann robot kit (right)

---

[20]SunFounder, *Smart Video Car Kit for Raspberry Pi*, online: https://www.sunfounder.com/products/smart-video-car-kit-for-raspberrypi.

10.6 Omni-Directional Robots

All the robots discussed in this chapter so far, with the exception of synchro-drive vehicles, have the same deficiency—they cannot drive in all possible directions. For this reason, these robots are called *non-holonomic*. In contrast, a *holonomic* or omni-directional robot is capable of driving in any direction. Most non-holonomic robots cannot drive in a direction perpendicular to their driven wheels. For example, a differential drive robot can drive forward/backward, in a curve, or turn on the spot, but it cannot drive sideways. The omni-directional robots introduced in this section, however, are capable of driving in any direction in a 2D plane.

10.6.1 Mecanum Wheels

The marvel behind the omni-directional drive design presented in this chapter is Mecanum wheels. This wheel design has been developed and patented by the Swedish company Mecanum AB with Bengt Ilon in 1973 (see article by Jonsson[21]), so it has been around for quite a while. Further details on Mecanum wheels and omni-directional drives can be found in papers by Carlisle;[22] Agullo, Cardona, Vivancos;[23] and Dickerson, Lapin.[24]

Figure 10.16: Mecanum wheel designs with rollers at 45°

[21]S. Jonsson, *New AGV with revolutionary movement*, in R. Hollier (Ed.), Automated Guided Vehicle Systems, IFS Publications, Bedford, 1987, pp. 345–353 (9).

[22]B. Carlisle, *An omni-directional mobile robot*, in B. Rooks (Ed.): Developments in Robotics 1983, IFS Publications, North-Holland, Amsterdam, 1983, pp. 79–87 (9).

[23]J. Agullo, S. Cardona, J. Vivancos, *Kinematics of vehicles with directional sliding wheels*, Mechanical Machine Theory, vol. 22, 1987, pp. 295–301 (7).

[24]S. Dickerson, B. Lapin, *Control of an omni-directional robotic vehicle with Mecanum wheels*, Proceedings of the National Telesystems Conference 1991, NTC'91, vol. 1, 1991, pp. 323–328 (6).

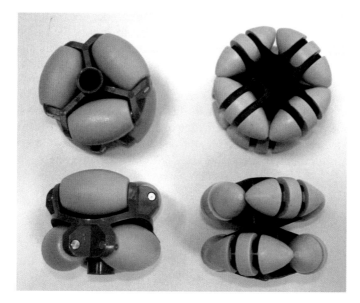

Figure 10.17: Mecanum wheel designs with rollers at 90°

There are a number of different Mecanum wheel variations; Figure 10.16 shows two of our designs. Each wheel's surface is covered with a number of free rolling cylinders. It is important to stress that the wheel hub is driven by a motor, but the rollers on the wheel surface are not. These are held in place by ball-bearings and can freely rotate about their axis. While the wheels in Figure 10.16 have the rollers at $\pm$ 45° and there is a left-hand and a right-hand versions of this wheel type, there are also Mecanum wheels with rollers set at 90° (Figure 10.17), and these do not require left-hand/right-hand versions.

A Mecanum-based robot can be constructed with either three or four independently driven Mecanum wheels. Vehicle designs with three Mecanum wheels require wheels with rollers set at 90° to the wheel axis, while the design we are following here is based on four Mecanum wheels and requires the rollers to be at an angle of $\pm$ 45° to the wheel axis. For the construction of a robot with four Mecanum wheels, two left-handed wheels (rollers at + 45° to the wheel axis) and two right-handed wheels (rollers at –45° to the wheel axis) are required (see Figure 10.18).

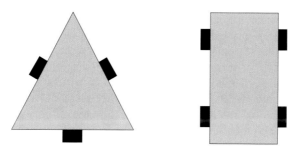

Figure 10.18: Three-wheel and four-wheel omni-directional vehicles

Although the rollers are freely rotating, this does not mean the robot can spin its wheels without moving. This would only be the case if the rollers were placed parallel to the wheel axis. However, the Mecanum wheels have the rollers placed at an angle (45° in Figure 10.16). Looking at an individual wheel (Figure 10.19, view from the bottom through a "glass floor"), the force generated by the wheel rotation acts on the ground through the one roller that has ground contact. At this roller, the force can be split in a vector parallel to the roller axis and a vector perpendicular to the roller axis. The force perpendicular to the roller axis will result in a small roller rotation, while the force parallel to the roller axis will exert a force on the wheel and thereby on the vehicle.

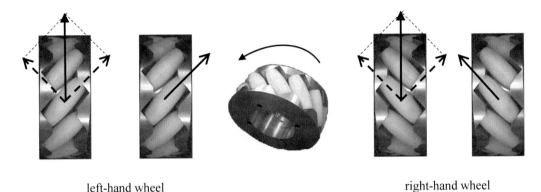

left-hand wheel
seen from below

right-hand wheel
seen from below

Figure 10.19: Mecanum principle, vector decomposition

Since Mecanum wheels do not appear individually, but here in a four-wheel assembly, the resulting wheel forces at 45° from each wheel have to be combined to determine the overall vehicle motion. If the two wheels as shown in Figure 10.19 are the robot's front wheels and both are rotated forward, then each of the two resulting 45° force vectors can be split into a forward and a sideways force. The two forward forces add up, while the two sideways forces (one to the left and one to the right) cancel each other out.

10.6.2 Omni-Directional Drive

Figure 10.20, left, shows the situation for the full robot with four independently driven Mecanum wheels. In the same situation as before, i.e., all four wheels being driven forward, we now have four vectors pointing forward that are added up and four vectors pointing sideways, two to the left and two to the right, that cancel each other out. Therefore, although the vehicle's chassis is subjected to additional perpendicular forces, the vehicle will simply drive straight forward.

In Figure 10.20, middle, assume wheels 1 and 4 are driven backward, and wheels 2 and 4 are driven forward. In this case, all forward/backward velocities cancel each other out, but the four vector components to the left add up and let the vehicle slide to the left.

The third case is shown in Figure 10.20, right. No vector decomposition is necessary in this case to reveal the overall vehicle motion. It can be clearly seen that the robot motion will be a clockwise rotation about its center.

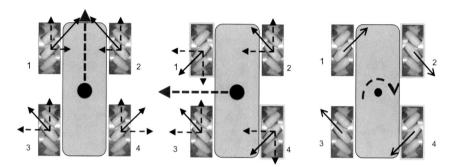

Figure 10.20: Mecanum principle for driving forward, sliding sideways and rotating; wheels without outline rotate forward, wheels with yellow outline rotate backward (seen from below)

The following list shows the basic motions, driving forward, driving sideways and turning on the spot, with their corresponding wheel directions (see Figure 10.21).

- Driving forward: all four wheels forward
- Driving backward: all four wheels backward
- Sliding left: 1, 4: backward;   2, 3: forward
- Sliding right: 1, 4: forward;   2. 3: backward
- Turning clockwise on the spot: 1, 3: forward;   2, 4: backward
- Turning counterclockwise: 1, 3: backward;   2, 4: forward

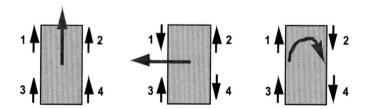

Figure 10.21: Kinematics of omni-directional robot

So far, we have only considered a Mecanum wheel spinning at full speed forward or backward. However, by varying the individual wheel speeds and by adding linear interpolations of basic movements, we can achieve driving directions along any vector in the 2D plane.

10.6.3 Omni-Directional Robot Design

We have developed three different Mecanum-based omni-directional robots: the demonstrator models Omni-1 (Figure 10.22, left), Omni-2 (Figure 10.22, right) and a full size omni-directional wheelchair (Figure 10.23).

Figure 10.22: Omni-1 and Omni-2

The first design, Omni-1, has the motor/wheel assembly tightly attached to the robot's chassis. Its Mecanum wheel design has rims that only leave a few millimeters clearance for the rollers. As a consequence, the robot can drive very well on hard surfaces, but it loses its omni-directional capabilities on softer surfaces like carpet. Here, the wheels will sink in a bit and the robot will then drive on the wheel rims, losing its capability to drive sideways.

Figure 10.23: Completed wheelchair and CAD design by Ben Woods, UWA

The deficiencies of Omni-1 led to the development of Omni-2. This robot has individual cantilever wheel suspensions with shock absorbers. This helps to navigate rougher terrain, since it will keep all four wheels on the ground. The robot also has a completely rimless Mecanum wheel design where rollers are held from the middle, which avoids sinking in and allows omni-directional driving on softer surfaces.

After experimenting with the scaled models, we built a full size omni-directional wheelchair using the more robust rimmed Mecanum wheel design. The wheelchair has been constructed for a payload of 100 kg and uses four powerful wheels controlled by an EyeBot embedded system through external power amplifiers. The wheelchair can be driven via a joystick with three degrees of freedom (*dof*), which are driving forward/backward, sliding sideways and rotating the wheelchair clockwise/counterclockwise—or any combination of these three commands.

The omni-directional wheelchair had been developed for severely handicapped drivers by Ben Woods[25] in 2006. Additional infrared distance sensors are used to keep a minimum distance from walls or obstacles, while an emergency switch will bring it to a full stop. Figure 10.23 shows Ben testing the wheelchair (left) and his CAD deign (right).

10.6.4 Omni-Drive Program

Operating omni-directional robots obviously requires an extended driving interface. The $v\omega$ routines for differential drive or Ackermann-steering robots are not sufficient since we also need to specify a vector for the driving direction in addition to a possible rotation. Also, for an omni-directional robot it is possible to drive along a vector and rotate at the same time, which has to be reflected by the software interface.

The code in Program 10.7 uses a low-level interface that sets each motor speed individually to achieve the basic omni-directional driving actions: forward/backward, sideways and turning on the spot.

Subroutine *Mdrive* sets each of the four motors (front left, front right, back left, back right) to their speed given by the associated parameter and holds it for 1.5 s (*OSWait* for 1'500 ms). After that it prints the updated encoder readings for each of the four wheels and then returns.

---

[25]B. Woods, *Omni-Directional Wheelchair*, Final Year Honours Thesis supervised by T. Bräunl, The University of Western Australia, Mechanical Eng., Oct. 2006, pp. (141).

Program 10.7: Omni-Directional Driving

```
#include "eyebot.h"
#define S 25 // Speed

void Mdrive(char* txt, int Fleft, int Fright, int Bleft, int
Bright)
{ LCDPrintf("%s\n", txt);
  MOTORDrive(1, Fleft);
  MOTORDrive(2, Fright);
  MOTORDrive(3, Bleft);
  MOTORDrive(4, Bright);
  OSWait(1500);
  LCDPrintf("Enc.%5d %5d\n %5d %5d\n", ENCODERRead(1),
    ENCODERRead(2), ENCODERRead(3), ENCODERRead(4));
}

int main ()
{ Mdrive("Forward",     S, S, S, S);
  Mdrive("Backward",   -S,-S,-S,-S);
  Mdrive("Left",       -S, S, S,-S);
  Mdrive("Right",       S,-S,-S, S);
  Mdrive("Left45",      0, S, S, 0);
  Mdrive("Right45",     S, 0, 0, S);
  Mdrive("Turn Spot L",-S, S,-S, S);
  Mdrive("Turn Spot R", S,-S, S,-S);
  nMdrive("Stop",       0, 0, 0, 0);
}
```

The *main* program demonstrates driving the standard maneuvers such as forward, backward and rotating left as a linear combination of wheel speeds. In this simple example, we only use a single fixed speed $\pm$ S or 0. More complex maneuvers can be driven with different speeds for each wheel and with wheel speeds varying over the duration of a driving pattern.

10.7 Drive Kinematics

In this chapter, we want to answer two fundamental questions:

(a) Given each wheel's movements,
    what is the vehicle's change in position and orientation (pose)?
(b) Which individual wheel speeds will be required,
    in order to achieve a desired change in vehicle pose?

The solution to question (a) is called *forward kinematics*, the solution to question (b) is called *inverse kinematics*. In the following sections, we will look at the kinematics of the three most prominent drive systems: Differential drive, Ackermann drive and Omni-directional drive.

10.7.1 Differential Drive Kinematics

In order to obtain a differential drive vehicle's current trajectory, we need to constantly monitor its shaft encoders (Figure 10.24).

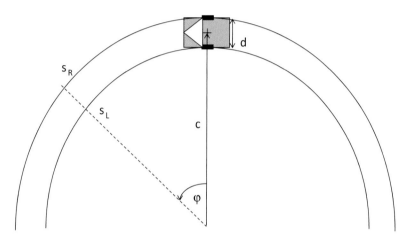

Figure 10.24: Trajectory calculation for differential drive

We know:

- r wheel radius
- d distance between driven wheels
- $ticks\_per\_rev$ number of encoder ticks for one full wheel revolution
- $ticks_L$ number of ticks during measurement in left encoder
- $ticks_R$ number of ticks during measurement in right encoder

First, we determine the values of s_L and s_R in meters, which are the distances traveled by the left and right wheels, respectively. Dividing the measured ticks by the number of ticks per revolution gives us the number of wheel revolutions. Multiplying this by the wheel circumference gives the traveled distance in meters:

$$s_L = 2\pi \cdot r \cdot \text{ticks}_L/\text{ticks\_per\_rev}$$
$$s_R = 2\pi \cdot r \cdot \text{ticks}_R/\text{ticks\_per\_rev}$$

We already know the distance the vehicle has traveled, i.e.,

$$s = (s_L + s_R)/2$$

This formula works for a robot driving forward, backward or turning on the spot. We still need to know the vehicle's rotation φ over the distance traveled. Assuming the vehicle follows a circular segment, we can define s_L and s_R as the traveled part of a full circle (φ in radians) multiplied by each wheel's turning radius. If the turning radius of the vehicle's center is c, then during a left turn, the turning radius of the right wheel is $c + d/2$, while the turning radius of the left wheel is $c - d/2$. Both circles have the same center.

$$s_R = \varphi \cdot (c + d/2)$$
$$s_L = \varphi \cdot (c - d/2)$$

Subtracting both equations eliminates c:

$$s_R - s_L = \varphi \cdot d$$

And finally solving for φ:

$$\varphi = (s_R - s_L)/d$$

Using wheel velocities $v_{L,R}$ instead of driving distances $s_{L,R}$ and using $\dot{\theta}_{L,R}$ as wheel rotations per second with radius r for left and right wheels, we get:

$$v_L = 2\pi r \cdot \dot{\theta}_L$$
$$v_R = 2\pi r \cdot \dot{\theta}_R$$

The formula specifying the velocities of a differential drive vehicle can now be expressed as a matrix. This is called the *forward kinematics*:

$$\begin{bmatrix} v \\ \omega \end{bmatrix} = 2\pi r \begin{bmatrix} 1/2 & 1/2 \\ -1/d & 1/d \end{bmatrix} \begin{bmatrix} \dot{\theta}_L \\ \dot{\theta}_R \end{bmatrix}$$

where

v	is the vehicle's linear speed (equals ds/dt or $\dot{s}$),
ω	is the vehicle's rotational speed (equals $d\theta/dt$ or $\dot{\theta}$),
$\dot{\theta}_{L,R}$	are the individual wheel speeds in revolutions per second,

r   is the wheel radius,

d   is the distance between the two driven wheels.

The *inverse kinematics* is derived from the previous formula, solving for the individual wheel speeds. It tells us the required wheel speeds for a desired vehicle motion (linear and rotational speed). We can find the inverse kinematics by inverting the 2 × 2 matrix of the forward kinematics:

$$\begin{bmatrix} \dot{\theta}_L \\ \dot{\theta}_R \end{bmatrix} = \frac{1}{2\pi r} \begin{bmatrix} 1 & -d/2 \\ 1 & d/2 \end{bmatrix} \begin{bmatrix} v \\ \omega \end{bmatrix}$$

10.7.2 Ackermann Drive Kinematics

If we consider the motion in a vehicle with Ackermann steering, then its front wheel motion is identical to the vehicle's forward motion s in the direction of the wheels with steering angle α (see Figure 10.25). The vehicle's movement vector s can be decomposed into a forward and a down component:

$$\text{forward} = s \cdot \cos \alpha$$
$$\text{down} = s \cdot \sin \alpha$$

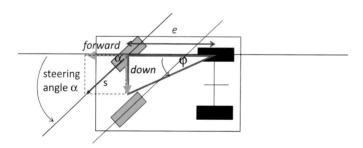

Figure 10.25: Motion of vehicle with Ackermann steering

If e denotes the distance between front and back wheels, then the overall vehicle rotation angle is:

$$\varphi = \text{down}/e$$

Note that this formula calculates φ directly and not sin (φ) as one might expect, since the front wheels follow the arc of a circle when turning. The calculation for the traveled distance and angle of a vehicle with Ackermann drive is shown in Figure 10.26, with:

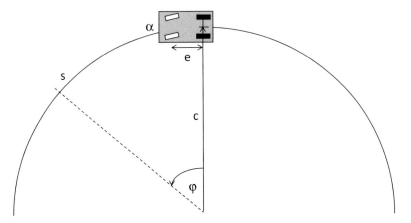

Figure 10.26: Trajectory calculation for Ackermann steering

α	steering angle,
e	distance between front and back wheels,
s_{front}	distance driven, measured at front wheels,
$\dot{\theta}$	driving wheel speed in revolutions per second,
s	total driven distance along arc,
φ	total vehicle rotation angle.

The trigonometric relationship between the vehicle's steering angle and overall movement is:

$$s = s_{front}$$
$$\varphi = s_{front} \cdot \sin(\alpha/e)$$

Expressing this relationship as velocities, we get:

$$v_{forward} = v_{motor} = 2\pi r \cdot \dot{\theta}$$
$$\omega = v_{motor} \cdot (\sin \alpha)/e$$

Therefore, the forward kinematics formula becomes relatively simple:

$$\begin{bmatrix} v \\ \omega \end{bmatrix} = 2\pi r \cdot \dot{\theta} \cdot \begin{bmatrix} 1 \\ \frac{\sin \alpha}{e} \end{bmatrix}$$

Note that this formula changes if the vehicle is rear-wheel drive and the wheel velocity is measured there. In this case, the *sin* function has to be replaced by the *tan* function.

10.7.3 Omni-Drive Kinematics

The *forward kinematics* for our four-wheel omni-drive system is a matrix formula that specifies which direction the robot will drive in (linear velocity v_x along the robot's center axis, v_y perpendicular to it) and what its rotational velocity ω will be for the *given* individual wheel speeds $\dot{\theta}_{FL}, \ldots, \dot{\theta}_{RB}$ and wheel distances d (left/right) and e (front/back):

$$\begin{bmatrix} v_x \\ v_y \\ \omega \end{bmatrix} = 2\pi r \begin{bmatrix} 1/4 & 1/4 & 1/4 & 1/4 \\ -1/4 & 1/4 & -1/4 & 1/4 \\ \frac{-1}{2(d+e)} & \frac{1}{2(d+e)} & \frac{-1}{2(d+e)} & \frac{1}{2(d+e)} \end{bmatrix} \cdot \begin{bmatrix} \dot{\theta}_{FL} \\ \dot{\theta}_{FR} \\ \dot{\theta}_{BL} \\ \dot{\theta}_{BR} \end{bmatrix}$$

with:

$\dot{\theta}_{FL}$, *etc.*	four individual wheel speeds in revolutions per second,
r	wheel radius,
d	distance between left and right wheel pairs,
e	distance between front and back wheel pairs,
v_x	vehicle velocity in forward direction,
v_y	vehicle velocity in sideways direction,
ω	vehicle rotational velocity.

The *inverse kinematics* is a matrix formula that specifies the required individual wheel speeds for given desired linear and angular velocities (v_x, v_y, ω) and can be derived by inverting the matrix of the forward kinematics. For details see Viboonchaicheep, Shimada, Kosaka,[26] 2003.

$$\begin{bmatrix} \dot{\theta}_{FL} \\ \dot{\theta}_{FR} \\ \dot{\theta}_{BL} \\ \dot{\theta}_{BR} \end{bmatrix} = \frac{1}{2\pi r} \begin{bmatrix} 1 & -1 & -(d+e)/2 \\ 1 & +1 & +(d+e)/2 \\ 1 & +1 & -(d+e)/2 \\ 1 & -1 & +(d+e)/2 \end{bmatrix} \cdot \begin{bmatrix} v_x \\ v_y \\ \omega \end{bmatrix}$$

---

[26]P. Viboonchaicheep, A. Shimada, Y. Kosaka, *Position rectification control for Mecanum wheeled omni-directional vehicles*, 29th Annual Conference of the IEEE Industrial Electronics Society, IECON'03, vol. 1, Nov. 2003, pp. 854–859 (6).

10.8 Tasks

1. Write a program to drive along a triangle (Differential drive or Ackermann steering).

2. Write a program to reach a waypoint at relative coordinates (x,y) as seen from the robot.

3. Write a program to reach a waypoint at relative coordinates (x,y) as seen from the robot and arrive at specified global orientation α.

4. Store a desired path as a sequence of waypoints in an array, then write a program for a robot to follow the sequence of waypoints. Whenever the robot is within an error margin of one waypoint, advance control to the next waypoint in the sequence.
   Start with a sine curve as the first waypoint sequence.

5. Implement (x,y,α) driving routines for an omni-directional robot using the inverse kinematics equation. Use EyeSim as the first step for testing (and if you do not have a real omni-directional robot).

6. Calculate kinematics equations (forward and inverse) for the 3-wheel omni-directional robot with rollers at 90° angles.

WALKING ROBOTS

Walking robots are an important alternative to driving robots, since the majority of the world's land area is unpaved. Although driving robots are more specialized and better adapted to flat surfaces—they can drive faster and navigate with higher precision—walking robots can be employed in more general environments. Walking robots follow nature by being able to navigate rough terrain, or even climb stairs or step over obstacles in a standard household situation, which would rule out most driving robots.

Robots with six or more legs have the advantage of stability. In a typical walking pattern of a six-legged robot, three legs are on the ground at all times, while three legs are moving. This gives static balance while walking, provided the robot's center of mass is within the triangle formed by the three legs on the ground. Four-legged robots are considerably harder to balance, but they are still fairly stable when compared to the dynamics of biped robots. Biped robots are the most difficult to balance, with only one leg on the ground and one leg in the air during walking. Static balance for biped robots can be achieved if the robot's feet are relatively large and the ground contact areas of both feet are overlapping. However, this is not the case in human-like android robots, which require dynamic balance for walking.

A collection of related research papers can be found in Rückert et al.[1] and Cho and Lee.[2]

Balancing robots have gained popularity after the introduction of the Segway vehicle;[3] however, many similar vehicles have been developed before. Most balancing robots are based on the inverted pendulum principle and have either wheels or legs. They can be studied in their own right or as a precursor for biped walking robots; for example, to experiment with individual sensors or

---

[1]U. Rückert, J. Sitte, U. Witkowski (Eds.), *Autonomous Minirobots for Research and Edutainment—AMiRE2001*, Proceedings of the 5th International Heinz Nixdorf Symposium, HNI-Verlagsschriftenreihe, no. 97, Univ. Paderborn, Oct. 2001.

[2]H. Cho, J.-J. Lee (Eds.) *Proceedings 2002 FIRA Robot World Congress*, Seoul, Korea, May 2002.

[3]Segway, *Welcome to the evolution in mobility*, http://www.segway.com, 2020.

© The Author(s), under exclusive license to Springer Nature Singapore Pte Ltd. 2022
T. Bräunl, *Embedded Robotics*,
https://doi.org/10.1007/978-981-16-0804-9_11

actuators. Inverted pendulum models have been used as the basis of a number of bipedal walking strategies: Caux et al.;[4] Kajita and Tani;[5] Ogasawara and Kawaji;[6] and Park and Kim.[7] The dynamics can be constrained to two dimensions and the cost of producing an inverted pendulum robot is relatively low, since it has a minimal number of moving parts.

11.1 Balancing Robots

The physical balancing robot is an inverted pendulum with two independently driven motors, to allow for balancing, as well as driving straight and turning (Figure 11.1). Tilt sensors, inclinometers, accelerometers, gyroscopes and digital cameras are used for experimenting with this robot and are discussed below.

Figure 11.1: BallyBot balancing robot

- Gyroscope
  Piezo-electric gyroscopes are designed for use in remote-controlled vehicles, such as model helicopters. The gyroscope modifies a servo control signal by an amount proportional to its measure of angular velocity. Instead of using the gyro to control a servo, we read back the gyro's modified servo output signal to obtain a measurement of

---

[4]S. Caux, E. Mateo, R. Zapata, *Balance of biped robots: special double-inverted pendulum*, IEEE International Conference on Systems, Man, and Cybernetics, 1998, pp. 3691–3696 (6).

[5]S. Kajita, K. Tani, *Experimental Study of Biped Dynamic Walking in the Linear Inverted Pendulum Mode*, IEEE Control Systems Magazine, vol. 16, no. 1, Feb. 1996, pp. 13–19 (7).

[6]K. Ogasawara, S. Kawaji, *Cooperative motion control for biped locomotion robots*, IEEE International Conference on Systems, Man, and Cybernetics, 1999, pp. 966–971 (6).

[7]J. Park, K. Kim, *Bipedal Robot Walking Using Gravity-Compensated Inverted Pendulum Mode and Computed Torque Control*, IEEE International Conference on Robotics and Automation, 1998, pp. 3528–3533 (6).

angular velocity. An estimate of angular displacement is obtained by integrating the velocity signal over time.

- Acceleration sensors
  These sensors output an analog signal, proportional to the acceleration in the direction of the sensor's axis of sensitivity. Mounting two acceleration sensors at 90° angles means that we can measure the translational acceleration experienced by the sensors in the plane through which the robot moves. Since gravity provides a significant component of this acceleration, we are able to estimate the orientation of the robot.

- Inclinometer
  An inclinometer is used to support the gyroscope. Although the inclinometer cannot be used alone because of its time lag, it can be used to reset the software integration of the gyroscope data when the robot is close to resting in an upright position.

- Inertial measurement unit (IMU)
  An inertial measurement unit typically combines 3 accelerometers and 3 inclinometers to cover all motion axes. Some models also include up to 3 magnetometers. This greatly simplifies system design compared to interfacing and preprocessing individual sensors.

- Digital camera
  Experiments have been conducted in using an artificial horizon or, more generally, the optical flow of the visual field to determine the robot's trajectory for balancing.

Variable	Description	Sensor
x	Position	Shaft encoders
v	Velocity	Differentiated encoder reading
θ	Angle	Integrated gyroscope reading
ω	Angular velocity	Gyroscope

Table 11.1 State variables

The PD control strategy selected for implementation on the physical robot requires the measurement of four state variables: $\{x, v, \theta, \omega\}$, see Table 11.1. An implementation relying on the gyroscope alone does not completely solve the problem of balancing the physical robot, remaining balanced on average for 5–15 s before falling over. This is an encouraging initial result, but it is still not a robust system. The system's balancing was greatly improved by adding an inclinometer to the robot. Although the robot was not able to balance with the inclinometer alone, because of inaccuracies and the time lag of the sensor, the combination of inclinometer and gyroscope proved to be the best solution. While the integrated data of the gyroscope gives accurate short-term orientation data, the inclinometer is used to recalibrate the robot's orientation value as well as the gyroscope's zero position at certain time intervals when the robot is moving at a low speed.

Software simulation can be used to investigate techniques for control systems that balance inverted pendulums. The first method investigated was an adaptive control system, based on a backpropagation neural network, which learns to balance the simulation with feedback limited to a single failure signal when the robot falls over. Disadvantages of this approach include the requirement for a large number of training cycles before satisfactory performance is obtained. Additionally, once the network has been trained, it is not possible to make quick manual changes to the operation of the controller. For these reasons, we selected a different control strategy for the physical robot.

An alternative approach is to use a simple PD control loop of the form:

$$u(k) = W \cdot X(k)$$

where:

u(k)   Horizontal force applied by motors to the ground,
X(k) kth measurement of the system state,
W   Weight vector applied to measured robot state.

Tuning of the control loop was performed manually, using the software simulation to observe the effect of modifying loop parameters. This approach quickly yielded a satisfactory solution in the software model and was selected for implementation on the physical robot (see Sutherland[8]).

A number of problems have been encountered with the sensors used. Over time the observed *zero velocity* signal received from the gyroscope can deviate (Figure 11.2). This means that not only does our estimate of the angular velocity become inaccurate, but since our estimate of the angle is the integrated signal, it becomes inaccurate as well.

The control system assumes that it is possible to accurately generate a horizontal force using the robot's motors. The force produced by the motors is related to the voltage applied, as well as the current shaft speed and friction. This relationship was experimentally determined and includes some simplification and generalization.

In certain situations, the robot needs to generate considerable horizontal force to maintain balance. On some surfaces this force can exceed the frictional force between the robot tires and the ground. When this happens, the robot loses track of its displacement, and the control loop no longer generates the correct output. This can be observed by sudden, unexpected changes in the robot displacement measurements.

---

[8]A. Sutherland, *Torso Driven Walking Algorithm for Dynamically Balanced Variable Speed Biped Robots*, Ph.D. Thesis, supervised by T. Bräunl, The University of Western Australia, June 2006, pp. (398), web: http://robotics.ee.uwa.edu.au/theses/2006-Biped-Suther-land-PhD.pdf.

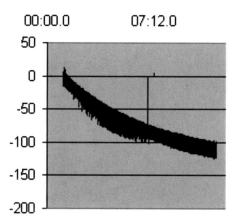

Figure 11.2: Measurement data revealing gyro drift

Program 11.1 is an excerpt from the balancing program. It shows the periodic timer routine for reading sensor values and updating the system state. Details of this control approach are described in Sutherland, Bräunl[9,10].

Program 11.1: Balance Timer Routine

```
void CGyro::TimerSample()
{ ...
  iAngVel = accreadX();
  if (iAngVel > -1)
  {
    iAngVel = iAngVel;
    // Get the elapsed time
    iTimeNow = OSGetCount();
    iElapsed = iTimeNow - g_iSampleTime;
    // Correct elapsed time if rolled over!
    if (iElapsed < 0) iElapsed += 0xFFFFFFFF; // ROLL OVER
    // Correct the angular velocity
    iAngVel -= g_iZeroVelocity;
    // Calculate angular displacement
    g_iAngle += (g_iAngularVelocity * iElapsed);
    g_iAngularVelocity = -iAngVel;
```

[9]A. Sutherland, T. Bräunl, *Learning to Balance an Unknown System*, Proceedings of the IEEE-RAS International Conference on Humanoid Robots, Humanoids 2001, Waseda University, Tokyo, Nov. 2001, pp. 385–391 (7).

[10]A. Sutherland, T. Bräunl, *An Experimental Platform for Researching Robot Balance*, 2002 FIRA Robot World Congress, Seoul, May 2002, pp. 14–19 (6).

```
g_iSampleTime = iTimeNow;
// Read inclinometer (drain residual values)
iRawADReading = OSGetAD(INCLINE_CHANNEL);
iRawADReading = OSGetAD(INCLINE_CHANNEL);
// If recording, and we have started...store data
if (g_iTimeLastCalibrated > 0)
  { ... /* re-calibrate sensor */
  }
}
// If correction factor remaining to apply, apply it!
if (g_iGyroAngleCorrection > 0)
{ g_iGyroAngleCorrection -= g_iGyroAngleCorrectionDelta;
  g_iAngle -= g_iGyroAngleCorrectionDelta;
}
}
```

A second two-wheel balancing robot had been built in a project by Ooi[11], Figure 11.3. Like the first robot it uses a gyroscope and inclinometer as sensors, but it employs a Kalman filter method for balancing (see Kalman[12] and Del Gobbo et al.[13]). A number of Kalman-based control algorithms have been implemented and compared with each other, including a pole-placement controller and a linear quadratic regulator (LQR); see Nakajima et al.[14] and Takahashi et al.[15]. An overview of Ooi's robot's control system is shown in Figure 11.4.

The robot also accepts driving commands from an infrared remote control, which are interpreted as a bias by the balance control system. They are used to drive the robot forward/backward or turn left/right on the spot.

[11]R. Ooi, *Balancing a Two-Wheeled Autonomous Robot*, B.E. Honours Thesis, The Univ. of Western Australia, Mechanical Eng., supervised by T. Bräunl, 2003, p. 56.

[12]R. Kalman *A New Approach to Linear Filtering and Prediction Problems*, Transactions of the ASME—Journal of Basic Engineering, Series D, vol. 82, 1960, pp. 35–45.

[13]D. Del Gobbo, M. Napolitano, P. Famouri, M. Innocenti, *Experimental application of extended Kalman filtering for sensor validation*, IEEE Transactions on Control Systems Technology, vol. 9, no. 2, 2001, pp. 376–380 (5).

[14]R. Nakajima, T. Tsubouchi, S. Yuta, E. Koyanagi, *A Development of a New Mechanism of an Autonomous Unicycle*, IEEE International Conference on Intelligent Robots and Systems, IROS 1997, vol. 2, 1997, pp. 906–912 (7).

[15]Y. Takahashi, N. Ishikawa, T. Hagiwara, *Inverse pendulum controlled two wheel drive system*, Proceedings of the 40th SICE Annual Conference, International Session Papers, SICE 2001, 2001, pp. 112–115 (4).

Figure 11.3: Balancing robot design by Ooi, UWA 2003

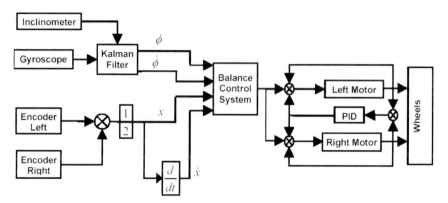

Figure 11.4: Kalman-based control system by Ooi, UWA 2003

11.2 Six-Legged Robots

Figure 11.5 shows two different six-legged robot designs. The *Crab* robot was built from scratch, while *Hexapod* utilizes walking mechanics from Lynxmotion[16] in combination with an EyeBot controller and additional sensors.

The two robots differ in their mechanical designs, which might not be obvious from the photos. Both robots use two servos per leg, to achieve leg lift (up/down) and leg swing (forward/backward) motion. However, Crab uses a mechanism that allows all servos to be firmly mounted on the robot's main chassis, while Hexapod only has the swing servos mounted to the robot body; the lift servos are mounted on small sub-assemblies, which are moved with each leg.

---

[16]Lynxmotion, *Lynxmotion – Imagine it. Build it. Control it.*, online: http://www.lynxmotion.com.

Figure 11.5: Crab six-legged walking robot and Hexapod robot, both Univ. Stuttgart

The second major difference is in sensor equipment. While Crab uses sonar sensors with a considerable amount of purpose-built electronics, Hexapod uses infrared PSD sensors for obstacle detection. These can be directly interfaced to the analog or digital input (depending on sensor type) of an embedded controller without any additional electronic circuitry.

Program 11.2 shows a very simple program generating a walking pattern for a six-legged robot. Since the same Raspberry Pi controller and the same RoBIOS operating system are used for driving and walking robots, the robot's hardware description table (HDT) has to be adapted to match the robot's physical appearance with corresponding actuator and sensor equipment.

Program 11.2: Six-Legged Gait Settings

```
#include "eyebot.h"
#define STEPS 6
#define MAXF 50
#define MAXU 60
#define CNTR 128
#define UP (CNTR+MAXU)
#define DN (CNTR-MAXU)
#define FW (CNTR-MAXF)
#define BK (CNTR+MAXF)

int GaitForward[STEPS][12]= {
//order: LFront, RFront, LMiddle, RMiddle, LRear, RRear
  {DN,FW, UP,BK, UP,BK, DN,FW, DN,FW, UP,BK},
  {DN,FW, DN,BK, DN,BK, DN,FW, DN,FW, DN,BK},
  {UD,FW, DN,BK, DN,BK, UP,FW, UP,FW, DN,BK},
  {UP,BK, DN,FW, DN,FW, UP,BK, UP,BK, DN,FW},
  {DN,BK, DN,FW, DN,FW, DN,BK, DN,BK, DN,FW},
  {DN,BK, UP,FW, UP,FW, DN,BK, DN,BK, UP,FW},
};
```

```
int GaitRight[STEPS][12]= {/* omitted for space */};
int GaitLeft [STEPS][12]= {/* omitted for space */};

int PosInit[12]=
  {CT,CT, CT,CT, CT,CT, CT,CT, CT,CT, CT,CT};
```

Data structures like *GaitForward* contain the actual positioning data for a gait. In this case it is six key frames for moving one full cycle for all legs. Function *gait* (see Program 11.3) then uses this data structure to step through these six individual key frame positions by subsequent calls of *move_joint*.

Function *move_joint* moves all the robot's 12 joints from one position to the next position using key frame averaging. For each of these iterations, new positions for all 12 leg joints are calculated and sent to the servos. Then a certain delay time is waited before the routine proceeds, in order to give the servos time to reach the specified positions.

Program 11.3: Walking Routines

```
void move_joint(int pos1[12], int pos2[12], int speed)
{ int i, servo, steps = 50;
  float size[12];
  for (servo=0; servo<NumServos; servo++)
    size[servo] = (float) (pos2[servo]-pos1[servo]) /
                  (float) steps;
  for (i=0;i<steps;i++)
  { for(servo=0; servo<NumServos; servo++)
      SERVOSet(servo, pos1[servo] + (int)((float)i *size[servo]));
    OSWait(10/speed);
  }
}

void gait(int g[][12], int speed)
{ int i;
  for (i=0; i<STEPS; i++)
    move_joint(g[i], g[i+1], speed);
}
```

11.3   Biped Robots

Finally, we get to robots that resemble what most people think of when hearing the term *robot*. These are biped walking robots, often also called *humanoid robots* or *android robots* because of their resemblance to human beings.

Figure 11.6: Johnny and Jack humanoid robots, E. Nicholls and J. Ng, UWA

Our first attempts at humanoid robot design were the two robots Johnny Walker and Jack Daniels, built in 1998 and named because of their struggle to maintain balance during walking (see Figure 11.6 by Nicholls[17] and Ng). Our goal was to achieve a humanoid robot design and control system with limited funds. We used servos as actuators, linked in an aluminum U-profile. Although servos are very easily interfaced to an embedded controller and do not require an explicit feedback loop, it is exactly this feature (built-in hardwired feedback loop and lack of external feedback) which causes most control problems. Without feedback sensors from the joints it is not possible to measure joint positions or joint torques.

These first-generation robots were equipped with foot switches (microswitches and binary infrared distance switches) and two-axis accelerometers in the hips. Like all of our other robots, both Johnny and Jack are completely autonomous robots, not requiring any umbilical cords or *remote brains*. Each robot carries an EyeBot controller as on-board intelligence and a set of rechargeable batteries for power.

The mechanical structure of Johnny has nine degrees of freedom (*dof*), four per leg plus one in the torso. Jack has eleven *dof*, two more for its arms. In each leg, three servos are used to bend the leg at the ankle, knee, and hip joints, all in the same plane. One servo is used to turn the leg in the hip, in order to allow the robot to turn. Both robots have an additional *dof* for bending the torso sideways as a counterweight. Jack is also equipped with arms, a single *dof* per arm enabling it to swing its arms, either for balance or for touching objects.

---

[17]E. Nicholls, *Bipedal Dynamic Walking in Robotics*, B.E. Honours Thesis, The Univ. of Western Australia, Electrical and Computer Eng., supervised by T. Bräunl, 1998.

A second-generation humanoid robot is Andy Droid (see Figure 11.7). This robot differs from the first-generation design in a number of ways, as described in Bräunl, Sutherland, Unkelbach:[18]

- Five *dof* per leg
  Allowing the robot to bend the leg and also to lean sideways.
- Lightweight design
  Using the minimum amount of aluminum and steel to reduce weight.
- Separate power supplies for controller and motors
  To eliminate incorrect sensor readings due to high currents and voltage fluctuations during walking.

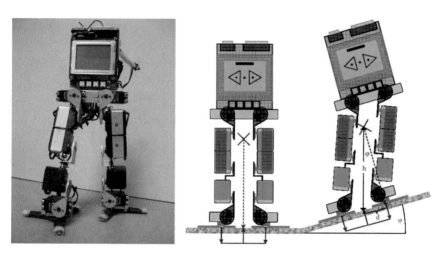

Figure 11.7: Andy Droid humanoid robot, J. Zimmermann, UWA / FH Koblenz

Figure 11.7, left, shows Andy without its arms and camera head, but with its second-generation foot design. Each foot consists of three adjustable toes, each equipped with a strain gauge. With this sensor feedback, the on-board controller can directly determine the robot's pressure point over the foot's support area and therefore immediately counteract an imbalance or adjust the walking gait parameters (Figure 11.7, right, by Zimmermann[19]).

Andy has a total of 13 *dof*, five per leg, one per arm, and one optional *dof* for the camera head. The robot's five *dof* per leg comprise three servos for bending the leg at the ankle, knee, and hips joints, all in the same plane (same as for Johnny). Two additional servos per leg are used to bend each leg sideways in the ankle and hip position, allowing the robot to lean sideways while keeping its torso level. There is no servo for turning a leg in the hip. The turning motion can still be achieved by different step lengths of left and right leg. One *dof* per

[18]T. Bräunl, A. Sutherland, A. Unkelbach, *Dynamic Balancing of a Humanoid Robot*, FIRA 1st Humanoid Robot Soccer Workshop (HuroSot), Daejeon Korea, Jan. 2002, pp. 19-23 (5).

[19]J. Zimmermann, *Balancing of a Biped Robot using Force Feedback*, Diploma Thesis, F.H. Koblenz/The Univ. of Western Australia, supervised by T. Bräunl, 2004.

arm allows swinging of the arms. Andy is 39 cm tall and weighs approximately 1 kg without batteries. For details see articles by Bräunl,[20] Montgomery,[21] Sutherland and Bräunl,[22] and Bräunl et al.[23]

Andy 2 (Figure 11.8) is the successor robot of *Andy Droid*. Instead of analog servos it uses digital servos that serve as both actuators and sensors. These digital servos are connected via RS232 and are daisy-chained, so a single serial port is sufficient to control all servos. Instead of pulse width modulation (PWM) signals, the digital servos receive commands as an ASCII sequence, including the individual servo number or a broadcast command. This reduces the computational load on the controller and generally simplifies operation. The digital servos can also act as sensors, returning position and electrical current data when being sent the appropriate command sequence.

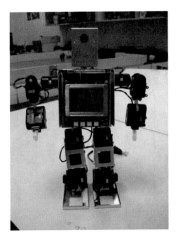

Figure 11.8: Andy 2 humanoid robot

Figure 11.9 visualizes feedback uploaded from the digital servos, showing the robot's joint positions and electrical currents (directly related to joint torque) during a walking gait, implemented by Harada.[24] High current (torque)

---

[20]T. Bräunl, *Design of Low-Cost Android Robots*, Proceedings of the First IEEE-RAS International Conference on Humanoid Robots, Humanoids 2000, MIT, Boston, Sept. 2000, pp. 1–6 (6).

[21]G. Montgomery, *Robo Crop—Inside our AI Labs*, Australian Personal Computer, Issue 274, Oct. 2001, pp. 80–92 (13).

[22]A. Sutherland, T. Bräunl, T. *Learning to Balance an Unknown System*, Proceedings of the IEEE-RAS International Conference on Humanoid Robots, Humanoids 2001, Waseda University, Tokyo, Nov. 2001, pp. 385–391 (7).

[23]T. Bräunl, A. Sutherland, A. Unkelbach, *Dynamic Balancing of a Humanoid Robot*, FIRA 1st Humanoid Robot Soccer Workshop (HuroSot), Daejeon Korea, Jan. 2002, pp. 19–23 (5).

[24]H. Harada, *Andy 2 Visualization Video*, 2006, online: http://robotics.ee.uwa.edu.au/eyebot5/mpg/walk-2leg/.

joints are color-coded, so problem areas like the robot's right hip servo in the figure can be easily detected.

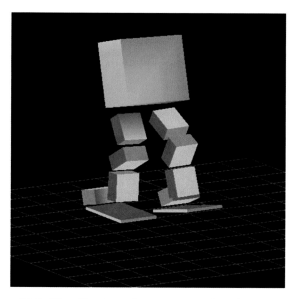

Figure 11.9: Visualization of servo sensor data by H. Harada, UWA/Hokkaido Univ.

A sample robot motion program without using any sensor feedback is shown in Program 11.4 (main) and Program 11.5 (subroutine *move*). This program for the Johnny/Jack servo arrangement demonstrates the robot's movements by letting it execute some squat exercises, continuously bending both knees and standing up straight again.

Program 11.4: Robot Gymnastics—Main

```
//RHipT, RHipB, RKnee, RAnkle, Torso, LAnkle. LKnee, LHipB, LHipT
int main()
{ int i, delay=2;
  int up [9] = {127,127,127,127,127,127,127,127,127};
  int down [9] = {127, 80,200, 80,127,200, 80,200,127};

  LCDMenu(" "," "," ","END");

  /* put servos in up position */
  for(i=0;i<9;i++) SERVOSet(i,up[i]);

  while (KEYRead() != KEY4) /* exercise until key press */
  { move(up,down, delay); /* move legs in bent pos.*/
```

```
    move(down,up, delay); /* move legs straight */
  }
  return 0;
}
```

Program 11.5: Robot Gymnastics—Move

```
void move(int old[], int new[], int delay)
{ int i,j; /* using int constant STEPS */
  float now[9], diff[9];

  for (j=0; j<9; j++) /* update all servo positions */
  { now[j] = (float) old[j];
    diff[j] = (float) (new[j]-old[j]) / (float) STEPS;
  }
  for (i=0; i<STEPS; i++) /* move servos to new pos.*/
  { for (j=0; j<9; j++)
    { now[j] += diff[j];
      SERVOSet(j, (int) now[j]);
    }
    OSWait(delay);
  }
}
```

The main program comprises two major steps:

- Setting all servos to the *up* position.
- Looping between *up* and *down* until keypress.

Robot configurations like *up* and *down* are stored as arrays with one integer value per *dof* of the robot (nine values in this example). They are passed as parameters to the subroutine *move*, which drives the robot servos to the desired positions by incrementally setting the servos to a number of intermediate positions.

Subroutine *move* uses local arrays for the current position (*now*) and the individual servo increment (*diff*). These values are calculated once. Then for a pre-determined constant number of steps, all servos are set to their next incremental position. An *OSWait* statement between loop iterations gives the servos some time to actually drive to their new positions.

11.4   Static Balance

There are two modes of walking:

- Static balance
  The robot's center of mass is at all times within the support area of its foot on the ground, or the combined support area of its two feet (convex hull), if both feet are on the ground.
- Dynamic balance
  The robot's center of mass may be outside the support area of its feet during a phase of its gait.

In this section we will concentrate on static balance, while dynamic balance is the topic of the following section. Our approach is to start with a semi-stable pre-programmed, but parameterized gait for a humanoid robot. Gait parameters are:

1. Step length
2. Height of leg lift
3. Walking speed
4. Leaning angle of torso in forward direction (constant)
5. Maximal leaning angle of torso sideways (variable, in sync with gait).

We will then update the gait parameters in real-time depending on the robot's sensors. Current sensor readings are compared with desired sensor readings at each time point in the gait. Differences between current and desired sensor readings will result in immediate parameter adaptations to the gait pattern. In order to get the right model parameters, we are conducting experiments with the BallyBot balancing robot as a testbed for acceleration, inclination and gyro sensors.

We constructed a gait generation tool (see Nicholls[25]), which is being used to generate gait sequences offline, which can subsequently be downloaded to the robot. This tool allows the independent setting of each *dof* for each time step and graphically displays the robot's attitude in three orthogonal views from the major axes (Figure 11.10). The gait generation tool also allows the playback of a designed gait sequence. However, it does not perform any dynamic analysis of the mechanical gait viability.

The first step toward walking is to achieve static balance. For this, we have the robot standing still, but use the acceleration sensors as a feedback with a software PI controller to the two hip joints. The robot is now actively standing straight. If pushed back, it will bend forward to counterbalance, and vice versa. Solving this isolated problem is similar to the inverted pendulum problem.

---

[25]E. Nicholls, *Bipedal Dynamic Walking in Robotics*, B.E. Honours Thesis, The Univ. of Western Australia, Electrical and Computer Eng., supervised by T. Bräunl, 1998.

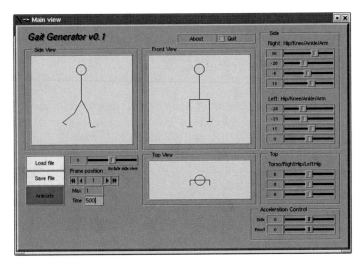

Figure 11.10: Gait generation tool

Real-world sensor data is never as clean as one would like. Figure 11.11 shows typical sensor readings from an inclinometer and foot switches for a walking experiment (Unkelbach[26]):

- The top curve shows the inclinometer data for the torso's side swing. The measured data is the absolute angle and does not require integration like the gyroscope.
- The two curves on the bottom show the foot switches for the right and left foot. First both feet are on the ground, then the left foot is lifted up and down, then the right foot is lifted up and down, and so on.

Program 11.6 demonstrates the use of sensor feedback for balancing a standing biped robot. In this example we control only a single axis (here forward/backward); however, the program could easily be extended to balance left/right as well as forward/backward by including a second sensor with another PI controller in the control loop.

The program's endless *while* loop starts with the reading of a new acceleration sensor value in the forward/backward direction. For a robot at rest, this value should be zero. Therefore, we can treat this value directly as an error value for a simple PI controller. All PI parameter values have to be determined experimentally.

[26]A. Unkelbach, *Analysis of sensor data for balancing and walking of a biped robot*, Project Thesis, Univ. Kaiserslautern/The Univ. of Western Australia, supervised by T. Bräunl and D. Henrich, 2002.

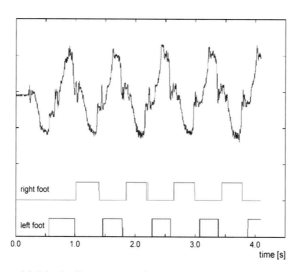

Figure 11.11: Inclinometer side swing and left/right foot switch data

Program 11.6: Balancing a Biped Robot

```c
void balance( void ) /* balance forward/backward */
{ int posture[9]= {127,127,127,127,127,127,127,127,127};
  float err, lastErr =0.0, adjustment, last_adj = 0.0;
  /* PID controller constants */
  float kP = 0.1, kI = 0.01;
  int i, delay = 10; // delay in ms

  while (1) /* endless loop */
  { /* read derivative. sensor signal = error */
    err = GetAccFB();
    /* adjust hip angles using PI controller */
    adjustment = last_adj
          + kP*( err - lastErr )
          + kI*((err + lastErr)/2));
    posture[lHipB] += adjustment;
    posture[rHipB] += adjustment;
    SetPosture(posture);
    lastErr = err;
    lastAdj = adjustment;
    OSWait(delay);
  }
}
```

11.5 Dynamic Balance

Walking gait patterns relying on static balance are not very efficient. They require large foot areas and only relatively slow gaits are possible, in order to keep dynamic forces low. Walking mechanisms with dynamic balance, on the other hand, allow the construction of robots with smaller feet, even feet that only have a single contact point, and can be used for much faster walking gaits or even running.

As has been defined in the previous section, dynamic balance means that at least during some phases of a robot's gait, its center of mass is not supported by its foot areas. Ignoring any dynamic forces and moments, this means that the robot would fall over if no counteraction is taken in real time. There are a number of different approaches to dynamic walking, which are discussed in the following.

11.5.1 Dynamic Walking Methods

In this section, we will discuss a number of different techniques for dynamic walking together with their sensor requirements.

1. Zero Moment Point (ZMP)
   Based on Fujimoto and Kawamura;[27] Goddard et al.;[28] Kajita et al.;[29] and Takanishi et al.[30]
   This is one of the standard methods for dynamic balance and is published in a number of articles. The implementation of this method requires the knowledge of all dynamic forces on the robot's body plus all torques between the robot's foot and ankle. This data can be determined by using accelerometers or gyroscopes on the robot's body plus pressure sensors in the robot's feet or torque sensors in the robot's ankles.
   With all contact forces and all dynamic forces on the robot known, it is possible to calculate the *zero moment point* (ZMP), which is the dynamic equivalent to the static center of mass. If the ZMP lies within the support area of the robot's foot (or both feet) on the ground, then the robot is in dynamic balance. Otherwise, corrective

---

[27]Y. Fujimoto, A. Kawamura, *Simulation of an autonomous biped walking robot including environmental force interaction*, IEEE Robotics and Automation Magazine, June 1998, pp. 33–42 (10).

[28]R. Goddard, Y. Zheng, H. Hemami, *Control of the heel-off to toe-off motion of a dynamic biped gait*, IEEE Transactions on Systems, Man, and Cybernetics, vol. 22, no. 1, 1992, pp. 92–102 (11).

[29]S. Kajita, T. Yamaura, A. Kobayashi, *Dynamic walking control of a biped robot along a potential energy conserving orbit*, IEEE Transactions on Robotics and Automation, Aug. 1992, pp. 431–438 (8).

[30]A. Takanishi, M. Ishida, Y. Yamazaki, I. Kato, *The realization of dynamic walking by the biped walking robot WL-10RD*, in ICAR'85, 1985, pp. 459–466 (8)

action has to be taken by changing the robot's body posture to prevent it from falling over.

2. Inverted Pendulum
   Based on Caux et al.;[31] Park and Kim;[32] Sutherland and Bräunl.[33]
   A biped walking robot can be modeled as an inverted pendulum. Dynamic balance can be achieved by constantly monitoring the robot's acceleration and adapting the corresponding leg movements.

3. Neural Networks
   Based on Miller;[34] Doerschuk et al.;[35] Kun and Miller[36]
   As for a number of other control problems, neural networks can be used to achieve dynamic balance. Of course, this approach still needs all the sensor feedback as in the other approaches.

4. Genetic Algorithms
   Based on Boeing and Bräunl.[37,38]
   A population of virtual robots is generated with initially random control settings. The best performing robots are reproduced using genetic algorithms for the next generation.
   This approach in practice requires a mechanics simulation system to evaluate each individual robot's performance and even then requires several CPU-days to evolve a good walking performance. The major issue here is the transferability of the simulation results back to the physical robot.

[31]S. Caux, E. Mateo, R. Zapata, *Balance of biped robots: special double-in- verted pendulum*, IEEE International Conference on Systems, Man, and Cybernetics, 1998, pp. 3691–3696 (6).

[32]J. Park, K. Kim, *Bipedal Robot Walking Using Gravity-Compensated In- verted Pendulum Mode and Computed Torque Control*, IEEE International Conference on Robotics and Automation, 1998, pp. 3528–3533 (6).

[33]A. Sutherland, T. Bräunl, *Learning to Balance an Unknown System*, Proceedings of the IEEE-RAS International Conference on Humanoid Robots, Humanoids 2001, Waseda University, Tokyo, Nov. 2001, pp. 385–391 (7).

[34]W. Miller III, *Real-time neural network control of a biped walking robot*, IEEE Control Systems, Feb. 1994, pp. 41–48 (8).

[35]P. Doerschuk, V. Nguyen, A. Li, *Neural network control of a three-link leg*, in Proceedings of the International Conference on Tools with Artificial Intelligence, 1995, pp. 278–281 (4).

[36]A. Kun, W. Miller III, *Adaptive dynamic balance of a biped using neural networks*, in Proceedings of the 1996 IEEE Intl. Conference on Robotics and Automation, Apr. 1996, pp. 240–245 (6).

[37]A. Boeing, T. Bräunl, *Evolving Splines: An alternative locomotion controller for a bipedal robot*, Seventh International Conference on Control, Automation, Robotics and Vision, ICARV 2002, CD-ROM, Singapore, Dec. 2002, pp. 1–5 (5).

[38]A. Boeing, T. Bräunl, *Evolving a Controller for Bipedal Locomotion*, Proceedings of the Second International Symposium on Autonomous Minirobots for Research and Edutainment, AMiRE 2003, Brisbane, Feb. 2003, pp. 43–52 (10).

5. PID Control
   Based on Bräunl;[39] Bräunl et al.[40]
   Classic PID control is used to control the robot's leaning front/back and left/right, similar to the case of static balance. However, here, we do not intend to make the robot stand up straight. Instead, in a teaching stage, we record the desired front and side lean of the robot's body during all phases of its gait. Later, when controlling the walking gait, we try to achieve this offset of front and side lean by using a PID controller. The following parameters can be set in a standard walking gate to achieve this leaning:
   - Step length
   - Height of leg lift
   - Walking speed
   - Amount of forward lean of torso
   - Maximal amount of side swing

6. Fuzzy Control
   Based on Lee et al.,[41,42]
   An adapted PID control can be used to replace the classic PID control by fuzzy logic for dynamic balance.

7. Artificial Horizon
   Based on Wicke.[43]
   This approach does not use any of the kinetics sensors of the other approaches, but a monocular grayscale camera. In the simple version, a black line on white ground (an *artificial horizon*) is placed in the visual field of the robot. We can then measure the robot's orientation by changes of the line's position and orientation in the image. For example, the line will move to the top if the robot is falling forward, it will be slanted at an angle if the robot is leaning left, and so on (Figure 11.12).

[39]T. Bräunl, *Design of Low-Cost Android Robots*, Proceedings of the First IEEE-RAS International Conference on Humanoid Robots, Humanoids 2000, MIT, Boston, Sept. 2000, pp. 1–6 (6).

[40]T. Bräunl, A. Sutherland, A. Unkelbach, *Dynamic Balancing of a Humanoid Robot*, FIRA 1st Humanoid Robot Soccer Workshop (HuroSot), Daejeon Korea, Jan. 2002, pp. 19–23 (5).

[41]C. Lee, A. Zaknich, T. Bräunl, *A Framework of Adaptive T-S type Rough-Fuzzy Inference Systems (ARFIS)*, FUZZ-IEEE 2008, 2008 World Congress on Computational Intelligence (WCCI), Ed. By G. Feng, 2008, pp. 567–574 (8).

[42]C. Lee, A. Zaknich, T. Bräunl, *An Adaptive T-S type Rough-Fuzzy Inference System (ARFIS) for Pattern Classification*, Annual Meeting of the North American Fuzzy Information Processing Society, IEEE NAFIPS, Eds. S. Rubin, M. Berthold, M. Reformat, San Diego, June 2007, p. 6.

[43]M. Wicke, *Bipedal Walking*, Project Thesis, Univ. Kaiserslautern / The Univ. of Western Australia, supervised by T. Bräunl, M. Kasper and E. von Puttkamer, 2001.

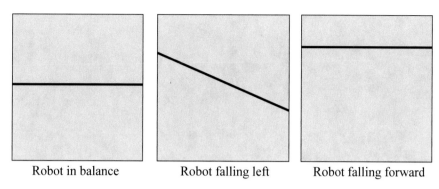

| Robot in balance | Robot falling left | Robot falling forward |

Figure 11.12: Artificial horizon

With a more powerful controller for image processing, the same principle can be applied even without the need for an artificial horizon. As long as there is enough texture in the background, general optical flow can be used to determine the robot's movements.

Figure 11.13 shows Johnny Walker during a walking cycle. Note the typical side swing of the torso to counterbalance the leg-lifting movement. This creates a large momentum around the robot's center of mass, which can cause problems with stability due to the limited accuracy of the servos used as actuators.

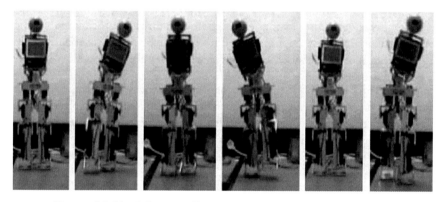

Figure 11.13: Johnny walking sequence

Figure 11.14 shows a similar walking sequence with Andy Droid. Here, the robot performs a much smoother and better controlled walking gait, since the mechanical design of the hip area allows a smoother shift of weight toward the side than in Johnny's case.

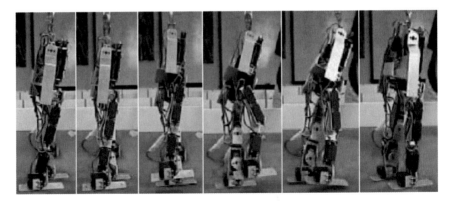

Figure 11.14: Andy walking sequence

11.5.2 Alternative Biped Designs

As has been mentioned before, servos have severe disadvantages for a number of reasons, most importantly because of their lack of external feedback. The construction of a biped robot with DC motors, encoders and end-switches, however, is much more expensive, requires additional motor driver electronics, and is considerably more demanding in software development. So instead of redesigning a biped robot by replacing servos with DC motors and keeping the same number of degrees of freedom, we decided to implement a minimal approach. Although Andy has 10 joints in both legs, it utilizes only three independent *dof*: bending each leg up and down, and leaning the whole body left or right. Therefore, it should be possible to build a robot that uses only three motors and uses mechanical gears or pulleys to achieve the articulated joint motion.

The CAD designs following this approach and the finished robot are shown in Figure 11.15, designed and built by Jungpakdee.[44] Each leg is driven by only one motor, while the mechanical arrangement lets the foot perform an ellipsoid curve for each motor revolution. The feet are only point contacts, so the robot has to keep moving continuously, in order to maintain dynamic balance. Only one motor is used for shifting a counterweight in the robot's torso sideways (the original drawing in Figure 11.15 specified two motors). Figure 11.16 shows the simulation of a dynamic walking sequence.

---

[44]K. Jungpakdee, *Design and construction of a minimal biped walking mechanism*, B.E. Honours Thesis, The Univ. of Western Australia, Dept. of Mechanical Eng., supervised by T. Bräunl and K. Miller, 2002.

Figure 11.15: Minimal biped design Rock Steady by Jungpakdee, UWA

Figure 11.16: Dynamic walking sequence by Jungpakdee, UWA

11.6 Tasks

1. Simulate one of the biped robot designs in ROS/Gazebo, e.g. a Nao robot.
   Write an application program to let it walk straight or turn.

2. Create a mechanical design of a biped walking robot and determine its possible movements. How many degrees of freedom are required? What will the motor actuator movement sequence be for the walking pattern?
   How many and which types of sensors do you plan to include in order to be able to walk reliably.

3. Simulate the biped design using a physics engine.

4. Build your real biped design and program a sensor-supported gait.

Autonomous Boats and Planes

I n this chapter, we will talk about autonomous boats, autonomous underwater vehicles (AUV), and also briefly about unmanned aerial vehicles (UAV), i.e., autonomous planes and drones.

12.1   Autonomous Boats

The design of an autonomous surface vessel or underwater vehicle requires one additional skill compared to the robot designs discussed previously: water tightness. Waterproof electrical connections are required for actuators and sensors for all autonomous boat types, while underwater vehicles have the additional challenge of increasing water pressure when diving.

Our interest in autonomous boats was ignited by an article titled "*Did a Solar-Powered Autonomous Boat Just Cross the Pacific Ocean?*" in *Make: magazine*.[1] The article describes the interesting journey of Damon McMillan's model boat SeaCharger, which successfully completes an autonomous journey from California to Hawaii. After welcoming the boat in Hawaii, McMillan—instead of taking the boat home—sets New Zealand as its new goal and it almost makes it there as well.

Figure 12.1 shows our first design, an autonomous solar raft, which carries a 100 W solar panel on top of three floating pipes. Two thruster motors can drive and steer the boat like a differential-drive robot without the need for a rudder. The solar panel provides enough energy to drive the boat during the day and to store excess energy in four lead–acid batteries for driving at night. All electronics have been included in the two outer pipes. The third pipe's role is simply to increase buoyancy.

---

[1]D. McMillian, *Did a Solar-Powered Autonomous Boat Just Cross the Pacific Ocean?*, Make:, 22 Aug. 2016, online: https://makezine.com/2016/08/22/solar-powered-autonomous-boat/.

T. Bräunl, *Embedded Robotics*,
https://doi.org/10.1007/978-981-16-0804-9_12

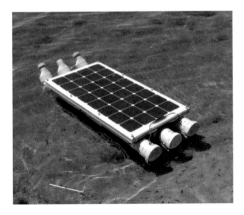

Figure 12.1: Autonomous solar raft by J. Borella, J. Hodge and A. Goldsworthy UWA

The *SolarRaft* uses two internal processors, a Raspberry Pi Zero for acquiring camera images and communication with the outside world, and an ArduPilot (an enhanced Arduino, see chapter on Arduino controller) for low level motor control. In addition, SolarRaft has the following sensors and communication devices:

- GPS and IMU linked to the ArduPilot
- Digital camera for recording and/or transmitting images
- Provisions for additional sensors, e.g., for water temperature, salinity.
- 3G modem for communication in coastal areas
  This can be simply exchanged for a satellite modem, however, the cost for transmitting data through these are prohibitive, while a 3G or 4G data plan is relatively cheap and sufficient for most rivers, lakes and islands close to the coast.

Figure 12.2 shows the layout of the electronics inside the main tube.

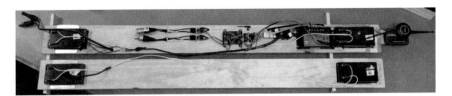

Figure 12.2: Autonomous solar raft electronics by J. Hodge and A. Goldsworthy UWA 2018

A redesign of the solar boat by Constant[2] resulted in *SolarCat*, a dual-hull catamaran using the same electrical and sensor design components, but in a more ergonomic hull. Figure 12.3 shows the mechanical redesign of the boat with the much more energy-efficient dual-hull concept. All electronic components of SolarCat are identical to the ones from SolarRaft.

Figure 12.3: SolarCat—autonomous solar boat design by P.-L. Constant, UWA 2020

Figure 12.4 shows the electronics that has been moved to a watertight case placed above the two hulls.

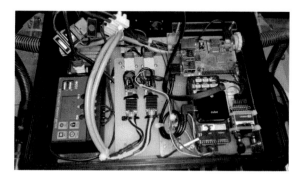

Figure 12.4: SolarCat electronics container by P.-L. Constant, UWA 2020

On the software side, SolarCat uses the MAVLink[3] communication protocol to link the autonomous boat to the ground station. Figure 12.5 shows the

---

[2]P.-L. Constant, *Autonomous Surface Vehicles*, Honours Thesis, UWA Electrical and Computer Eng., supervised by T. Bräunl, 2020.

[3]MAVLink, *MAVLink Developer Guide*, online: https://mavlink.io/en/.

MAVLink route planner for an autonomous trip around Rottnest Island, off the coast of Western Australia.

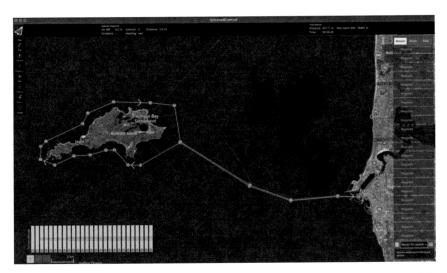

Figure 12.5: MAVLink route planning for a trip around Rottnest Island by P.-L. Constant, UWA 2020

Quite a number of international competitions have been established lately to let autonomous boats compete with each other. These will definitely stimulate student interest in this research area:

- RoboBoat   https://roboboat.org
- Microtransat Challenge   https://www.microtransat.org
- SailBot   https://www.sailbot.org
- RobotX Challenge   https://robotx.org
- AUVSI   https://www.auvsi.org

12.2   Autonomous Underwater Vehicles

Unlike many other areas of mobile robots, AUVs have an immediate application area conducting various sub-sea surveillance and manipulation tasks for the resources industry. In the following, we want to concentrate on intelligent control and not on general engineering tasks such as constructing AUVs that can go to great depths, as there are industrial remotely operated vehicle (ROV) solutions available that have solved these problems.

While most autonomous mobile robot applications can also use wireless communication to a host station, this is a lot harder for an AUV. Once submerged, none of the standard communication methods work; Bluetooth or WLAN only operates up to a water depth of about 50 cm. The only wireless communication method available is sonar with a very low data rate, but unfortunately, these systems have been designed for the open ocean and can usually not cope with signal reflections as they occur when using them in a pool. So unless some wire-bound communication method is used, AUV applications have to be truly autonomous.

The dynamic model of an AUV describes the AUV's motions as a result of its shape, mass distribution, forces/torques exerted by the AUV's motors and external forces/torques (e.g., ocean currents). Since we are operating at relatively low speeds, we can disregard the Coriolis force and present a simplified dynamic model by Gonzalez[4] with:

M   mass and inertia matrix
v   linear and angular velocity vector
D   hydrodynamic damping matrix
G   gravitational and buoyancy vector
τ   force and torque vector (AUV motors and eternal forces/torques)

D can be further simplified as a diagonal matrix with zero entries for y-axis movements (AUV can only move forward/backward along x-axis, and dive/surface along z-axis, but not move sideways), and zero entries for rotations about x and y (AUV can actively rotate only about z, while its self-righting movement greatly eliminates rotations about x and y).

G is non-zero only in its z component, which is the sum of the AUV's gravity and buoyancy vectors.

τ is the product of the force vector combining all of an AUV's motors with a pose matrix that defines each motor's position and orientation, based on the AUV's local coordinate system.

12.2.1 Mako AUV

Mako (Figure 12.6) was designed from scratch as a dual PVC hull containing all electronics and batteries, linked by an aluminum frame and propelled by four trolling motors, two of which are for active diving. The advantages of this design over competing proposals are shown in Bräunl et al.:[5]

---

[4]L. Gonzalez, *Design, Modelling and Control of an Autonomous Underwater Vehicle*, B.E. Honours Thesis, The University of Western Australia, Electrical and Computer Eng., supervised by T. Bräunl, 2004.

[5]T. Bräunl, A. Boeing, L. Gonzales, A. Koestler, M. Nguyen, J. Petitt, *The Autonomous Underwater Vehicle Initiative—Project Mako*, 2004 IEEE Conference on Robotics, Automation, and Mechatronics (IEEE-RAM), Dec. 2004, Singapore, pp. 446–451 (6).

- Ease in machining and construction due to its simple structure
- Relative ease in ensuring watertight integrity because of the lack of rotating mechanical devices such as bow planes and rudders
- Substantial internal space owing to the existence of two hulls
- High modularity due to the relative ease with which components can be attached to the skeletal frame
- Cost-effectiveness because of the availability and use of common materials and components
- Relative ease in software control implementation when compared to using a ballast tank and single thruster system
- Ease in submerging with two vertical thrusters
- Static stability due to the separation of the centers of mass and buoyancy, and dynamic stability due to the alignment of thrusters.

Figure 12.6: Autonomous submarine Mako by Gonzalez, UWA

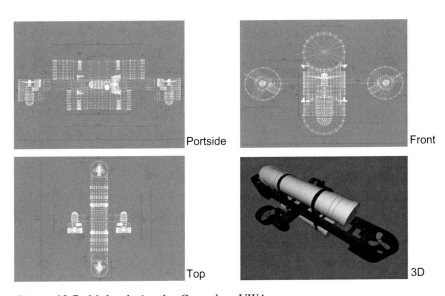

Figure 12.7: Mako design by Gonzalez, UWA

Simplicity and modularity were key goals in both the mechanical and electrical system designs. With the vehicle not intended for use below 5 m depth, pressure did not pose a major problem. The Mako AUV measures 1.34 m long, 64.5 cm wide and 46 cm tall (Figure 12.7).

The vehicle comprises two watertight PVC hulls mounted to a supporting aluminum skeletal frame. Two thrusters are mounted on the port and starboard sides of the vehicle for longitudinal movement, while two others are mounted vertically on the bow and stern for depth control. The Mako's vertical thruster diving system is not power conservative, however, when a comparison is made with ballast systems that involve complex mechanical devices, the advantages such as precision and simplicity that come with using these two thrusters far outweigh those of a ballast system.

Propulsion is provided by four modified 12 V, 7 A trolling motors that allow horizontal and vertical movement of the vehicle. These motors were chosen for their small size and the fact that they are intended for underwater use; a feature that minimized construction complexity substantially and provided watertight integrity (Figure 12.8).

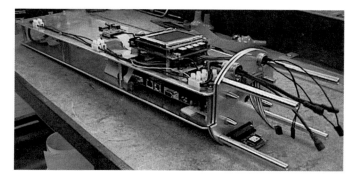

Figure 12.8: Electronics and controller setup inside Mako's top hull

The starboard and port motors provide both forward and reverse movement while the stern and bow motors provide depth control in both downward and upward directions. Roll is passively controlled by the vehicle's innate righting moment (Figure 12.9). The top hull contains mostly air besides light electronics equipment, the bottom hull contains heavy batteries. Therefore, mainly a buoyancy force pulls the top cylinder up and gravity pulls the bottom cylinder down. If for whatever reason, the AUV rolls as in Figure 12.9, right, these two forces ensure that the AUV will right itself.

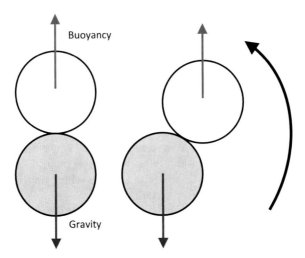

Figure 12.9: Self-righting moment of AUV

Overall, this provides the vehicle with 4 *dof* that can be actively controlled, which provide an ample range of motion suited to accomplishing a wide range of tasks.

The control system of the Mako is run by a Raspberry Pi embedded controller. The Pi's is controlling the AUV's movement through its four thrusters and its sensors and has also sufficient processing power for the vision system.

Motor controllers designed and built specifically for the thrusters provide both speed and direction control. Each motor controller interfaces with the Pi via two servo ports. Due to the high current used by the thrusters, each motor controller produces a large amount of heat. To keep the temperature inside the hull from rising too high and damaging electronic components, a heat sink attached to the motor controller circuit on the outer hull was devised. Hence, the water continuously cools the heat sink and allows the temperature inside the hull to remain at an acceptable level.

The sonar/navigation system utilizes an array of Navman Depth2100 echo sounders, operating at 200 kHz. One of these sensors is facing down and thereby providing an effective depth sensor (assuming the pool depth is known), while the other three sensors are used as distance sensors pointing forward, left and right. An auxiliary control board based on a PIC controller has been designed by Alfirevich[6] to multiplex the four sonars and connect them to the main controller.

---

[6]E. Alfirevich, *Depth and Position Sensing for an Autonomous Underwater Vehicle*, B.E. Honours Thesis, The University of Western Australia, Electrical and Computer Eng., supervised by T. Bräunl, 2005.

Figure 12.10: Mako in operation

A digital magnetometer module provides for yaw or heading control. A simple water detector circuit is connected to an analog input to detect a possible hull breach. Two probes run along the bottom of each hull, which allow for the location (upper or lower hull) of the leak to be known. The software periodically monitors whether the hull integrity of the vehicle has been compromised, and if so, immediately surfaces the vehicle. Another analog input is used for a power monitor that will ensure that the system voltage remains at an acceptable level. Figure 12.10 shows the Mako in operation.

12.2.2 USAL AUV

The USAL AUV uses a commercial ROV as a basis, which was heavily modified and extended (Figure 12.11). All original electronics were taken out and replaced by an embedded controller (Figure 12.12). The hull was split and extended by a trolling motor for active diving, which allows the AUV to hover, while the original ROV had to use active rudder control during a forward motion for diving, see Gerl[7] and Drtil.[8] Figure 12.12 shows USAL's complete electronics subsystem.

Figure 12.11: Autonomous submarine USAL by B. Gerl, UWA/TU Munich

---

[7]B. Gerl, *Development of an Autonomous Underwater Vehicle in an Interdisciplinary Context*, Diploma Thesis, The University of Western Australia and Technical Univ. München, Electrical and Computer Eng., supervised by T. Bräunl, 2006.

[8]M. Drtil, *Electronics and Sensor Design of an Autonomous Underwater Vehicle*, Diploma Thesis, The University of Western Australia and FH Koblenz, Electrical and Computer Eng., supervised by T. Bräunl, 2006.

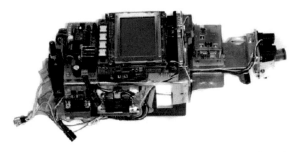

Figure 12.12: USAL controller and sensors by Gerl and Drtil, UWA/TUM/FH Koblenz

For simplicity and cost reasons, we decided to use infrared PSD sensors for the USAL instead of the echo sounders used on the Mako. Since the front part of the hull was made out of clear Perspex, we were able to place the PSD sensors inside the AUV hull, so we did not have to worry about waterproofing sensors and cabling. Figure 12.13 shows the results of measurements conducted by Drtil, using this sensor setup in air (through the hull), and in different grades of water quality. Assuming good water quality, as can be expected in a swimming pool, the sensor setup returns reliable results up to a distance of about 1.1 m, which is sufficient for using it as a collision avoidance sensor, but too short for using it as a navigation aid in a large pool.

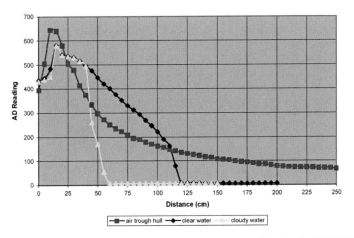

Figure 12.13: Underwater PSD measurement by Drtil, UWA/FH Koblenz

The USAL system overview is shown in Figure 12.14. Numerous sensors are connected to the on-board controller. These include a digital camera, four analog PSD infrared distance sensors, a digital compass, a three-axis solid-state accelerometer and a depth pressure sensor. The Bluetooth wireless communication system can only be used when the AUV has surfaced or is diving close to the surface. The energy control subsystem contains voltage regulators and level converters, additional voltage and leakage sensors, as well as motor drivers for the stern main driving motor, the rudder servo, the diving trolling motor and the bow thruster pump.

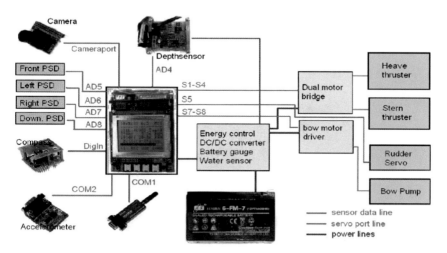

Figure 12.14: USAL system overview by Drtil, UWA/FH Koblenz

Figure 12.15 shows the arrangement of the three thrusters and the stern rudder, together with a typical turning movement of the USAL.

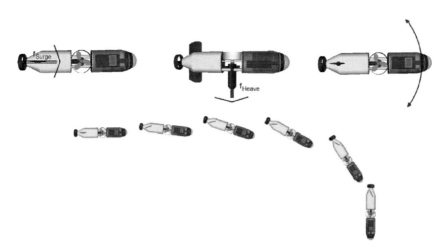

Figure 12.15: USAL thrusters and rudder turning maneuver by Drtil, UWA/FH Koblenz

12.2.3 BlueROV2

BlueROV2 (see Figure 12.16) is a commercially built platform for remotely operated vehicle (ROV) research. The ROV has 6 thrusters, which let it maneuver easily in all directions and orientations. A Raspberry Pi is used an internal controller with a front-facing camera. Space is available for adding additional sensors, especially an IMU.

BlueROV2 is operated remotely from a PC through a long data cable, but still requires its local battery onboard the ROV. The cable is strong enough to pull the ROV back in case of any problems driving back.

Figure 12.16: BlueROV2

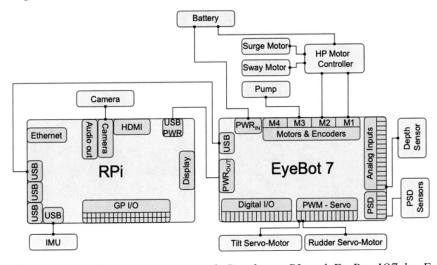

Figure 12.17: Control structure with Raspberry PI and EyeBot IO7 by F. Hidalgo, UWA

Hidalgo[9] used the Raspberry Pi in combination with the EyeBot IO7 interface board for the BlueROV2 (see Figure 12.17). Some results of his research into underwater navigation using ORB-SLAM2 on the ROV are demonstrated in Figure 12.18.

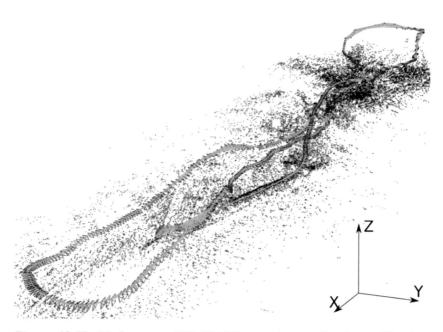

Figure 12.18: Underwater ORB-SLAM2 experiments for Omeo Wreck at Coogee Beach, Western Australia, by F. Hidalgo, UWA 2019

12.2.4 AUV Tasks

Rather than starting with the complex competition tasks for surface vessels and AUVs, we suggest to start with a number of simpler training tasks as shown in Figure 12.19 and outlined below.

---

[9]F. Hidalgo, *Simultaneous Localization and Mapping in Underwater Robots*, Ph.D. thesis, supervised by T. Bräunl and A. Boeing, UWA 2019.

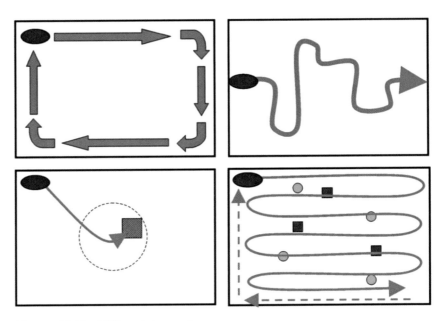

Figure 12.19: AUV training tasks

1. Wall Following
   The AUV is placed close to a corner of the pool and has to follow the pool wall without touching it. The AUV should perform one lap around the pool, return to the starting position, and then stop.

2. Pipeline Following
   A plastic pipe is placed along the bottom of the pool, starting on one side of the pool and terminating on the opposite side. The pipe is made out of straight pieces and 90° angles.
   The AUV is placed over the start of the pipe on one side of the pool and has to follow the pipe on the ground until the opposite wall has been reached.

3. Target Finding
   The AUV has to locate a target plate with a distinctive texture that is placed at a random position within a 3m diameter from the center of the pool.

4. Object Mapping
   A number of simple objects (balls or boxes of distinctive color) are placed at the bottom of the pool, distributed over the whole pool area. The AUV has to survey the whole pool area, e.g., by diving along a sweeping pattern, and record all objects found at the bottom of the pool. Finally, the AUV has to return to its start corner and upload the coordinates of all objects found.

These tasks can be solved initially in simulation with the free EyeSim[10,11] system:

https://robotics.ee.uwa.edu.au/eyesim/.

Figure 12.20 shows an example setup for AUV Mako in EyeSim.

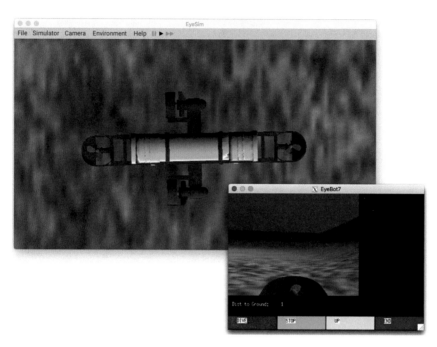

Figure 12.20: UV Mako in EyeSim VR simulator

A simple program for actuating the Mako's dive motors is shown in Program 12.1.

Program 12.1: Mako Dive Operation

```
#include "eyebot.h"
#define LEFT 1 // Thruster IDs.
#define FRONT 2
#define RIGHT 3
#define BACK 4
#define PSD_DOWN 6 // PSD direction

void dive(int speed)
{ MOTORDrive(FRONT, speed);
```

[10]T. Bräunl, *EyeSim VR—EyeBot Mobile Robot Simulator*, online: https://robotics.ee.uwa.edu.au/eyesim/.

[11]T. Bräunl, *Robot Adventures in Python and C*, Springer Verlag 2020.

```
  MOTORDrive(BACK, speed);
}

int main()
{ BYTE img[QVGA_SIZE];
  char key;

  LCDMenu("DIVE", "STOP", "UP", "END");
  CAMInit(QVGA);
  do { LCDSetPrintf(19,0, "Dist to Ground:%6d\n",
       PSDGet(PSD_DOWN));
       CAMGet(img);
       LCDImage(img);

       switch(key=KEYRead())
       { case KEY1: dive(-100); break;
         case KEY2: dive( 0); break;
         case KEY3: dive(+100); break;
       }
     } while (key != KEY4);
  return 0;
}
```

12.3   Unmanned Aerial Vehicles (UAVs)

The term UAV comprises autonomous fixed-wing aircraft (planes) as well as autonomous multi-rotor vehicles (drones). Although we have designed a number of autonomous planes and drones over the years at UWA, this section will be rather brief, as the development in this area advances with incredible speed. New and more powerful systems are developed by industry in every single year.

Figure 12.21 shows the first autonomous plane we developed at UWA with the implemented control system shown in Figure 12.22.

Figure 12.21: Autonomous fixed-wing aircraft, C. Croft and T. Bräunl, UWA

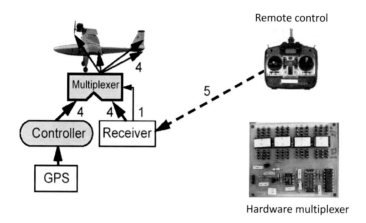

Hardware multiplexer

Figure 12.22: Control system for autonomous plane

The system design shows the plane controlled by either the embedded controller or by the receiver of the remote control. The idea is to allow the transmitter to gain back control of the plane in case something goes wrong in the autonomous system. Rather than letting the embedded controller read and interpret the receiver signal in software, we decided to build a (more trusted) hardware multiplexer, which can be directly switched back and forth by a dedicated channel on the remote control's transmitter. That way the operator can easily switch between manual and autonomous control of the plane.

After years of working with petrol and later electric model fixed-wing planes, we finally started working with multi-rotor drones as shown in Figure 12.23. All drones carry "the usual" sensors that are being used for autonomous vehicles, which are GNSS receiver, IMU and Lidar. One of the applications is to detect and follow a mobile target—in this case a model car—as shown in Figure 12.23, right.

Figure 12.23 Autonomous multi-rotor vehicle and target ground vehicle

Figure 12.24 shows an application where the drone autonomously surrounds an object or a building at various height levels before it returns to its starting point. The generated image sequence is then being used to reconstruct a 3D model of the object (see Mohanty[12]).

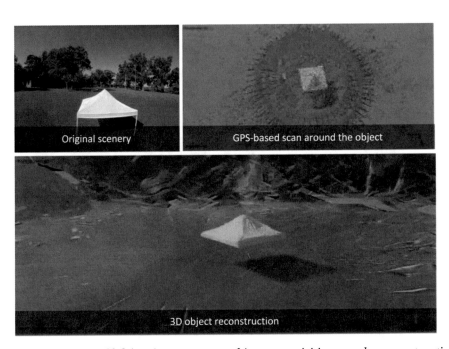

Figure 12.24: Autonomous object acquisition and reconstruction by M. Mohanty, UWA

---

[12]M. Mohanty, *Environmental Mapping in 2D/3D with the use of a Multirotor Unmanned Aerial Vehicle,* Bachelor of Engineering Thesis, supervised by T. Bräunl and C. Croft, UWA, Oct. 2015.

12.4 Tasks

1. Using EyeSim, implement the wall following task described in this chapter (for submarine Mako or USAL).

2. Using EyeSim, implement the pipeline following task described in this chapter (for submarine Mako or USAL).

3. Using EyeSim, implement the target finding task described in this chapter (for submarine Mako or USAL).

4. Using EyeSim, implement the object mapping task described in this chapter (for submarine Mako or USAL).

5. Modify a model boat for autonomous control by replacing its receiver with an embedded processor. Add a distance sensor and an orientation sensor.
   Implement a program that lets the boat cross a small moat driving straight, then detect the opposite quay and stop in time before it collides, turn exactly 180°, then finally drive back to the starting point and stop there.

6. Build your own real autonomous boat with GPS and Wi-Fi or 3G modem.
   Set up a collision-free path for the boat to follow and test it on a small pond where it is easy to recover the boat in case of any problems.

ROBOT MANIPULATORS

13

The main focus on the robotics side of this book is on mobile robots. However, we also want to give a brief introduction to the area of stationary manipulators, as they still form the vast majority of all commercial robot systems. Traditional applications of robot manipulators are spot welding and spray painting (Figures 13.1 and 13.2), especially in the automotive industry, as well as packaging and filling tasks, e.g., in the chemical and pharmaceutical industry. Robot manipulators can work in hazardous environments (e.g., nuclear power plants) and can conduct a variety of tasks from simple repetitive movements to complex sensor-based assemblies.

Figure 13.1: Industrial application and individual manipulator. Photo courtesy of Kuka Systems GmbH

© The Author(s), under exclusive license to Springer Nature Singapore Pte Ltd. 2022
T. Bräunl, *Embedded Robotics*,
https://doi.org/10.1007/978-981-16-0804-9_13

Figure 13.2: Spray painting cabin. Photo courtesy of ABB Robotics

13.1   Homogeneous Coordinates

Any design work involving robot manipulators is based on kinematics, and *homogeneous coordinates* are a necessary prerequisite for this. With *pose,* we describe the combination of a position and orientation in 3D space.

In traditional three-dimensional geometry, the addition of 3×1 vectors is used for translation operations and the multiplication with 3×3 matrices is used for rotation operations. This works quite well for simple applications; however, for robot manipulators we often have longer chains of translations and rotations, such as the following:

$$\text{Trans}(x_1, y_1, z_1) \rightarrow \text{Rot}(x, 90) \rightarrow \text{Trans}(x_2, y_2, z_2) \rightarrow \text{Rot}(z, -45)$$

So, applying this kinematic chain to point p in 3D-space would the following result:

$$p' = (((p + \text{Trans}(x_1, y_1, z_1)) \cdot \text{Rot}(x, 90)) + \text{Trans}(x_2, y_2, z_2)) \cdot \text{Rot}(z, -45)$$

Unfortunately, when using standard 3D geometry operations, there is no way to simplify this expression in a way so that we can apply a single operation to a point in order to transform it from start to destination. This may become a performance issue if there are many points that require the same transformation.

3.1 Homogeneous Coordinates

Homogeneous coordinates (introduced by Möbius[1] in 1827!) solve this problem very elegantly; almost two centuries after their discovery they are today's standard in geometry calculations for robot manipulators.

Homogeneous coordinates extend both translations and rotations by a fourth coordinate (scaling factor), which is set to constant 1 for our purposes, and they use the same transformation format of a 4×4 matrix for rotation as well as for translation. So each homogeneous 4×4 transformation contains a rotational part followed by a translational part. Either part can be zero (or rather the identity transformation).

For example, a 3D translation Trans (1, 2, 3) now becomes:

$$\begin{bmatrix} 1 & 0 & 0 & 1 \\ 0 & 1 & 0 & 2 \\ 0 & 0 & 1 & 3 \\ 0 & 0 & 0 & 1 \end{bmatrix}$$

Note that the top-left 3×3 submatrix is the identity matrix (three ones along the main diagonal) and three zeros and a one in the bottom row. This means there is no rotational part. In robotics applications, the bottom row *always* reads (0, 0, 0, 1). Other values could be used for scaling, but we do not need that.

A 90° rotation about the *x*-axis will then become:

$$\begin{bmatrix} \mathbf{1} & \mathbf{0} & \mathbf{0} & 0 \\ \mathbf{0} & \mathbf{0} & \mathbf{-1} & 0 \\ \mathbf{0} & \mathbf{1} & \mathbf{0} & 0 \\ 0 & 0 & 0 & 1 \end{bmatrix}$$

The 3×3 matrix of the rotation is maintained as the top-left submatrix, while the translation vector in the right column is (0, 0, 0), so no translation is being performed. A bottom row of (0, 0, 0, 1) is added as before.

So this gives us the following 4×4 matrices for a general translation along x, y, z and for general rotations about the main axes x, y, z, respectively:

$$\mathrm{Trans}\left(v_x, v_y, v_z\right) = \begin{bmatrix} 1 & 0 & 0 & v_x \\ 0 & 1 & 0 & v_y \\ 0 & 0 & 1 & v_z \\ 0 & 0 & 0 & 1 \end{bmatrix} \quad \mathrm{Rot}(x, \theta) = \begin{bmatrix} 1 & 0 & 0 & 0 \\ 0 & \cos\theta & -\sin\theta & 0 \\ 0 & \sin\theta & \cos\theta & 0 \\ 0 & 0 & 0 & 1 \end{bmatrix}$$

$$\mathrm{Rot}(y, \theta) = \begin{bmatrix} \cos\theta & 0 & \sin\theta & 0 \\ 0 & 1 & 0 & 0 \\ -\sin\theta & 0 & \cos\theta & 0 \\ 0 & 0 & 0 & 1 \end{bmatrix} \quad \mathrm{Rot}(z, \theta) = \begin{bmatrix} \cos\theta & -\sin\theta & 0 & 0 \\ \sin\theta & \cos\theta & 0 & 0 \\ 0 & 0 & 1 & 0 \\ 0 & 0 & 0 & 1 \end{bmatrix}$$

---

[1] A. Möbius, *Der barycentrische Calcl: ein neues Hilfsmittel zur analytischen Behandlung der Geometrie*, Crelle's Journal, Leipzig, 1827.

4 × 4 matrices are also a convenient tool to describe the 3D position and orientation (*pose*) of an object, e.g., a robot's end-effector (*hand*), in a single data structure.

13.2 Manipulator Kinematics

The standard textbooks on Kinematics are by Paul[2] and Craig,[3] and we will only scratch the surface of this topic. Each manipulator is made up of a number of links (metal bars) and joints (motor-actuated hinges or pivots). A typical manipulator has a base, an end-effector (e.g., a gripper, welding or spray-nozzle attachment) and six motorized joints with links between them. These allow the robot to reach any 3D position and orientation (pose). Each actuated joint is called a degree of freedom, so a manipulator with six joints is a six-degree-of-freedom manipulator, or in short: 6 dof.

We start by introducing the standard manipulator joints and their unambiguous drawing norm. There are three basic types or manipulator joints (see Figure 13.3):

- rotational joints with the rotation axis along the link
- rotational joints with the rotation axis perpendicular to the link (hinge joints) and
- prismatic joints (telescopic joints).

All other joints, e.g., a more complex ball joint, can be described as a combination of these basic types.

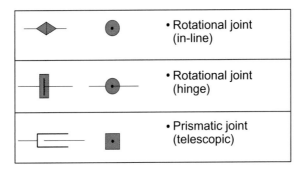

Figure 13.3: Side view and top view of basic manipulator joint types

[2]N. Paul, *Robot Manipulators: Mathematics, Programming, and Control*, MIT Press, Cambridge MA, 1981.

[3]J. Craig, *Introduction to Robotics*, 3rd Ed., Addison-Wesley, Reading MA, 2003.

3.2 Manipulator Kinematics

The Puma 560 robot from Unimation/Stäubli is a standard 6-dof manipulator that is frequently used as a model in textbooks. Figure 13.4 shows a simulation screenshot in RoboSim[4] of this robot and its conceptual drawing, labeling its joints $\theta 1, \ldots, \theta 6$.

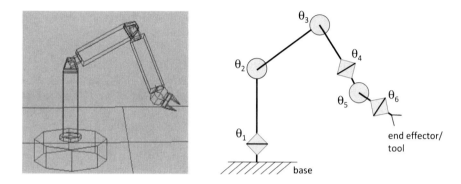

Figure 13.4: Puma 560 simulation and conceptual drawing

Manipulator kinematics deals with answering the following two basic questions for a specified manipulator geometry:

1. For a given set of joint values, what is the manipulator's end-effector's pose?
   This will be answered by *forward kinematics*.
2. What joint values are required to reach a desired end-effector pose?
   This will be answered by *inverse kinematics*.

There are of course a number of additional and more complex questions in manipulator kinematics. Below are a few examples, but these are beyond the scope of this chapter.

- How can I make the end-effector move in a straight line?
  (Motion equations)
- How can I make the end-effector apply a certain force or torque to an object? (Jacobi matrix)

---

[4]T. Bräunl, *RoboSim—A Simple 6-DOF Robot Manipulator Simulation System*, 1999, online: https://robotics.ee.uwa.edu.au/robosim/.

13.2.1 Forward Kinematics

The forward kinematics of a manipulator describes its transformations from the manipulator's base via subsequent joints to the end-effector. Each partial transformation, i.e., from joint to joint, can be described by a 4×4 homogeneous matrix, and we need to multiply all 4×4 matrices (six matrices for a 6 dof manipulator) in order to get a single 4×4 matrix that describes the full manipulator transformation from base to end-effector.

We can use individual 4×4 matrices to describe the transition from joint to joint for simple manipulators, but Denavit and Hartenberg have developed a standard called the *Denavit–Hartenberg Notation*[5] (or short DH) that allows a standardized description of *any* joint configuration using four 4×4 transformations to go from one joint to the next.

When going from joint i-1 to joint i, we assume that the joint axis for revolutionary joints is aligned with the local z-axis. We then perform two translations and two rotations as follows:

1. $\text{Rot}(x_{i-1}, \alpha_{i-1})$
    Rotation about x_{i-1} to align new z_i-axis
2. $\text{Trans}(a_{i-1}, 0, 0)$
    Translation along x_{i-1} to position new z_i-axis
3. $\text{Rot}(z_i, \theta_i)$
    Actual *variable* joint rotation ($\pm$max. range) about new z_i-axis
4. $\text{Trans}(0, 0, d_i)$
    Translation along new z_i to position new x_i-axis

So the transformation from one joint to the next can be written as a product of these four individual transformations and is then reduced to a single 4×4 transformation matrix. The indices on the bottom and top of transformation symbol T denote starting from joint i-1, going to joint i.

The four transformations written down explicitly and multiplied out as a single 4×4 matrix are:

$$_i^{i-1}T = \text{Rot}(x, \alpha_{i-1}) \cdot \text{Trans}(a_{i-1}, 0, 0) \cdot \text{Rot}(z, \theta_i) \cdot \text{Trans}(0, 0, d_i)$$

$$_i^{i-1}T = \begin{bmatrix} \cos\theta_i & -\sin\theta_i & 0 & a_{i-1} \\ \sin\theta_i \cdot \cos\alpha_{i-1} & \cos\theta_i \cdot \cos\alpha_{i-1} & -\sin\alpha_{i-1} & -\sin\alpha_{i-1} \cdot d_i \\ \sin\theta_i \cdot \sin\alpha_{i-1} & \cos\theta_i \cdot \sin\alpha_{i-1} & \cos\alpha_{i-1} & \cos\alpha_{i-1} \cdot d_i \\ 0 & 0 & 0 & 1 \end{bmatrix}$$

As a basic example for this, see Figure 13.5. It shows a planar (2D) manipulator with three joints. The first joint coincides with the manipulator's base, while there is an unspecified end-effector or tool attached after joint three.

---

[5]Wikipedia: *Denavit-Hartenberg parameters*, online: https://en.wikipedia.org/wiki/Denavit-Hartenberg_parameters.

3.2 Manipulator Kinematics

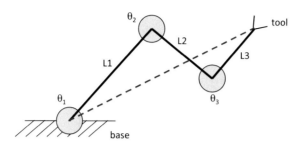

Figure 13.5: 3-dof manipulator example

We can now fill in the Denavit–Hartenberg table for this sample manipulator by placing individual local coordinate systems in each joint and finding the four parameters each for the transition from joint to joint (see Figure 13.6). Note that right-handed coordinate systems are used at all times.

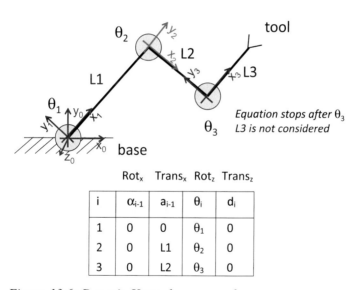

i	α_{i-1}	a_{i-1}	θ_i	d_i
1	0	0	θ_1	0
2	0	L1	θ_2	0
3	0	L2	θ_3	0

Figure 13.6: Denavit–Hartenberg example

The overall manipulator transformation (going from base until just after joint three) is:

$$^{0}_{3}\text{T} =^{0}_{1} \text{T} \cdot^{1}_{2} \text{T} \cdot^{2}_{3} \text{T}$$

The final link (here L3) is usually not included in the DH notation as it would change the manipulator's equation with every new end-effector or tool being used. This final step is a translation-only and can be simply added later separately.

We also know all individual transition matrices from i-1 to i, respectively. They are all translations along the x-axis followed by rotations about θ_i, as outlined by the DH notation–or by directly deriving the transformations, which will be easier for this simple example. For reading clarity, we are now using the abbreviations s and c for sin and cos.

$$
{}^0_1T = \begin{bmatrix} c\theta_1 & -s\theta_1 & 0 & 0 \\ s\theta_1 & c\theta_1 & 0 & 0 \\ 0 & 0 & 1 & 0 \\ 0 & 0 & 0 & 1 \end{bmatrix} \quad {}^1_2T = \begin{bmatrix} c\theta_2 & -s\theta_2 & 0 & L_1 \\ s\theta_2 & c\theta_2 & 0 & 0 \\ 0 & 0 & 1 & 0 \\ 0 & 0 & 0 & 1 \end{bmatrix} \quad {}^2_3T = \begin{bmatrix} c\theta_3 & -s\theta_3 & 0 & L_2 \\ s\theta_3 & c\theta_3 & 0 & 0 \\ 0 & 0 & 1 & 0 \\ 0 & 0 & 0 & 1 \end{bmatrix}
$$

As each transformation is a 4×4 matrix, we can now multiply them together to find the complete manipulator transition from base to tool—something we could not have done without homogeneous coordinates. As further abbreviations, we use here s_i for $sin(\theta_i)$, c_{123} for $cos(\theta_1 + \theta_2 + \theta_3)$ and so on.

$$
{}^0_3T = \begin{bmatrix} c_{123} & -s_{123} & 0 & L_1 \cdot c_1 + L_2 \cdot c_{12} \\ s_1 & c_1 & 0 & L_1 \cdot s_1 + L_2 \cdot s_{12} \\ 0 & 0 & 1 & 0 \\ 0 & 0 & 0 & 1 \end{bmatrix}
$$

This transformation can be applied to a point p on the tool tip of the manipulator (relative to joint number three), and it will return its 3D pose. If the manipulator moves, we do not have to recalculate the 4×4 matrix, we only have to update the matrix's joint angles (here: θ_1, θ_2, θ_3).

If we assume the sample configuration $\theta_1 = 0°$, $\theta_2 = 90°$, $\theta_3 = -90°$, then we can calculate the end-effector position L3 as shown in Figure 13.7 (the top index of point ${}^0P$ indicates the relative coordinate system used, so ${}^0P$ is using the global base coordinate system).

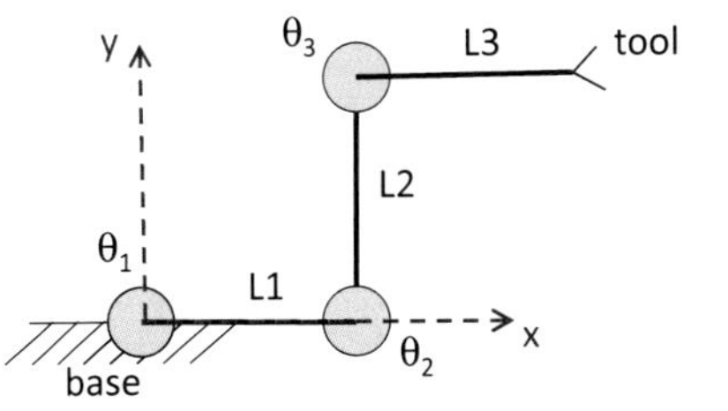

with $\theta_1 = 0°$, $\theta_2 = 90°$, $\theta_3 = -90°$

$$
{}^0P = {}^0_3T \cdot {}^3P = \begin{bmatrix} 1 & 0 & 0 & L1 \\ 0 & 1 & 0 & L2 \\ 0 & 0 & 1 & 0 \\ 0 & 0 & 0 & 1 \end{bmatrix} \cdot \begin{bmatrix} L3 \\ 0 \\ 0 \\ 1 \end{bmatrix} = \begin{bmatrix} L1 + L3 \\ L2 \\ 0 \\ 1 \end{bmatrix}
$$

Figure 13.7: Sample configuration

3.2.2 Inverse Kinematics

Performing the inverse kinematics calculation proves to be much more complex than the forward kinematics, and there has been at least anecdotal evidence of manipulator manufacturers changing their mechanical design in order to simplify the inverse kinematics solution. On the other hand, inverse kinematics is the more important task of the two, as we usually need to know how to get to a desired manipulator pose, rather than just checking where the manipulator would go for a given set of joint angles.

Unfortunately, for any given manipulator there is no general or simple way of deriving the inverse kinematics equation. The algebraic solution for the Puma 560 stretches over five pages in Craig,[6] and there are multiple solutions for most goal poses as the manipulator does have some mechanical redundancies.

The use of *numeric methods*[7] may be a good alternative for complex manipulator designs. These iterative methods use the inverse of the Jacobi matrix to derive approximate solutions that have sufficient accuracy for any practical application. The derivation of a closed symbolic solution will then not be required.

13.3 Manipulator Simulation

There are a number of robot manipulator simulators available either as public domain systems or as commercial products:

- RoboSim[8] (Figure 13.4)
  A very basic open-source wireframe manipulator simulator, developed at Univ. Stuttgart and UWA. RoboSim is based on Java and runs under all operating systems.

- EyeSim-VR[9] (Figure 13.8)
  A versatile simulator for mobile robots and robot manipulators, freely available from UWA. EyeSim runs natively under Windows, MacOS and Linux.

---

[6]J. Craig, *Introduction to Robotics*, 3rd Ed., Addison-Wesley, Reading MA, 2003.

[7]Wikipedia, *Inverse Kinematics*, online: https://en.wikipedia.org/wiki/Inverse_kinematics.

[8]T. Bräunl, R. Pollak, J. Schützner, *RoboSim – A Simple 6-DOF Robot Manipulator Simulation System*, online: https://robotics.ee.uwa.edu.au/robosim/.

[9]T. Bräunl, *EyeSim VR – EyeBot Mobile Robot Simulator*, online: https://robotics.ee.uwa.edu.au/eyesim/.

- RoboDK[10] (Figure 13.10)

  Commercial system for robot manipulator simulation; free academic licenses are available upon request. RoboDK runs under Windows, MacOS, Ubuntu Linux, as well as Raspberry Pi, Android and iOS.

- ROS OpenManipulator[11] and Gazebo

  OpenManipulator is a package of the overall open-source ROS project. Gazebo is the simulation environment for ROS. ROS is runs under Ubuntu Linux.

All manipulator manufacturers provide simulation systems for application planning with their products and in many cases specialized programming environments as well.

In EyeSim, we use the servo command *SERVOSet* for a simple setting of manipulator joint angles and prismatic lengths. Program 13.1 shows a simple repetitive manipulator movement. In the main *while*-loop, the user can select between joint 1 to joint 6 by pressing the *KEY1* button and then either increment or decrement the angle by a servo value of 10, using buttons *KEY2* and *KEY3*. Note that servo values are in byte range [0, 255] and not in degrees [0, 360]. Pressing *KEY4* will terminate the program. With this program, it is easy to familiarize oneself with the UR5 robot's joint locations and manipulator movements.

Program 13.1: Manipulator Movement Program in EyeSim in C

```c
#include "eyebot.h"

int main ()
{ int angles[6] = {0,0,0,0,0,0};
  int joint = 0, done = false;

  LCDMenu("JOINT","+","-","END");
  while (!done)
  { LCDSetPrintf(0,0, "Joint %d - Angles %3d %3d %3d %3d %3d %3d",
    joint+1, angles[0],angles[1],angles[2],
        angles[3],angles[4],angles[5]);
    switch (KEYGet())
    { case KEY1: joint = (joint+1)%6; break;
      case KEY2: angles[joint] +=10; break;
      case KEY3: angles[joint] -=10; break;
```

[10]RoboDK, *Simulate Robot Applications*, online: https://robodk.com.

[11]ROS, *ROS Robots OpenManipulator*, online: https://robots.ros.org/openmanipulator/.

3.3 Manipulator Simulation

```
    case KEY4: done = true;
    }
    SERVOSet(joint+1, angles[joint]); // drive to new angle
    }
}
```

In order to run the simulation, a SIM file helps to simplify the setup significantly. The setup shown in Program 13.2 uses the standard box environment, so no specific environment file is needed. The UR5 robot is introduced with the line:

```
UR5 1000 1000 0 joints.x
```

This places the manipulator at (x, y)-coordinates [1'000 mm, 1'000 mm] without any rotation (0°) and will automatically start the executable program *joints*.x for the UR5. The two lines following the *Objects* comment place a small soccer ball into the simulation environment at position [500 mm, 1'000 mm], also without rotation. The outcome of this setup and demonstration program is shown in Figure 13.8

Program 13.2: Simulation Environment.sim File for Manipulator

```
# Standard box world

# Robots
UR5 1000 1000 0 joints.x

# Objects
object smallball/smallsoccerball.esObj
smallball 500 1000 0
```

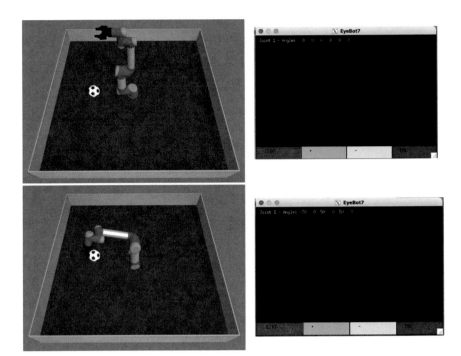

Figure 13.8: Manipulator program simulation in EyeSim

13.4 Teaching and Programming

The next step after simulation for setting up a real robot is usually a process called *teaching*, in which the operator manually drives the robot in a certain configuration and then stores this pose by pressing a button on the teach pad controls. These poses are then subsequently used as reference points in a robot program, e.g., for repeatedly moving to certain poses for a pick-and-place task. Figure 13.9 shows the typical manipulator teach pads for the UR5 and Nachi robots.

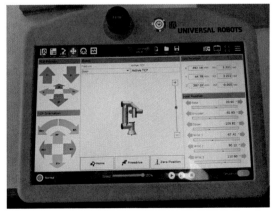

Figure 13.9: Manipulator teach pads for UR5 (left) and Nachi (right)

There is a large and growing number of programming languages and library packages available for robot manipulators. There are procedural languages like the traditional AL,[12] VAL-II[13] and even functional languages. On the other end of the spectrum are library packages for linking with C/C++, C# or Python. The tendency clearly goes away from specialized robotics languages toward libraries that can be called from standard programming languages.

Application programs can either be directly loaded onto the manipulator's control processor, or they can run on a remote processor and communicate with the manipulator's processor via sockets. This is especially then required if the robot's movements depend on the evaluation of external sensors.

Many robot manipulator users try to avoid all programming and use integrated tools such as Rhino3D[14] and Grasshoppper3D,[15,16] instead. For an introduction into the workflow using these tools, see the article by Donovan.[17]

13.5 Industrial Manipulators

In this section, we show a couple of robot manipulators that we work with at UWA. These are UR5, a robot for small payloads up to 5 kg and for educational use, and Nachi ST133TF, a full-size industrial robot with a payload of 133 kg that is being used in the production industry.

13.5.1 Universal Robots UR5

Universal Robots[18] is a Danish company that produces 6 *dof* robots with 3, 5, 10 and 16 kg payloads that are consequently named UR3, UR5, UR10 and UR16. These robots are very popular both in industry and in education, as they are safe to work in the same area as people (collaborative robots or *cobots*). These robots move with a limited torque and will stop immediate if a collision —e.g., with a person—is being detected. Therefore, no safety cage is required around them, which is a huge saving when setting up a robot application.

---

[12]S. Mujtaba, R. Goldman, *AL Users' Manual*, 3rd Ed., Stanford Dept. of Computer Science, report no. STAN-CS-81–889, Dec. 1981.

[13]B. Shimano, C. Geschke, C. Spalding, *VAL-II: A new robot control system for automatic manufacturing*, IEEE International Conference on Robotics and Automation, March 1984, pp. 278–292 (15).

[14]Rhinoceros, *Rhino3D*, online: https://www.rhino3d.com.

[15]Grasshopper3D, *Grasshopper – Algorithmic Modeling for Rhino*, online: https://www.grasshopper3d.com

[16]Rhinoceros, *Grasshopper – New in Rhino 6*, online: https://www.rhino3d.com/6/new/grasshopper/.

[17]J. Donovan, *Using Rhino and Grasshopper*, online: https://www.designrobotics.net/robot-fabrication-using-rhino-and-grasshopper/.

[18]Universal Robots, *Flexible Automation for Manufacturers of All Sizes*, online: https://www.universal-robots.com.

Figure 13.10: Real UR5 robot (left) and simulation in RoboDK (right)

Robot UR5 has a payload of 5 kg and weighs 18.4 kg without end-effector. Figure 13.10 shows a UR5 robot mounted on a trolley and the same robot in the RoboDK simulation system. The gripper and camera unit on the robot are from Canadian company Robotiq.[19] RoboDK can also be used to simulate more complex scenarios. Figure 13.11 shows an example setup where two robots cooperate in moving parts from a palette via a conveyor belt.

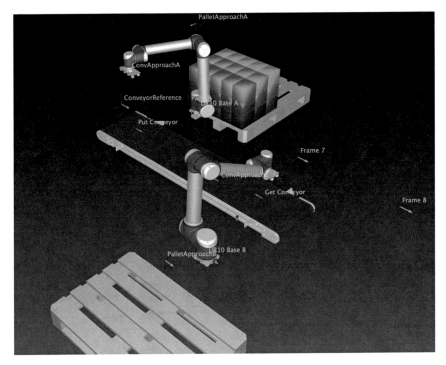

Figure 13.11: Two cooperating UR5 robots in RoboDK simulation

---

[19]Robotiq, *Robotiq: Start Production Faster*, 2020, online: https://robotiq.com.

The UR robots can be programmed in various ways with either Python or C++ interfaces, with and without the ROS[20] (Robot Operating System) interface. ROS is currently the most powerful tool for the programming of either robot manipulators or mobile robots. It is a free open-source system that has models of most commercial and research robots as well as for numerous sensors and actuators. Its strongest points are the available advanced algorithms for perception, path planning and mapping, especially simultaneous localization and mapping (SLAM).

The biggest drawback of ROS is its dependence on the Ubuntu Linux system, even requiring special release versions of it. ROS is not trivial to install, and it has a significant learning curve until one can use it productively. It is a great system for experienced programmers, but a steep step for newcomers.

13.5.2 Nachi ST133TF

According to Statista,[21] the world's largest companies producing industrial robot manipulators are:

1. ABB (Switzerland)
2. Omon (Japan)
3. Fanuc (Japan)
4. Kawasaki Robotics (Japan)
5. KUKA (Germany)
6. Yaskawa (Japan).

Other large robot manufacturers include Siemens (Germany) and Nachi (Japan), which shows the dominance of the countries Japan, Switzerland and Germany in the industrial robotics sector.

At UWA, we use a Nachi ST133TF robot, donated from Perth company Fleetwood RV. Figure 13.12 shows a photograph of the robot.

---

[20]ROS, *Powering the world's robots*, online: https://www.ros.org.

[21]Statista, *Leading companies in the global industrial robot market in 2017, based on revenue from industrial robot sales*, 2020, online: https://www.statista.com/statistics/257177/global-industrial-robot-market-share-by-company/.

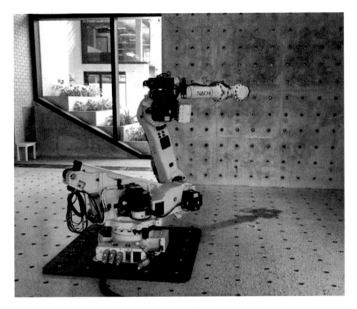

Figure 13.12: Nachi ST133TF robot

The Nachi is a large and powerful industrial robot with a payload of 133 kg and a weight of 1'070 kg without end-effector. Operating this robot requires a number of safety features, including a secure cage, which can be built from a metal fence or a light curtain (see Figure 13.13). No person is allowed inside the robot's operating cage while it is in motion, which is being ensured through automated shutdown systems, triggered by automated gate latches or light curtains.

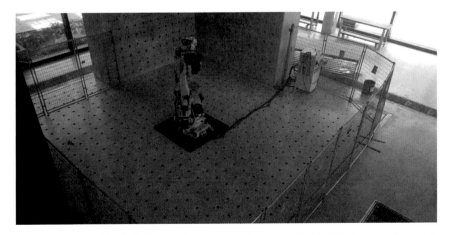

Figure 13.13 Robot safety cell at the UWA EZONE Engineering Hub

3.6 Tasks

1. Familiarize yourself with the UR5 manipulator in either EyeSim, RoboDK or ROS/Gazebo simulation. Program a sequence of different movements.

2. Program the simulated robot to do a repetitive task. Grasp a cylindrical object (can) from the bottom of a feeder, then place it back at the top of the feeder.

3. Combine two manipulators and a driving robot in simulation.
   The first manipulator places a box object from a feeder onto a mobile robot, which then drives to the second manipulator and stops there. The second manipulator then unloads the box and puts it into a collection area. The empty robot then drives back to the first manipulator where it gets loaded again. These three independent programs need to be synchronized with each other.

4. Program a manipulator to move its end-effector along a straight line.

5. Program a manipulator to spay-paint (fill) a square area on a canvas. Graphically visualize the end-effector's movement.

6. Calculate the required manipulator joint torques, in order to let the end-effector exert a certain linear force at a predefined angle.

7. If you have access to a real manipulator robot and a camera, implement a pick-and-place task. Use the camera to identify position and orientation of color-coded objects to be selected in a rectangular area that is reachable for the manipulator. Then use the manipulator to grasp the object and place it in a designated area, e.g. into a bucket.

PART III
ROBOT SOFTWARE

14

LOCALIZATION AND NAVIGATION

L ocalization and navigation are the two most important tasks for mobile robots. We want to know where we are at any point in time, and we need to be able to make a plan for how to reach the goal destination. Of course, these two problems are not isolated from each other, but rather closely linked. If a robot does not know its exact position at the start of a planned trajectory, it will encounter problems in reaching its destination.

In the past, a variety of separate algorithmic approaches was developed for localization, navigation and mapping. However, today probabilistic methods that minimize uncertainty are applied to the whole problem complex at once, especially simultaneous localization and mapping (SLAM).

In this chapter, we will look at navigation algorithms that operate with or without maps. A navigation algorithm without a map like *DistBug* can be used in a continuously changing environment or if a path has to be traveled only once and therefore does not necessarily have to be optimal. If a map is provided, then algorithms like *Dijkstra* or *A** can be applied to find the shortest path offline before the robot starts driving. Navigation algorithms without maps operate in direct interaction with the robot's sensors while driving. Navigation algorithms with maps require a nodal distance graph that has to be either provided or needs to be extracted from the environment with an algorithm like the *Quadtree* method.

14.1 Localization

One of the central problems for driving robots is localization. For many application scenarios, we need to know a robot's position and orientation (pose) at all times. For example, a cleaning robot needs to make sure it covers the whole floor area without repeating lanes or getting lost, or an office delivery

© The Author(s), under exclusive license to Springer Nature Singapore Pte Ltd. 2022
T. Bräunl, *Embedded Robotics*,
https://doi.org/10.1007/978-981-16-0804-9_14

robot needs to be able to navigate a building floor and needs to know its pose relative to its starting point. This is a nontrivial problem in the absence of global sensors.

The localization problem can be solved by using a Global Positioning System. In an outdoor setting this could be a Global Navigation Satellite System (GNSS) such as GPS. In an indoor setting, a global sensor network with infrared, sonar, laser or radio beacons can be employed.

14.1.1 Radio Beacons

Let us assume a driving environment that has a number of synchronized radio beacons that are sending out radio or sonar signals at the same regular time intervals, but at different (distinguishable) frequencies (see Figure 14.1). By receiving signals from two or three different beacons, the robot can determine its local position from the time difference of the signals' arrival times.

Using two beacons can narrow down the robot position to two possibilities, since two circles have two intersection points. For example, if the two signals arrive at exactly the same time, the robot is located in the middle between the two transmitters. If, say, the left beacon's signal arrives before the right one, then the robot is closer to the left beacon by a distance proportional to the time difference. Using local position coherence, this may already be sufficient for global positioning. However, to be able to determine a 2D position without local sensors, three beacons are required.

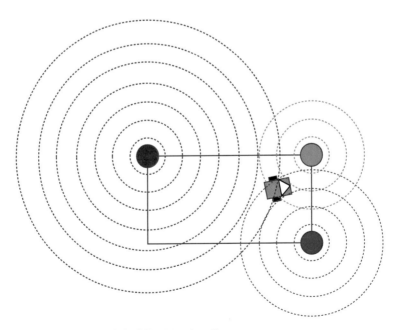

Figure 14.1: Global Positioning System

Only the robot's position can be determined by this method, not its orientation. The orientation has to be deduced from the change in position (difference between two subsequent positions), which is exactly the method employed for satellite-based GNSS, or from an additional compass sensor.

Using global sensors is in many cases not possible because of restrictions in the robot environment, or not desired because it limits the autonomy of a mobile robot (such as overhead or global vision systems for robot soccer in some competitions). On the other hand, in some cases, it is possible to convert a system with global sensors as in Figure 14.1 to one with local sensors. For example, if the sonar sensors can be mounted on the robot and the beacons are converted to reflective markers, then we have an autonomous robot with local sensors.

14.1.2 Light Beacons

Another idea is to use light-emitting homing beacons instead of radio or sonar beacons—the equivalent of a lighthouse. With two light beacons with different, distinguishable colors, the robot can determine its position at the intersection of the lines from the beacons at the measured angle. The advantage of this method is that the robot can determine its position and orientation. However, in order to do so, the robot has either to perform a 360° rotation, or it needs to possess an omni-directional vision system that allows it to determine the angle of each recognized light beacon.

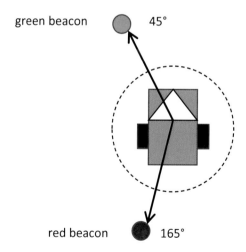

Figure 14.2: Beacon measurements on robot

For example, after doing a 360° scan in Figure 14.2, the robot knows it sees a green beacon at an angle of 45° and a red beacon at an angle of 165° in its local coordinate system.

We then need to fit these two vectors into the robot's environment using the known fixed beacon positions. But since we do not know the robot's distance from either of the beacons, we cannot determine which of the many possible positions is correct (see Figure 14.3, left). Knowing only two beacon angles is not sufficient for localization. If the robot in addition knows its global orientation, for example, by using an on-board compass, or if three beacons are being used, then correct localization is possible (see Figure 14.3, right).

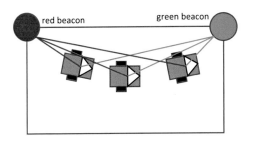
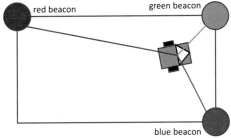

Figure 14.3: Positioning using light beacons

14.1.3 Dead Reckoning

In many cases, driving robots have to rely on their wheel encoders alone for short-term localization and can update their true position and orientation from time to time, for example, when reaching a certain waypoint. So-called *dead reckoning* is the standard localization method under these circumstances. Dead reckoning is a nautical term from the 1700s when ships did not have modern navigation equipment and had to rely on vector-adding their course segments to establish their current position.

Dead reckoning can be described as local polar coordinates or—more practically—as turtle graphics geometry. As can be seen in Figure 14.4, it is required to know the robot's starting position and orientation. For all subsequent driving actions (straight sections, curves or rotations on the spot), the robot's current position is updated as per the feedback from the wheel encoders.

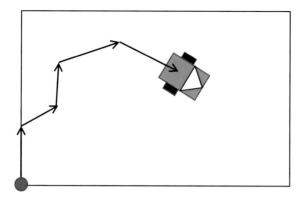

Figure 14.4: Dead reckoning

Obviously, this method has severe limitations when applied over a longer time. All inaccuracies due to sensor error or wheel slippage will add up over time. Especially bad are errors in orientation, because over time they have the largest effect on position accuracy. This is why an on-board compass is very valuable in the absence of global sensors. It makes use of the Earth's magnetic field to determine a robot's absolute orientation. Even simple digital compass modules work indoors and outdoors and are accurate to about 1°.

14.2 Environment Representation

We have seen how a robot can drive a certain distance or turn about a certain angle in its *local coordinate system*. For many applications, however, it is important to first establish a map (in an unknown environment) or to plan a path (in a known environment). These path points are usually specified in *global* or *world coordinates*.

Translating local robot coordinates to global world coordinates is a 2D transformation that requires a translation and a rotation, in order to match the two coordinate systems (Figure 14.5).

Assume the robot has the global position $[r_x, r_y]$ and has global orientation φ. If it senses an object at local coordinates $[o_x', o_y']$, then the global coordinates $[o_x, o_y]$ of the object can be calculated as follows:

$$[o_x, o_y] = \mathrm{Trans}(r_x, r_y) \cdot \mathrm{Rot}(\varphi) \cdot [o_x', o_y']$$

For example, the marked position in Figure 14.5 has local coordinates [3, 1], so it is ahead of the robot. The robot's global position is [5, 3] and its orientation is 120°. Then the global object position can be calculated as:

Localization and Navigation

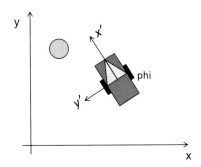

Figure 14.5: Global and local coordinate systems

$$[o_x, o_y] = \text{Trans}(5,3) \cdot \text{Rot}(120°) \cdot [3,1]$$
$$= \text{Trans}(5,3) \cdot [-2.37, 2.11]$$
$$= [2.63, 5.11]$$

Coordinate transformations such as this one can be greatly simplified by using *homogeneous coordinates*. As already shown for robot manipulators in an earlier chapter, arbitrary long 3D transformation sequences can be summarized in a single 4×4 matrix (see Craig[1] for details). In the 2D case above, a 3×3 homogeneous matrix is sufficient. This matrix has the rotation part about the z-axis in its top left 2×2 section and the (x,y)-translation in its right 2×1 column. The bottom row always gets the vector $[0, 0, 1]$.

$$\begin{bmatrix} o_x \\ o_y \\ 1 \end{bmatrix} = \begin{bmatrix} 1 & 0 & 5 \\ 0 & 1 & 3 \\ 0 & 0 & 1 \end{bmatrix} \cdot \begin{bmatrix} \cos\alpha & -\sin\alpha & 0 \\ \sin\alpha & \cos\alpha & 0 \\ 0 & 0 & 1 \end{bmatrix} \cdot \begin{bmatrix} 3 \\ 1 \\ 1 \end{bmatrix}$$

$$\begin{bmatrix} o_x \\ o_y \\ 1 \end{bmatrix} = \begin{bmatrix} \cos\alpha & -\sin\alpha & 5 \\ \sin\alpha & \cos\alpha & 3 \\ 0 & 0 & 1 \end{bmatrix} \cdot \begin{bmatrix} 3 \\ 1 \\ 1 \end{bmatrix}$$

for a = 30° this comes to:

$$\begin{bmatrix} o_x \\ o_y \\ 1 \end{bmatrix} = \begin{bmatrix} -0.5 & -0.87 & 5 \\ 0.87 & -0.5 & 3 \\ 0 & 0 & 1 \end{bmatrix} \cdot \begin{bmatrix} 3 \\ 1 \\ 1 \end{bmatrix} = \begin{bmatrix} 2.63 \\ 5.11 \\ 1 \end{bmatrix}$$

---

[1] J. Craig, *Introduction to Robotics – Mechanics and Control*, 2nd Ed., Addison-Wesley, Reading MA, 1989.

Navigation, however, is much more than just driving to a certain specified location—it all depends on the particular task to be solved. For example: Are the destination points known or do they have to be searched? Are the dimensions of the driving environment known? Are all objects in the environment known? Are objects moving or stationary? and so on.

There are a number of well-known navigation algorithms, which we will briefly discuss in the following. However, some of them are of a more theoretical nature and do not closely match the real problems encountered in practical navigation scenarios. For example, some of the shortest path algorithms require a set of node positions and full information about their distances, which in many practical applications are not available. See also Arkin[2] for more details.

The two standard representations of a 2D environment are *Configuration Space* and *Occupancy Grid*. In configuration space, we are given the dimensions of the environment plus the coordinates of all obstacles, e.g., walls, represented by line segments. In an occupancy grid, the environment is specified at a certain resolution with individual pixels either representing *free space* (white pixels) or an *obstacle* (black pixels). These two formats can easily be transformed into each other. For transforming a configuration space into an occupancy grid, we can "print" the obstacle coordinates on a canvas data structure of the desired resolution. For transforming an occupancy grid into a configuration space, we can extract obstacle line segment information by combining neighboring obstacle pixels into individual line segments.

While many navigation algorithms work directly on the environment description (configuration space or occupancy grid), some algorithms, such as Dijkstra and A*, require a distance graph as input (see Figure 14.6). A distance graph is an environment description at a higher level. It does not contain the full environment information, but it allows for an efficient initial path planning step (e.g., from room to room) that can be subsequently refined.

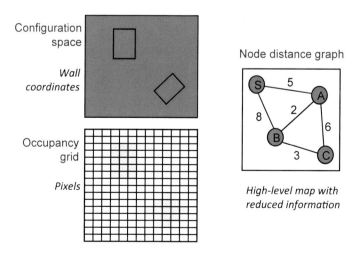

Figure 14.6: Basic environment representations

---

[2]R. Arkin, *Behavior-Based Robotics*, MIT Press, Cambridge MA, 1998.

A distance graph contains only a few individually identified node positions from the environment and their relative distances. Neither of these two basic environment formats leads directly to a distance graph, so we are interested in algorithms that can automatically derive a distance graph.

The brute force solution to this problem would be starting with an occupancy grid and treating each pixel of the grid as a node in the distance graph. If the given environment is in configuration space, it can easily be converted to occupancy grid by "printing it" as mentioned before.

However, this approach has a number of problems. First, the number of nodes in the resulting distance graph will be huge, so this will not be computationally feasible for larger environments or finer grids. Second, path planning in such a graph will result in suboptimal paths, as neighboring pixels have been transformed into neighboring graph nodes and therefore only support turning angles that are multiples of $\pm 45°$ (when using eight-nearest neighbors) of multiples of $\pm 90°$ (when using four-nearest neighbors).

14.3 Quadtree

Using a quadtree can generate a node graph with a low number of nodes and without any turning angle restriction. To generate a quadtree, the given environment in either configuration space or occupancy grid format is recursively divided into four quadrants. If a quadrant is either completely empty (no obstacles) or completely covered by an obstacle, it becomes a terminal node, also called a *leaf*. Those quadrant nodes that contain a mix of free space and obstacles will be further divided in the next recursive step (see Figure 14.7).

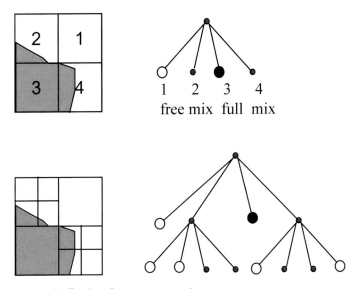

Figure 14.7: Quadtree construction

This procedure continues until either all nodes are terminal or until a maximum resolution is reached.

All free nodes of the quadtree (or more precisely their center positions) can now be used as nodes in the distance graph. We construct a complete graph by linking each free node with each other free node, then eliminate those edges for which the linked two nodes cannot be connected through a collision-free line, because of a blocking obstacle (e.g., lines c–e and b–e in Figure 14.8). For the remaining edges, we determine their relative distances by measuring in the original environment and enter these values into the distance graph (Figure 14.8, right). As the final path planning step, we can then use, e.g., the A* algorithm on the distance graph.

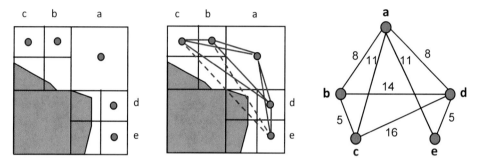

Figure 14.8: Distance graph construction from quadtree

Program 14.2 shows the recursive generation of a quadtree. The subroutine *quad* will be called with the coordinates of the leftmost corner of the area and its size, e.g., for an area of 1024×1024 pixels:

```
quad(0,0, 1024)
```

Our implementation requires a square area with the size being a power of two, as this simplifies the area subdivision. The program first checks through two nested loops whether the given area is completely free (*allFree*) or completely occupied (*allOcc*) and sets these variables accordingly. If the area is completely free, then the coordinates of this node are printed for subsequent use. Otherwise, if not the whole area is part of an obstacle and (redundantly) if the area size is larger than 1×1, then quad calls itself recursively for each of the four quadrants of the given area, as shown in the example in Figure 14.9.

Program. 14.1: Recursive Quadtree Construction

```
void quad(int x, int y, int size) // start pos + size
{ bool allFree=true, allOcc=true;
  for (int i=x; i<x+size; i++)
   for (int j=y; j<y+size; j++)
    if (field[i][j]) allFree=false; //at least 1 occ.
            else allOcc =false; //at least 1 free
   if (allFree) printf("free %d %d %d\n", x, y, size);
      else if (!allOcc && (size>1))
      { int s2 = size/2;
        quad(x, y, s2);
        quad(x+s2, y, s2);
        quad(x, y+s2, s2);
        quad(x+s2, y+s2, s2);
      }
}
```

Start: x=0, y=0, size=4

Loop: check all pixels

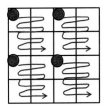

Recursive calls:
x=0, y=0, size=2
x=2, y=0, size=2
x=0, y=2, size=2
x=2, y=2, size=2

Figure 14.9: Quadtree decomposition example

So, summarizing the algorithm again:

1. Recursive function for Quadtree decomposition
2. Make mid-points of all free squares to nodes
3. For all node pairs (a,b) calculate the Euclidian distance:
   $d(a,b) = \sqrt{[(a_x-b_x)^2 + (a_y-b_y)^2]}$
4. Delete all lines (a,b) that cross through any occupied or mixed square.

Steps 1 + 2 are part of Program 14.1, step 3 is trivial (see Figure 14.8), but step 4 is quite complex and requires applying an algorithm such as Liang–Barsky[3] or Cohen–Sutherland[4].

We need to check two conditions as visualized in Figure 14.10:

a.   Are all box coordinates on the same side of the line segment?
     → If yes, there is no intersection

b.   Else, do the box and line shadows intersect in *x and y*?
     → if yes, then there is an intersection, if no, then there is not.

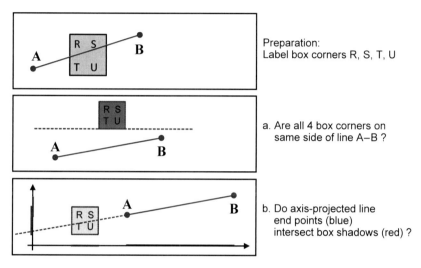

Figure 14.10: Line-box intersection algorithm

We can use the simplified method by Alejo[5] to check these two conditions:

a.   Are all 4 corners of box RSTU on same side of line of line A–B?
     For each corner point $P \in \{R, S, T, U\}$ calculate then line equation and then check against A and B:
     $F(P) = (B_y - A_y)*P_x + (A_x - B_x)*P_y + (B_x*A_y - A_x*B_y)$
     - $F(P) = 0$ means point P is on the line A–B.
     - $F(P) > 0$ means point P is above line A–B.
     - $F(P) < 0$ means point P is below line A–B.

[3]Wikipedia, *Liang-Barsky algorithm*, online: https://en.wikipedia.org/wiki/Liang-Barsky_algorithm.

[4]Wikipedia, *Cohen-Sutherland algorithm*, online: https://en.wikipedia.org/wiki/Cohen-Sutherland_algorithm.

[5]Alejo, Stack Overflow, *How to test if a line segment intersects an axis-aligned rectangle in 2D?*, online: https://stackoverflow.com/questions/99353/how-to-test-if-a-line-segment-intersects-an-axis-aligned-rectange-in-2d.

b. Else project line endpoints on x-axis and check if shadow intersects square shadow in *x-* **and** *y*-axis:
- if $(A_x > U_x$ and $B_x > U_x) \rightarrow$ **No** intersection,
- if $(A_x < R_x$ and $B_x < R_x) \rightarrow$ **No** intersection,
- if $(A_y > U_y$ and $B_y > U_y) \rightarrow$ **No** intersection,
- if $(A_y < R_y$ and $B_y < R_y) \rightarrow$ **No** intersection,
- else $\rightarrow$ There **is an** intersection.

This algorithm can be further improved by using a framed quadtree, as shown in Yahja et al.[6] The following sections discuss more evolved methods for automatically generating a distance graph.

14.4 Visibility Graph

The visibility graph method uses a different idea to identify node positions for the distance graph. While the Quadtree method identifies points with some free space around them, the visibility graph method uses corner points of obstacles instead.

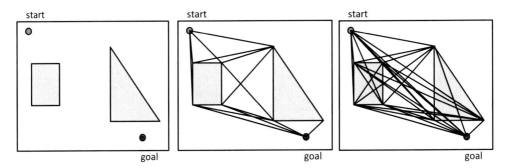

Figure 14.11: Selecting nodes for visibility graph

If the environment is represented as a configuration space, then we already have the polygon description of all obstacles. We simply collect a list of all start and end points of obstacle border lines, plus the robot's start and goal position.

As shown in Figure 14.11, center, we then construct a complete graph by linking every node position of every other one. Finally, we delete all the lines that intersect an obstacle, leaving only those lines that allow us to drive from one node to another in a direct line (Figure 14.11, right).

One problem of this approach is that it allows lines to pass very closely to an obstacle, so this would only work for a theoretical robot with a zero diameter. However, this problem can be easily solved by virtually enlarging each obstacle by half of the robot's diameter before applying the algorithm (see Figure 14.12).

---

[6]A. Yahja, A. Stentz, S. Singh, B. Brummit, *Framed-Quadtree Path Planning for Mobile Robots Operating in Sparse Environments*, IEEE Conference on Robotics and Automation (ICRA), Leuven Belgium, May 1998, pp. (6).

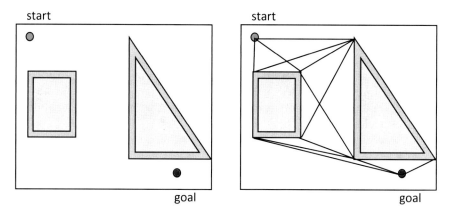

Figure 14.12: Enlarging obstacles by half of robot diameter

More advanced versions of the visibility graph algorithm can be used to also include the robot's orientation for driving, which is especially important for robots that are not of cylindrical shape, as shown by Bicchi et al.[7]. For non-cylindrical robots (non-circular in their 2D projection) there may exist possible paths that require a change of orientation in order to drive through a narrow passageway between two obstacles. This more complex problem has become known as the *piano mover's problem* (see Hopcroft et al.[8] and Figure 14.13).

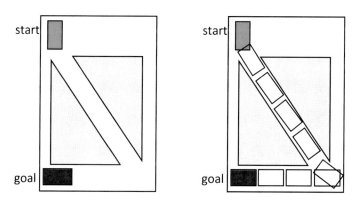

Figure 14.13: Piano mover's problem

[7]A. Bicchi, G. Casalino, C. Santilli, *Planning Shortest Bounded-Curvature Paths*, Journal of Intelligent and Robotic Systems, vol. 16, no. 4, Aug. 1996, pp. 387–405 (9).

[8]J. Hopcroft, J. Schwartz, M. Sharir, *Planning, geometry, and complexity of robot motion*, Ablex Publishing, Norwood NJ, 1987.

14.5 Voronoi Diagram and Brushfire Algorithm

A Voronoi diagram is another method for extracting distance node information from a given 2D environment. The work by Voronoi[9], Dirichlet[10] and Delaunay[11] dates back more than a century ago. The principle way the algorithm works is by constructing a skeleton of points with minimal distances to obstacles and walls.

We can define a Voronoi diagram as follows:

F free space in environment , i.e., white pixels in a binary image

F' occupied space, i.e., black pixels in binary image

$b \in F'$ is basis point for $p \in F$

         $\Leftrightarrow b$ has minimal distance to p (no other point in F' is closer to p)

Voronoi diagram = { p ∈ F | p has at least two basis points }

Figure 14.14 demonstrates the relationship between basis points and Voronoi points. Only free space points with at least two basis points qualify as Voronoi points.

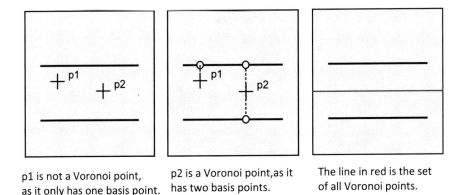

p1 is not a Voronoi point, as it only has one basis point.   p2 is a Voronoi point, as it has two basis points.   The line in red is the set of all Voronoi points.

Figure 14.14: Basis points and Voronoi points

Figure 14.15 shows the set of Voronoi points for a closed box. The Voronoi points span a minimal distance skeleton of the given box structure.

[9]G. Voronoi, *Nouvelles applications des parametres continus – la theorie des formes quadratiques*, Journal für die Reine und Angewandte Mathematik, vol. 133, 1907, pp. 97–178 (82).

[10]G. Dirichlet, *Über die Reduktion der positiven quadratischen Formen mit drei unbestimmten ganzen Zahlen*, Journal für die Reine und Angewandte Mathematik, vol. 40, 1850, pp. 209–227 (19).

[11]B. Delaunay, *Sur la sphere vide*, Izvestia Akademii Nauk SSSR, Otdelenie Matematicheskikh i Estestvennykh Nauk, vol. 7, 1934, pp. 793–800 (8).

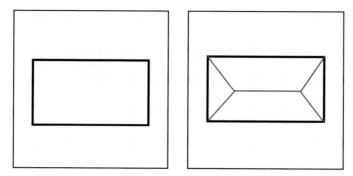

Figure 14.15: Box share and Voronoi points in red

If we have a Voronoi diagram, we can use the end points of all Voronoi lines as nodes to construct the distance graph. However, deriving all Voronoi points is not as easy as the simple definition may suggest. The brute force approach of checking every single pixel in the image for its Voronoi property would take a very long time, as this would involve two nested loops over all pixels. Much more efficient methods for determining Voronoi points are the Delauney triangulation and the Brushfire algorithm, which we will describe in detail below.

14.5.1 Delaunay Triangulation

The Delaunay triangulation tries to construct a Voronoi diagram with much less computational effort. We start with the definition of a Delaunay triangle:

q_1, q_2, q_3 ∈ F' form a Delaunay triangle

⇔ there is point p ∈ F that is equidistant to all q_1, q_2, q_3 and no other point in F' is nearer to p.

Point p in a Delaunay triangle is a Voronoi point.
All corner points in F' are also Voronoi points.

This means: Voronoi point p is the center of a free space circle touching obstacle or boundary points q_1, q_2, q_3 without any other obstacle point inside it.

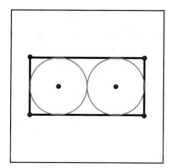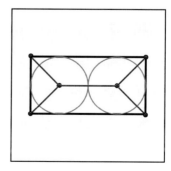

Figure 14.16: Delaunay triangulation example

Figure 14.16, left, shows an example of two Voronoi points inside circles plus four corner Voronoi points. Voronoi points can be used directly as graph nodes for a navigation algorithm or they can be joined with their nearest neighbors through straight lines to form a complete Voronoi diagram, as in Figure 14.16, right.

14.5.2 Brushfire Algorithm

The Brushfire algorithm by Lengyel et al.[12] is a discrete graphics algorithm for generating Voronoi diagrams on an occupancy grid (using 1 for occupied, 0 for free). The algorithm steps are defined as follows:

1. Identify each obstacle and each border with a unique label (color).
2. Iterate i starting at 2 until no more changes occur:

   a. If a free pixel is a neighbor to a labeled pixel or a border, then label the pixel with 'i' in the same color. (Use 4-nearest or 8-nearest neighborhood)

   b. If a free pixel is being *overwritten twice or more in different colors* by this procedure, then make it a Voronoi point.

   c. If a pixel and its top or right neighbor were both overwritten with 'i' in different colors by this procedure, then make this pixel a Voronoi point.

The border labels (or "colors") are slowly moving in from the sides toward the center, so in that sense Brushfire is a type of flood-fill algorithm.

Figure 14.17 shows the step-by-step execution of the Brushfire algorithm and the resulting Voronoi diagram for a sample environment.

---

[12]J. Lengyel, M. Reichert, B. Donald, D. Greenberg, *Real-time robot motion planning using rasterizing computer graphics hardware*, Proceedings of ACM SIGGRAPH 90, Computer Graphics vol. 24, no. 4, 1990, 327–335 (9).

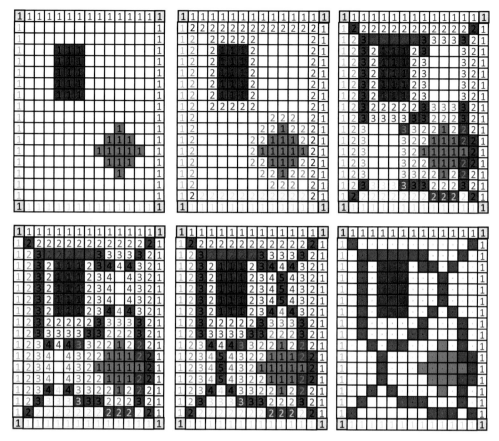

Figure 14.17: Brushfire algorithm execution and resulting Voronoi diagram

14.6   Potential Field Method

References   Arbib, House,[13] Koren, Borenstein,[14] Borenstein, Everett, Feng[15]
Description   Global map generation algorithm with virtual forces.
Required   Start and goal position, positions of all obstacles and walls.
Algorithm   Generate a map with virtual attracting and repelling forces. Start
               point, obstacles, and walls are repelling, goal is attracting; force
               strength is inverse to object distance; robot simply follows force field.

---

[13]M. Arbib, D. *House, Depth and Detours: An Essay on Visually Guided Behavior*, in M. Arbib, A. Hanson (Eds.), Vision, Brain and Cooperative Computation, MIT Press, Cambridge MA, 1987, pp. 129–163 (35).

[14]Y. Koren, J. Borenstein, *Potential Field Methods and Their Inherent Limitations for Mobile Robot Navigation*, Proceedings of the IEEE Conference on Robotics and Automation, Sacramento CA, April 1991, pp. 1398–1404 (7).

[15]J. Borenstein, H. Everett, L. Feng, *Navigating Mobile Robots: Sensors and Techniques*, AK Peters, Wellesley MA, 1998.

Example Figure 14.18 shows an example with repelling forces from obstacles and walls, plus a superimposed general field direction from start to goal Figure 14.19 exemplifies the potential field generation steps in the form of 3D surface plots. A ball placed on the start point of this surface would roll toward the goal point, which demonstrates the derived driving path of a robot. The 3D surface on the left only represents the force vector field between start and goal as a height difference, as well as repelling walls. The 3D surface on the right has the repelling forces for two obstacles added.

Problem The robot can get stuck in local minima. In this case, the robot has reached a spot with zero force (or a level potential), where repelling and attracting forces cancel each other out. So the robot will stop and never reach the goal.

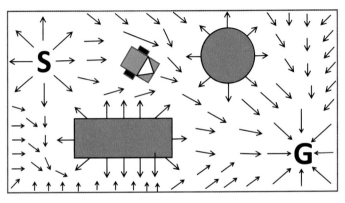

Figure 14.18: Potential field

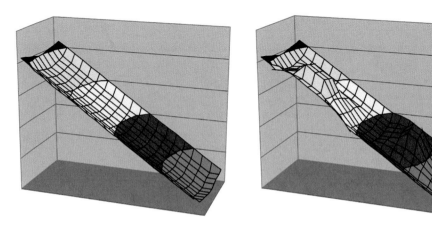

Figure 14.19: Potential fields as 3D surfaces

14.7   Wandering Standpoint Algorithm

Reference	von Puttkamer[16]
Description	Local path planning algorithm.
Requires	Local distance sensor.
Algorithm	Try to reach goal from start in direct line. When encountering an obstacle, measure avoidance angle for turning left and for turning right, then turn to smaller angle. Continue with boundary following around the object until goal direction is clear again.
Example	Figure 14.20 shows the subsequent robot positions from Start through 1...6 to Goal. The goal is not directly reachable from the start point. Therefore, the robot switches to boundary-following mode until at point 1 it can drive again unobstructed toward the goal. At point 2, another obstacle has been reached, so the robot once again switches to boundary-following mode. Finally at point 6, the goal is directly reachable in a straight line without further obstacles.
Problem	The algorithm can lead to an endless loop for extreme obstacle placements. In this case, the robot keeps driving, but never reaches the goal.

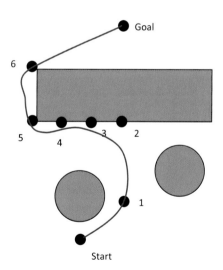

Figure 14.20: Wandering standpoint

---

[16]E. von Puttkamer, *Autonome Mobile Roboter*, Lecture notes, Univ. Kaiserslautern, Fachbereich Informatik, 2000.

14.8   Bug Algorithm Family

References	Bug1 and Bug2: Lumelsky, Stepanov[17] 1986
	DistBug: Kamon, Rivlin[18] 1997
	For a summary and comparison of Bug algorithms see Ng, Bräunl[19] 2007.
Description	Local planning algorithm that guarantees convergence and will find a path if one exists or report that goal is unreachable.
Required	Own position (odometry), goal position, and touch sensor (for Bug1 and Bug2) or distance sensor data (for DistBug).
Algorithm	<u>Bug1:</u> Drive straight toward the goal until an obstacle is hit (hit point). Then do boundary following while recording the shortest distance to goal (leave point). When hit point is reached again, drive to leave point and continue algorithm from there.
	<u>Bug2:</u> Using an imaginary straight line M from start to goal, follow M line until an obstacle is hit (hit point). Then follow the boundary until a point on M is reached that is closer to the goal (leave point). Continue the algorithm from here.
	<u>DistBug:</u> Drive straight toward the goal until an obstacle is hit (hit point). Then follow the boundary while recording the shortest distance to goal. If the goal is visible or if there is sufficient free space toward the goal, continue the algorithm from here (leave point). However, if the robot returns to the previous hit point, then the goal is unreachable.

Below is our algorithmic version of DistBug, adapted from the original version.

Constant:	STEP	min. distance of two leave points, e.g. 1cm
Variables:	P	current robot position (x, y)
	G	goal position (x, y)
	Hit	location where current obstacle was first hit
	Min_dist	minimal distance to goal during boundary following

[17]V. Lumelsky, A. Stepanov, *Dynamic Path Planning for a Mobile Automaton with Limited Information on the Environment*, IEEE Transactions on Automatic Control, vol. 31, Nov. 1986, pp. 1058—1063 (6).

[18]I. Kamon, E. Rivlin, *Sensory-Based Motion Planning with Global Proofs*, IEEE Transactions on Robotics and Automation, vol. 13, no. 6, Dec. 1997, pp. 814–822 (9).

[19]J. Ng, T. Bräunl, *Performance Comparison of Bug Navigation Algorithms*, Journal of Intelligent and Robotic Systems, Springer-Verlag, no. 50, 2007, pp. 73–84 (12).

1. *Main* **program**

```
int main()
{while (1)
  { "drive towards goal" // non-blocking, program continues while driving
    if (P==G)                          { "success"; exit(0); }
    if ("obstacle collision") { Hit = P; follow(); }
  }
}
```

2. **Subroutine** *follow*

```
void follow ()
  {    Min_dist = ; // init
       Turn(left); // to align with wall
       while (1)
       { "drive following obstacle boundary"; // non-blocking, continue
         program
           D = dist(P, G) // air-line distance from current position to goal
           F = free(P, G) // space in direction of goal, e.g. PSD measurement
           if (D < Min_dist) Min_dist = D;

           if ((F ≥ D) or (D-F ≥ Min_dist - STEP)) then return;
           // goal or closer point reachable
           if (P == Hit) then { "goal unreachablez"; exit(1); }
       }
  }
```

Problem: Although this algorithm has nice theoretical properties, it is not very usable in practice, as the positioning accuracy and sensor distance required for the success of the algorithm are usually not achievable. Most variations of the DistBug algorithm suffer from a lack of robustness against noise in sensor readings and robot driving/positioning.

Figure 14.21 shows two standard DistBug examples, here simulated with the EyeSim system. In the example on the left-hand side, the robot starts in the main program loop, driving forward toward the goal, until it hits the U-shaped obstacle. A hit point is recorded, and subroutine *follow* is called. After a left turn, the robot follows the boundary around the left leg, at first getting further away from the goal, then getting closer and closer. Eventually, the free space in goal direction will be greater or equal to the remaining goal distance ($\rightarrow$ leave point). Then the boundary follow subroutine returns to the main program, and the robot will for the second time drive directly toward the goal. This time the goal is reached, and the algorithm terminates.

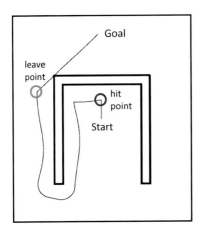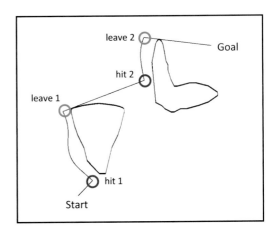

Figure 14.21: Distbug examples, J. Ng, UWA

Figure 14.21, right, shows another example. The robot will stop boundary following at the first leave point, because its sensors have detected that it can reach a point closer to the goal than before. After reaching the second hit point, boundary following is called a second time, until at the second leave point the robot can drive directly to the goal.

To point out the differences between the two algorithms, we show the execution of the algorithms Bug1 and Bug2 (Figure 14.22) in the same environment as Fig. 14.21, right.

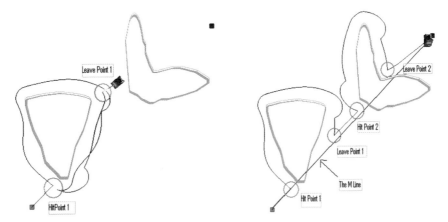

Figure 14.22: Bug1 (left) and Bug2 (right) examples, J. Ng, UWA

Figure 14.23 shows two more examples that further demonstrate the Dist-Bug algorithm. In Figure 14.23, left, the goal is inside the E-shaped obstacle and cannot be reached. The robot first drives straight toward the goal, hits the obstacle and records the hit point, then starts boundary following. After completion of a full circle around the obstacle, the robot returns to the hit point, which is the termination condition for an unreachable goal.

Figure 14.23, right, shows a more complex environment. After the hit point has been reached, the robot surrounds almost the whole obstacle until it finds the entry to the maze-like structure. It continues boundary following until the goal is directly reachable from the leave point.

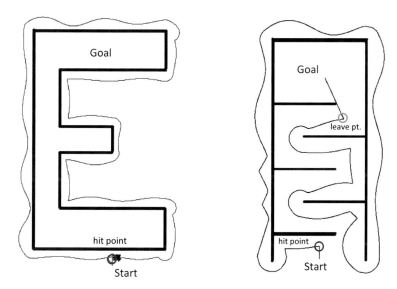

Figure 14.23: Complex DistBug examples, Ng, UWA

14.9   Dijkstra's Algorithm

Reference   Dijkstra[20]

Description Algorithm for computing all shortest paths from a given starting node in a fully connected graph. Time complexity for e edges and v nodes is $O(e + v^2)$ for a naive implementation, which can be reduced to $O(e + v \cdot \log v)$.

Distances between neighboring nodes are given as edge(n, m).

[20]E. Dijkstra, *A note on two problems in connexion with graphs*, Numerische Mathematik, Springer-Verlag, Heidelberg, vol. 1, pp. 269–271 (3), 1959.

Required   Relative distances (non-negative) between all nodes.

Algorithm   While all previous algorithms worked directly on the environment data, Dijkstra (and also A* below) requires a distance graph to be constructed first. Start *ready set* with start node. In loop select node with shortest distance in every step, then compute distances to all of its neighbors and store shortest paths with corresponding path predecessors. Add current node to *ready set*; loop finishes when all nodes are included.

1. Init

```
Set start distance to 0:              dist[s] = 0
Set all other distances to infinite:  dist[i] = ∞ (for i≠s)
Clear ready set:                      Ready = { }
```

2. Loop until all nodes are in Ready set

```
Select node n with shortest known distance that is not in Ready set
Ready = Ready + {n}
for "each neighbor node m of n":
  if (dist[n]+edge(n,m) < dist[m]) // then shorter path found
    {dist[m] = dist[n] + edge(n,m); //record new shortest distance
     pre[m] = n;                    // record predecessor node
    }
```

As an example, consider the nodes and distances in Figure 14.24. On the left-hand side is the distance graph, on the right-hand side is the table with the shortest distances found so far and the immediate path predecessors that lead to this distance.

In the beginning (initialization step), we only know that start node S is reachable with distance 0 (by definition). The distances to all other nodes are infinite at this stage and we do not have a path predecessor recorded yet. Proceeding from step 0 to step 1, we have to select the node with the shortest distance from all nodes that are not yet included in the Ready set. Since Ready is still empty, we have to look at all nodes. Clearly S has the shortest distance (0), compared to infinite distances for all other nodes.

For step 1, S is now included into the Ready set and the distances and path predecessors (equal to S) for all its neighbors are being updated. Since S is neighbor to nodes a, c and d, the distances for these three nodes are being updated and their path predecessor is being set to S.

When moving to step 2, we have to select the node with the shortest path among a, b, c, d, as S is already in the Ready set. Among these, node c has the shortest path (5). The table is updated for all neighbors of c, which are S, a, b and d. New shorter distances are found for a, b and d, each entering c as their immediate path predecessor.

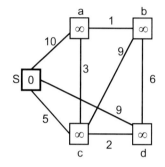

From s to:	S	a	b	c	d
Distance	0	∞	∞	∞	∞
Predecessor	-	-	-	-	-

Step 0: Init list, no predecessors

Ready = {}

From s to:	**S**	a	b	c	d
Distance	0	10	∞	5	9
Predecessor	-	S	-	S	S

Step 1: Closest node is S, add to Ready

Update distances and pred. to all neighbors of S

Ready = {S}

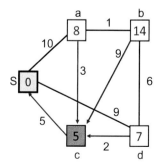

From s to:	**S**	a	b	**c**	d
Distance	0	1̶0̶ 8	14	5	9̶ 7
Predecessor	-	S̶ c	c	S	S̶ c

Step 2 : Next closest node is c, add to Ready

Update distances and pred for a and d

Ready = {S, c}

Figure 14.24: Dijkstra's algorithm example

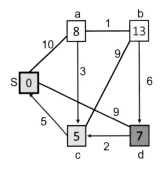

From s to:	**S**	a	b		**c**	**d**
Distance	0	8	1̶4̶	13	5	7
Predecessor	-	c	c̶	d	S	c

Step 3 : Next closest node is *d,* add to Ready

Update distances and pred. for *b*

Ready = {*S, c, d*}

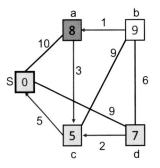

From s to:	**S**	**a**	b		**c**	**d**
Distance	0	8	1̶3̶	9	5	7
Predecessor	-	c	d̶	a	S	c

Step 4 : Next closest node is *a,* add to Ready

Update distances and pred. for *b*

Ready = {*S, a, c, d*}

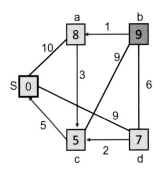

From s to:	**S**	**a**	**b**	**c**	**d**
Distance	0	8	9	5	7
Predecessor	-	c	a	S	c

Step 5 : Closest node is *b* add to Ready

Check all neighbors of *s*

Ready = {*S, a, b, c, d*} _complete!_

Figure 14.24: (continued)

In the following steps 3 through 5, the algorithm's loop is repeated, until finally, all nodes are included in the Ready set and the algorithm terminates. The table now contains the shortest path from the start node *S* to each of the other nodes, as well as the path predecessor for each node, allowing us to reconstruct the shortest path.

Figure 14.25 shows how to construct the shortest path from each node's predecessor. For finding the shortest path between *S* and *b*, we already know the shortest distance (9), but we have to reconstruct the shortest path backwards from *b*, by following the predecessors:

$$\text{pre}[b] = a$$
$$\rightarrow \text{pre}[a] = c$$
$$\rightarrow \text{pre}[c] = S$$

Therefore, the shortest path is: $S \rightarrow c \rightarrow a \rightarrow b$

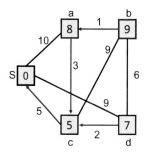

From s to:	S	a	b	c	d
Distance	0	8	9	5	7
Predecessor	-	c	a	S	c

Example: Find shortest path S →b

dist[b] = 9

pre[b] = a

  ↳ pre[a] = c

    ↳ pre[c] = S

Shortest path: $S \rightarrow c \rightarrow a \rightarrow b$ (total length is 9)

Figure 14.25: Determine shortest path

14.10   A* Algorithm

Reference	Hart, Nilsson, Raphael[21] 1968
Description	A* or A-Star is a heuristic algorithm for computing the shortest path from one given start node to one given goal node. The average time complexity is $O(k \cdot \log_k v)$ for v nodes with branching factor k, but it can also be quadratic in the worst case.
Required	Distance graph with relative distance information between all nodes plus lower bound of distance to goal from each node (e.g., air-line or linear distance).
Algorithm	Maintain sorted list of paths to goal, in every step expand only the currently shortest path by adding adjacent node with shortest distance, including estimate of remaining distance to goal.

An Example is shown in Figure 14.26. Each node has a lower-bound distance to the goal (e.g., the Euclidean distance from a Global Positioning System).

[21]P. Hart, N. Nilsson, B. Raphael, *A Formal Basis for the Heuristic Determination of Minimum Cost Paths*, IEEE Transactions on Systems and Cybernetics, vol. SSC-4, no. 2, 1968, pp. 100–107 (8).

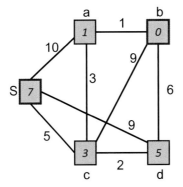

Node values are now lower
bound distances to goal *b*
(e.g. linear distances)

Edge values are distances
between neighboring nodes

Figure 14.26: A* example

For the first step, there are three choices:

1. {S, a} with min. length 10 + 1 = 11
2. **{S, c} with min. length 5 + 3 = 8** *(shortest path so far)*
3. {S, d} with min. length 9 + 5 = 14

Using a *best-first* algorithm, we explore the shortest estimated path first: We *remove* {S, c} and replace it with all possible extensions of this path with one additional step. All other paths remain unchanged. So the path list for our next step will be:

1. {S, a}   with min. length 10 + 1 = 11
2.1. **{S, c, a} with min. length 5 + 3 + 1 = 9** *(shortest path so far)*
2.2. {S, c, b} with min. length 5 + 9 + 0 = 14
2.3. {S, c, d} with min. length 5 + 2 + 5 = 12
3. {S, d}   with min. length 9 + 5 = 14

As it turns out, the currently shortest partial path is {S, c, a}, which will now be replaced by its expansions with one step further. As it turns out, only one such extension exists: {S, c, a, b}. As before, all other paths remain unchanged:

1. {S, a}   with min. length 10 + 1 = 11
2.1.1. {S, c, a, b} with min. length 5 + 3 + 1 + 0 = 9 *(shortest path so far)*
2.2. {S, c, b}   with min. length 5 + 9 + 0 = 14
2.3. {S, c, d}   with min. length 5 + 2 + 5 = 12
3. {S, d}   with min. length 9 + 5 = 14

The shortest path in this current iteration is now *{S, c, a, b}*, and this path already reaches the goal node *b*. Since we now have a shortest path that reaches the goal, any other possible path to goal that we have not yet explored must be longer, as all calculated partial distances are *minimal* distances, so if anything, the second shortest partial path {S, c, d} can only be 12 or higher. We therefore have found the **shortest path {S, c, a, b}** and our algorithm terminates.

This algorithm may look complex since there seems to be the need to store incomplete paths and their lengths at various places. However, using a recursive best-first search implementation can solve this problem in an elegant way without the need for explicit path storing. The quality of the lower-bound goal distance from each node greatly influences the timing complexity of the algorithm. The closer the given lower bound is to the true distance, the shorter the execution time.

14.11   Probabilistic Localization

All robot motions and sensor measurements are affected by a certain degree of noise. The aim of probabilistic localization is to provide the best possible estimate of the robot's current configuration based on all previous data and their associated distribution functions. The final estimate will be a probability distribution because of the inherent uncertainty as shown by Choset et al.[22] (Figure 14.27).

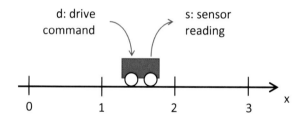

Figure 14.27: Uncertainty in actual position

As an example, assume a robot is driving in a straight line along the x-axis, starting at the true position $x = 0$. The robot executes driving commands with distance d, where d is an integer, and it receives sensor data from its on-board Global Positioning System s, where s is also an integer. The values for d and $\Delta s = s - s'$ (current position measurement minus position measurement before executing driving command) may differ from the true position $\Delta x = x - x'$.

The robot's driving accuracy from an arbitrary starting position has to be established by extensive experimental measurements and can then be expressed by a probability mass function (PMF), e.g.,

$$p(\Delta x = d - 1) = 0.2; \quad p(\Delta x = d) = 0.6; \quad p(\Delta x = d + 1) = 0.2$$

[22]H. Choset, K Lynch, S. Hutchinson, G. Kantor, W. Burgard, L. Kavraki, S. Thrun. *Principles of Robot Motion: Theory, Algorithms, and Implementations*, MIT Press, Cambridge MA, 2005.

Note that in this example, the robot's true position can only deviate by plus or minus one unit (e.g., cm); all position data values are discrete.

In a similar way, the accuracy of the robot's position sensor has to be established by measurements before it can be expressed as a PMF. In our example, there will again only be a possible deviation from the true position by plus or minus one unit:

$$p(x = s - 1) = 0.1; \quad p(x = s) = 0.8; \quad p(x = s + 1) = 0.1$$

Assuming the robot has executed a driving command with $d = 2$ and after completion of this command, its local sensor reports its position as $s = 2$. The probabilities for its actual position x are as follows, with n as normalization factor:

$$p(x = 1) = n \cdot p(s = 2 \,|\, x = 1) \cdot p(x = 1 \,|\, d = 2, x' = 0) \cdot p(x' = 0)$$
$$= n \cdot 0.1 \cdot 0.2 \cdot 1 = 0.02n$$
$$p(x = 2) = n \cdot p(s = 2 \,|\, x = 2) \cdot p(x = 2 \,|\, d = 2, x' = 0) \cdot p(x' = 0)$$
$$= n \cdot 0.8 \cdot 0.6 \cdot 1 = 0.48n$$
$$p(x = 3) = n \cdot p(s = 2 \,|\, x = 3) \cdot p(x = 3 \,|\, d = 2, x' = 0) \cdot p(x' = 0)$$
$$= n \cdot 0.1 \cdot 0.2 \cdot 1 = 0.02n$$

Positions 1, 2 and 3 are the only ones the robot can be at after a driving command with distance 2, since our PMF has probability 0 for all deviations greater than plus or minus one. Therefore, the three probabilities must add up to one. We can use this fact to determine the normalization factor n:

$$0.02n + 0.48n + 0.02n = 1$$
$$\Rightarrow n = 1.92$$

Now we can calculate the probabilities for the three positions, which reflect the *robot's belief*:

$$p(x = 1) = 0.04;$$
$$p(x = 2) = 0.92;$$
$$p(x = 3) = 0.04.$$

So the robot is most likely to be in position 2, but it remembers all probabilities at this stage.

Continuing with the example, let us assume the robot executes a second driving command, this time with $d = 1$, but after execution its sensor still reports $s = 2$. The robot will now recalculate its position belief according to the conditional probabilities, with x denoting the robot's true position after driving and x' before driving:

4.11 Probabilistic Localization

$p(x = 1) = n \cdot p(s = 2 \mid x = 1)$
$$[p(x = 1 \mid d = 1, x' = 1) \cdot p(x' = 1)$$
$$+ p(x = 1 \mid d = 1, x' = 2) \cdot p(x' = 2)$$
$$+ p(x = 1 \mid d = 1, x' = 3) \cdot p(x' = 3)]$$
$$= n \cdot 0.1 \cdot (0.2 \cdot 0.04 + 0 \cdot 0.92 + 0 \cdot 0.04)$$
$$= 0.0008n$$

$p(x = 2) = n \cdot p(s = 2 \mid x = 2)$
$$[p(x = 2 \mid d = 1, x' = 1) \cdot p(x' = 1)$$
$$+ p(x = 2 \mid d = 1, x' = 2) \cdot p(x' = 2)$$
$$+ p(x = 2 \mid d = 1, x' = 3) \cdot p(x' = 3)]$$
$$= n \cdot 0.8 \cdot (0.6 \cdot 0.04 + 0.2 \cdot 0.92 + 0 \cdot 0.04)$$
$$= 0.1664n$$

$p(x = 3) = n \cdot p(s = 2 \mid x = 3)$
$$[p(x = 3 \mid d = 1, x' = 1) \cdot p(x' = 1)$$
$$+ p(x = 3 \mid d = 1, x' = 2) \cdot p(x' = 2)$$
$$+ p(x = 3 \mid d = 1, x' = 3) \cdot p(x' = 3)]$$
$$= n \cdot 0.1 \cdot (0.2 \cdot 0.04 + 0.6 \cdot 0.92 + 0.2 \cdot 0.04)$$

Note that only states $x = 1$, 2 and 3 were computed since the robot's true position can only differ from the sensor reading by one. Next, the probabilities are normalized to 1.

$$0.0008n + 0.1664n + 0.0568n = 1$$
$$\Rightarrow \quad n = 4.46$$
$$\Rightarrow \quad p(x = 1) = 0.0036$$
$$p(x = 2) = 0.743$$
$$p(x = 3) = 0.254$$

These final probabilities are reasonable because the robot's sensor is more accurate than its driving, hence $p(x = 2) > p(x = 3)$. Also, there is a very small chance the robot is in position 1, and indeed this is represented in its belief.

The biggest problem with this approach is that the configuration space must be discrete. That is, the robot's position can only be represented discretely. A simple technique to overcome this is to set the discrete representation to the minimum resolution of the driving commands and sensors, e.g., if we may not expect driving or sensors to be more accurate than 1 cm, we can then express all distances in 1 cm increments. This will, however, result in a large number of measurements and a large number of discrete distances with individual probabilities.

The *particle filter* technique can be used to address this problem and will allow the use of non-discrete configuration spaces. The key idea in particle filters is to represent the robot's belief as a set of N particles, collectively known as M. Each particle consists of a robot configuration x and a weight $w \in [0, 1]$.

After driving, the robot updates the jth particle's configuration x_j by first sampling the PDF (probability density function) of $p(x_j \mid d, x_{j'})$; typically a Gaussian distribution. After that, the robot assigns a new weight $w_j = p(s \mid x_j)$ for the jth particle. Then, weight normalization occurs such that the sum of all weights equals one. Finally, resampling occurs such that only the most likely particles remain, e.g., by using the algorithm from Choset et al.[23]:

```
M = { }
R = rand(0, 1/N)
c = w[0]
i = 0
for (j=0; j<N; j++)
{ u = R + j/N
  while (u > c)
  { i = i + 1
    c = c + w[i]
  }
  M = M + { (x[i], 1/N) } // add particle to set
}
```

Like in the previous example, we assume the robot starts at $x = 0$, but this time the PDF for driving is a uniform distribution specified by:

$$p(\Delta x = d + b) = \begin{cases} 1 & for\ b \in [-0.5, +0.5] \\ 0 & otherwise \end{cases}$$

while the sensor PDF is specified by:

$$p(x = s + b) = \begin{cases} 4 + 16b & for\ b \in [-0.25, 0] \\ 4 - 16b & for\ b \in [0, 0.25] \\ 0 & otherwise \end{cases}$$

[23]H. Choset, K Lynch, S. Hutchinson, G. Kantor, W. Burgard, L. Kavraki, S. Thrun. *Principles of Robot Motion: Theory, Algorithms, and Implementations*, MIT Press, Cambridge MA, 2005.

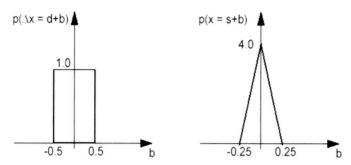

Figure 14.28: Probability density functions

The PDF for $x' = 0$ and $d = 2$ is shown in Figure 14.28, left, the PDF for $s = 2$ is shown in Figure 14.28, right.

Assuming the initial configuration $x = 0$ is known with absolute certainty and our system consists of four particles (this is a very small number; in practice around 10'000 particles are used). Then the initial set is given by:

$$M = \{(0, 0.25), (0, 0.25), (0, 0.25), (0, 0.25)\}$$

Now the robot is given a driving command $d = 2$ and after completion its sensors report the position as $s = 2$. The robot first updates the configuration of each particle by sampling the PDF in Figure 14.28, left, four times. One possible result of sampling is: 1.6, 1.8, 2.2 and 2.1. Hence, M is updated to:

$$M = \{(1.6, 0.25), (1.8, 0.25), (2.2, 0.25), (2.1, 0.25)\}$$

Now the weights for the particles are updated according to the PDF shown in Figure 14.28, right. This results in:

$$p(x = 1.6) = 0.0$$
$$p(x = 1.8) = 0.8$$
$$p(x = 2.2) = 0.8$$
$$p(x = 2.1) = 2.4$$

Therefore, M is updated to:

$$M = \{(1.6, 0), (1.8, 0.8), (2.2, 0.8), (2.1, 2.4)\}$$

After that, the weights are normalized to add up to one. This gives:

$$M = \{(1.6, 0), (1.8, 0.2), (2.2, 0.2), (2.1, 0.6)\}$$

Finally, the resampling procedure is applied with R = 0.1. The new M will then be:

$$M = \{(1.8, 0.25), (2.2, 0.25), (2.1, 0.25), (2.1, 0.25)\}$$

Note that the particle value 2.1 occurs twice because it is the most likely, while 1.6 drops out. If we need to know the robot's position estimate P at any time, we can simply calculate the weighted sum of all particles. In the example, this comes to:

$$P = 1.8 \cdot 0.25 + 2.2 \cdot 0.25 + 2.1 \cdot 0.25 + 2.1 \cdot 0.25 = 2.05$$

14.12 SLAM

The most successful method for solving this problem is a statistical method called simultaneous localization and mapping, or *SLAM* for short, see Durrant-Whyte, Bailey.[24],[25] While in conventional methods, positioning and orientation errors accumulate over time and make the whole map obsolete, SLAM combines a Kalman filter with a particle filter to generate a robust map that can deal with sensor inaccuracies.

Figure 14.29: Lidar scan of complex environment without SLAM in EyeSim, by Frewin, UWA

While the robot manages to complete the complex maze environment in Figure 14.29, left, the generated map in Figure 14.29, right, clearly shows the adverse effect of drift in positioning in the absence of SLAM. The perceived angles from odometry deviate too much from the correct 90° turns and as a consequence, some of the maze corridors overwrite previously recorded parts of the map when returning to the starting position at the lower left corner.

---

[24]H. Durrant-Whyte, T. Bailey, *Simultaneous Localisation and Mapping (SLAM): Part I*, IEEE Robotics & Automation Magazine, vol. 13, no. 2, June 2006, pp. 99–110.
[25]T. Bailey, H. Durrant-Whyte, *Simultaneous Localisation and Mapping (SLAM),: Part II*, IEEE Robotics & Automation Magazine, vol. 13, no. 3, Sep. 2006, pp. 108–117.

SLAM is an important part of ROS,[26] the open-source *Robot Operating System*, originally developed by Willow Garage. However, there are also implementations of SLAM outside of ROS, such as BreezySLAM from S. Levy,[27] which provides interfaces for most popular programming languages, including C/C++, Python and Java. There is even a minimal implementation by the name of tinySLAM (Steux, El Hamzaoui [28]), which implemented the core algorithm in less than 200 lines of C-code. The full source code is listed in this paper and on GitHub.[29] Some websites also refer to the same system as CoreSLAM or OpenSLAM.

Program. 14.2: SLAM Main Loop, Adapted from Levy

```
while (exploring):
   slam.update(LIDARGet())
   x, y, theta = slam.getpos(scan)
slam.getmap(mapbytes)
```

The Python algorithm steps by Levy are shown in Program 14.2. In a loop, the algorithm does an update of its internal state by using the latest Lidar scan. It then updates the robot's pose (*x*, *y*, *theta*). Finally, the generated map can be exported from the system.

While most SLAM algorithms are based on Lidar scan data, e.g., Google Cartographer,[30] there are also software packages for *visual navigation*, using SLAM for camera image data, such as ORB-SLAM2.[31]

Figure 14.30 shows the complete campus map of UWA, generated by Yuchen Du[32] using the Google Cartographer package under ROS. The map was generated with the *n*UWA*y* autonomous shuttle bus, fusing its eight on-board Lidar sensors.

---

[26]ROS, *Robot Operating System*, online; https://www.ros.org.

[27]S. Levy, *BreezySLAM*, Washington and Lee University, Lexington, VA, 2019, online: https://github.com/simondlevy/BreezySLAM.

[28]B. Steux, O. El Hamzaoui, *tinySLAM: a SLAM Algorithm in less than 200 lines C-Language Program,* 11th Int. Conf. on Control, Automation, Robotics and Vision (ICARV), Singapore, 7–10th December 2010, pp. 1975–1979 (5).

[29]OpenSLAM/tinySlam, online: https://github.com/OpenSLAM-org/openslam_tinyslam.

[30]Google, *Cartographer ROS Intewgration*, online: https://google-cartographer-ros.readthedocs.io/en/latest/.

[31]R. Mur-Artal, J. Tardos, *ORB-SLAM2: an Open-Source SLAM System for Monocular, Stereo and RGB-D Cameras*, online: https://arxiv.org/abs/1610.06475.

[32]Y. Du, *LiDAR-based Simultaneous Localization and Mapping System for nUWAy*, Master thesis, supervised by T. Bräunl, UWA 2020, online https://robotics.ee.uwa.edu.au/theses/2020-REV-ShuttleNavigation-Du.pdf.

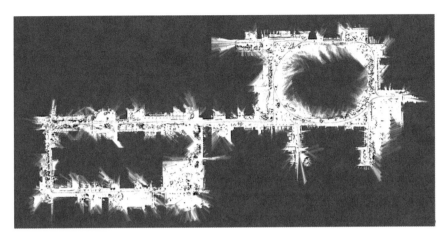

Figure 14.30: Cartographer SLAM mapping of UWA campus by Yuchen Du, REV/UWA

Finally, Figure 14.31 shows the localization of the REV *n*UWA*y* shuttle bus on the UWA campus, implemented by Yuchen Du. Google Cartographer conducts localization by using a SLAM algorithm to find the most likely vehicle pose in the previously generated map.

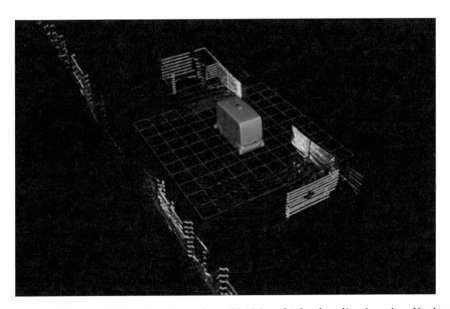

Figure 14.31: Cartographer SLAM vehicle localization by Yuchen Du, REV/UWA

4.13 Tasks

1. Implement dead reckoning on a simulated or a real robot.

2. Determine the accuracy of the driving and turning movements.

3. How long can you drive before the angular inaccuracy is greater than 30°?

4. Implement the light beacon localization for the arrangement shown in this chapter.

5. As a first step, implement it as an independent program (outside of a simulation system). The algorithm should determine the robot's position from three given angles of beacons at known locations.

6. Implement a conversion program that translates a configuration space into an occupancy grid. The configuration space data is given as a list of wall/obstacle segments in the form (x_1, y_1, x_2, y_2).

7. Implement the Wandering Standpoint algorithm for a mobile robot.

8. Implement the DistBug algorithm for a mobile robot.

9. Implement the Quadtree decomposition of a given occupancy grid.

10. Implement the Brushfire algorithm for a given occupancy grid.

11. Implement the Dijkstra algorithm for a given node distance graph.

12. Implement the A* algorithm for a given node distance graph.

13. Explore BreezySLAM with EyeSim or the Cartographer SLAM package in ROS. How much can you improve localization accuracy with SLAM in the presence of sensor and actuator noise?

15 MAZE NAVIGATION

Mobile robot competitions have been around for well over 40 years, with the Micro Mouse Contest being the first of its kind in 1977. These competitions have inspired generations of students, researchers and hobbyists alike, while consuming vast amounts of funding, time and effort. Competitions provide a goal together with an objective performance measure, while extensive media coverage allows participants to present their work to a wider forum.

As the competition robots evolved over the years, becoming faster and smarter, so did the competitions themselves. Today, interest has shifted from the "mostly solved" maze contest toward more complex problems such as robot soccer (FIRA[1] and RoboCup[2]), robot rescue scenarios (ELROB,[3] RoboCup Rescue[4]) or traffic scenarios (Carolo-Cup[5]).

15.1 Micro Mouse Contest

"The stage was set. A crowd of spectators, mainly engineers, were there. So were reporters from the Wall Street Journal, the New York Times, other publications, and television. All waited in expectancy as Spectrum's Mystery Mouse Maze was unveiled. Then the color television cameras of CBS and NBC began to roll; the moment would be recreated that evening for viewers of the

---

[1]FIRA, *Welcome to FIRA RoboWorld Cup*, online: http://www.firaworldcup.org.

[2]RoboCup, *Welcome to RoboCup*, https://www.robocup.org.

[3]ELROB, *ELROB – The European Land Robot Trial*, online: https://www.elrob.org.

[4]Wikipedia, *Rescue Robot League*, online: https://en.wikipedia.org/wiki/Rescue_Robot_League.

[5]Carolo-Cup, online: https://wiki.ifr.ing.tu-bs.de/carolocup/en/carolo-cup.

© The Author(s), under exclusive license to Springer Nature Singapore Pte Ltd. 2022
T. Bräunl, *Embedded Robotics*,
https://doi.org/10.1007/978-981-16-0804-9_15

Walter Cronkite and John Chancellor-David Brinkley news shows" Roger Allan.[6]

This report from the first "Amazing Micro-Mouse Maze Contest" demonstrates the enormous media interest in the first mobile robot competition in New York in 1977. The academic response was overwhelming. Over 6'000 entries followed the announcement of Donald Christiansen,[7] who originally suggested the contest in 1977.

Figure 15.1: Maze from London Micro Mouse Contest in EyeSim

The task is for a robot *mouse* to drive from the start to the goal in the fastest time. Rules have changed somewhat over the years, in order to allow exploration of the whole maze and then to compute the shortest path, while also counting exploration time at a reduced factor.

The first mice constructed were rather simple—some of them did not even contain a microprocessor as a controller, but were simple *wall huggers*, which would find the goal by always following the left wall (or always the right wall). A few of these robots scored even higher than some of the intelligent mice, which were mechanically slower.

John Billingsley[8] made the Micro Mouse Contest popular in Europe in 1982 and called for the first rule change: starting in a corner, the goal should be in the center and not in another corner, to eliminate wall huggers. From then on, more intelligent behavior was required to solve a maze (Figure 15.1). Virtually all university robotics labs at that time were building micromice in one form or

---

[6]R. Allan, *The amazing micromice: see how they won*, IEEE Spectrum, Sept. 1979, vol. 16, no. 9, pp. 62–65 (4).

[7]D. Christiansen, *Announcing the amazing micromouse maze contest*, IEEE Spectrum, vol. 14, no. 5, May 1977, p. 27 (1).

[8]J. Billingsley, *Micromouse - Mighty mice battle in Europe, Practical Computing*, Dec. 1982, pp. 130–131 (2).

another; a real micromouse craze was going around the world. All of a sudden, researchers and students had a goal and could share ideas with a large number of colleagues who were working on exactly the same problem.

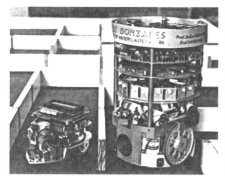 

Figure 15.2: Micromice of Hinkel and von Puttkamer, Univ. Kaiserslautern (left); micromouse of Gordon Wyeth, University of Queensland (right)

Micromouse technology evolved quite a bit over the years, as did the running time. A typical sensor arrangement was to use three sensors to detect any walls in front, to the left and to the right of the mouse. Early mice used simple microswitches as touch sensors, while later sonar, infrared and optical sensors became popular (see Hinkel[9] in Figure 15.2, left). The Micromouse of the University of Queensland, Australia (see in Figure 15.2, right) uses three extended arms with several infrared sensors each for reliable wall distance measurement.

While the mouse's size is restricted by the maze's wall distance, smaller and lighter mice have the advantage of higher acceleration/deceleration and therefore higher speed. Even smaller mice became able to drive in a straight diagonal line instead of going through a sequence of left/right turns, which exists in most mazes. As sensors and processors have improved significantly over the last years, the remaining possible improvement are mainly on the mechanics side (see Bräunl[10]).

---

[9]R. Hinkel, Low-Level Vision in an Autonomous Mobile Robot, EUROMICRO 1987; 13th Symposium on Microprocessing and Microprogramming, Portsmouth England, Sept. 1987, pp. 135–139 (5).

[10]T. Bräunl, Research Relevance of Mobile Robot Competitions, IEEE Robotics and Automation Magazine, vol. 6, no. 4, Dec. 1999, pp. 32–37 (6).

15.2 Maze Exploration Algorithms

For maze exploration, we will develop two algorithms: a simple iterative procedure that follows the left wall of the maze (*wall hugger*), and a slightly more complex recursive procedure to explore an unrestricted maze.

15.2.1 Wall Following

Our first approach for the exploration part of the problem is to always follow the left wall. For example, if a robot comes to an intersection with several open sides, it follows the leftmost path. Program 15.1 shows the implementation of this approach. The start square is assumed to be at position (0,0), the four directions north, west, south and east are encoded as integers 0, 1, 2, 3.

Program 15.1: Explore-Left

```
do
{ // 1. Check walls
  Ffree = PSDGet(PSD_FRONT) > THRES;
  Lfree = PSDGet(PSD_LEFT) > THRES;
  Rfree = PSDGet(PSD_RIGHT) > THRES;
  LCDSetPrintf(4,0, "L%d F%d R%d", Lfree, Ffree, Rfree);

  // 2. Rotations
  if (Lfree) { VWTurn(+90, 90); VWWait(); } // Turn left
    else if (Ffree) { } // NO turn
      else if (Rfree) { VWTurn(-90, 90); VWWait(); } // Turn right
        else { VWTurn(180, 90); VWWait(); } // Dead end

  // Driving straight 1 square
  VWStraight(360, 360); VWWait(); // in an ideal world ...
} while (KEYRead() != KEY4);
```

This procedure is quite simple and comprises only a few lines of code. It contains a single *do*-loop that runs until the stop key is being pressed. As an addition, the program could keep track of the goal coordinates and stop when they are reached. In each iteration, it is determined by reading the robot's infrared sensors whether a wall exists on the front, left- or right-hand side (Boolean variables *Ffree, Lfree* and *Rfree*. The robot then selects the *leftmost* direction for its further journey. That is, if possible, it will always drive left, if not it will try driving straight, and only if the other two directions are blocked will it try to drive right. If none of the three directions are free, the robot will turn on the spot and go back one square, since it has obviously arrived at a dead end.

This simple and elegant algorithm works very well for *benign* mazes. However, as mentioned earlier, it will not work for competition mazes that have the goal in the center without connecting walls. The goal in the middle of the maze in Figure 15.3 cannot be reached by a wall-following robot. The recursive algorithm shown in the following section, however, will be able to cope with arbitrary mazes.

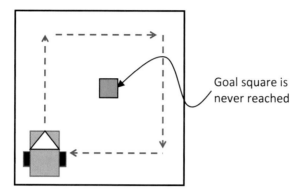

Figure 15.3: Problem for wall following

There is, however, another real problem. The algorithm as written in Program 15.1 would only work in a simulation of a perfect world without any error settings, and certainly not in the real world. This is because each rotation command over 90° will not perfectly rotate the robot. It may end up at 89° or 91° and this error will accumulate over time. Likewise, the driving straight operation may not be perfectly straight and not exactly of the right distance either. As EyeSim uses a physics simulation system, the approach of Program 15.1 would fail even there. We will solve this problem later in this chapter (see Section on *real robots*).

15.2.2 Recursive Exploration

The algorithm for full maze exploration guarantees that each reachable square in the maze will be visited, independent of the maze construction. This, of course, requires us to generate an internal representation of the maze and to maintain a bit field for marking whether a particular square has already been visited. Our new algorithm is structured in several stages for exploration and navigation:

1. Explore the whole maze:
   Starting at the start square, visit all reachable squares in the maze, then return to the start square by using a recursive algorithm.
2. Compute the shortest distance from the start square to any other square by using a *flood fill* algorithm.

3. Allow the user to enter the coordinates of a desired destination square, then determine the shortest driving path by reversing the path in the distance array from the destination to the start square.

The difference between the wall-following algorithm and this recursive exploration of all paths is sketched in Figure 15.4. While the wall-following algorithm only takes a single path, the recursive algorithm explores all possible paths subsequently. Of course, this requires some bookkeeping of squares already visited to avoid an infinite loop.

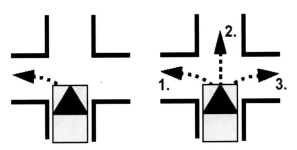

Figure 15.4: Left wall following versus recursive exploration

Program 15.2: Explore

```
void explore()
{ int front_open, left_open, right_open, old_dir;

  mark[rob_y][rob_x] = 1; /* set mark */
  PSDGet(psd_left), PSDGet(psd_right));
  front_open = PSDGet(psd_front) > THRES;
  left_open = PSDGet(psd_left) > THRES;
  right_open = PSDGet(psd_right) > THRES;
  maze_entry(rob_x,rob_y,rob_dir, front_open);
  maze_entry(rob_x,rob_y,(rob_dir+1)%4, left_open);
  maze_entry(rob_x,rob_y,(rob_dir+3)%4, right_open);
  old_dir = rob_dir;

  if (front_open && unmarked(rob_y,rob_x,old_dir))
    { go_to(old_dir); /* go 1 forward */
      explore(); /* recursive call */
      go_to(old_dir+2); /* go 1 back */
    }
  if (left_open && unmarked(rob_y,rob_x,old_dir+1))
    { go_to(old_dir+1); /* go 1 left */
      explore(); /* recursive call */
      go_to(old_dir-1); /* go 1 right */
    }
```

5.2 Maze Exploration Algorithms

```
if (right_open && unmarked(rob_y,rob_x,old_dir-1))
  { go_to(old_dir-1); /* go 1 right */
    explore(); /* recursive call */
    go_to(old_dir+1); /* go 1 left */
  }
}
```

Program 15.2 shows an excerpt from the central recursive function *explore*. Similar to before, we determine whether there are walls in front and to the left and right of the current square. However, we also mark the current square as visited (data structure *mark*) and enter any walls found into our internal representation using auxiliary function *maze_entry*. Next, we have a maximum of three recursive calls, depending on whether the directions *front*, *left*, or *right* are open (no wall) and the next square in this direction has not been visited before. If this is the case, the robot will drive into the next square and procedure *explore* will be called recursively. Upon termination of this call, the robot will return to the previous square. Overall, this will result in the robot exploring the whole maze and returning to the start square upon completion of the algorithm.

A possible extension of this algorithm is to check in every iteration if all surrounding walls of a new, previously unvisited square are already known (e.g., if all surrounding squares have been visited). In that case, it is not required for the robot to actually visit this square. The trip can be saved, and the internal database can be updated.

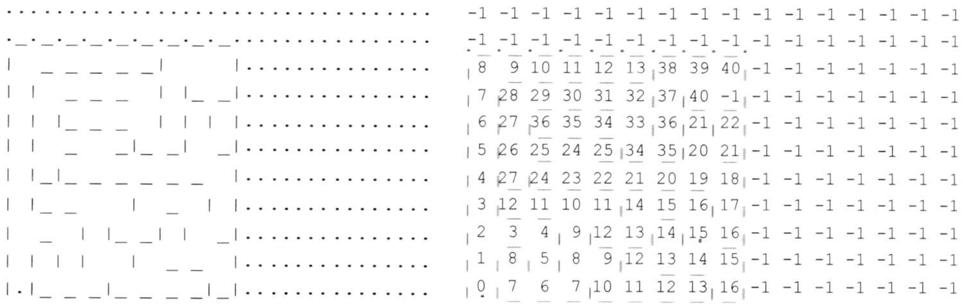

```
....................................    -1 -1 -1 -1 -1 -1 -1 -1 -1 -1 -1 -1 -1 -1 -1 -1
._._._._._._._._._._._............      -1 -1 -1 -1 -1 -1 -1 -1 -1 -1 -1 -1 -1 -1 -1 -1
|  _ _ _ _ _|    |............          | 8  9 10 11 12 13|38 39 40|-1 -1 -1 -1 -1 -1 -1
| |  _ _ _  | |_ _|............         | 7 28 29 30 31 32|37|40 -1|-1 -1 -1 -1 -1 -1 -1
| | |_ _ _  | | | |............         | 6 27|36 35 34 33|36|21 22|-1 -1 -1 -1 -1 -1 -1
| |  _ _|_ _|  _|............           | 5 26 25 24 25|34 35|20 21|-1 -1 -1 -1 -1 -1 -1
| |_|_ _ _ _ _   |............          | 4 27|24 23 22 21 20 19 18|-1 -1 -1 -1 -1 -1 -1
| |_ _   |  _ | |............           | 3 12 11 10 11|14 15 16|17|-1 -1 -1 -1 -1 -1 -1
|  _ | |_|  |   _|............          | 2  3  4| 9|12 13|14|15 16|-1 -1 -1 -1 -1 -1 -1
| | | |   |  _ _  |............         | 1  8| 5| 8  9|12 13 14 15|-1 -1 -1 -1 -1 -1 -1
|.|_ _ _|_ _ _ _|_|............         | 0  7  6  7|10 11 12 13|16|-1 -1 -1 -1 -1 -1 -1
```

Figure 15.5: Maze algorithm output

We have now completed the first step of the algorithm. The result can be seen in the top of Figure 15.5. We now know for each square whether it can be reached from the start square or not, and we know all walls for each reachable square.

In the second step, we want to find the minimum distance of each maze square from the start square. Figure 15.5, bottom, shows the shortest distances for each point in the maze from the start point. A value of −1 indicates a position that cannot be reached (for example outside the maze boundaries). We are using a flood fill algorithm to generate this data (see Program 15.3).

Program 15.3: Flood Fill Algorithm

```
int shortest_path(int goal_y, int goal_x)
{ int i,j,iter;

  for (i=0; i<MAZESIZE; i++) for (j=0; j<MAZESIZE; j++)
  { map [i][j] = -1; /* init */
    nmap[i][j] = -1;
  }
  map [0][0] = 0;
  nmap[0][0] = 0;
  iter=0;

  do
  { iter++;
    for (i=0; i<MAZESIZE; i++) for (j=0; j<MAZESIZE; j++)
    { if (map[i][j] == -1)
      { if (i>0)
          if (!wall[i][j][0] && map[i-1][j] != -1)
            nmap[i][j] = map[i-1][j] + 1;
        if (i<MAZESIZE-1)
          if (!wall[i+1][j][0] && map[i+1][j] != -1)
            nmap[i][j] = map[i+1][j] + 1;
        if (j>0)
          if (!wall[i][j][1] && map[i][j-1] != -1)
            nmap[i][j] = map[i][j-1] + 1;
        if (j<MAZESIZE-1)
          if (!wall[i][j+1][1] && map[i][j+1] != -1)
            nmap[i][j] = map[i][j+1] + 1;
      }
    }
    for (i=0; i<MAZESIZE; i++) for (j=0; j<MAZESIZE; j++)
      map[i][j] = nmap[i][j]; /* copy back */

  } while (map[goal_y][goal_x]==-1 && iter<(MAZESIZE*MAZESIZE));
  return map[goal_y][goal_x];
}
```

5.2 Maze Exploration Algorithms

The program uses the 2D integer array *map* for recording the distances, plus a copy *nmap*, which is used and copied back after each iteration. In the beginning, each square (array element) is marked as unreachable (−1), except for the start square [0,0], which can be reached in zero steps. Then, we run a *while*-loop as long as at least one map entry changes. Since each change reduces the number of unknown squares (value −1) by at least one, the upper bound of loop iterations is the total number of squares ($MAZESIZE^2$). In each iteration, we use two nested *for*-loops to examine all unknown squares. If there is a path (no wall to north, south, east or west; e.g., *!wall[i][j][0]*) to a known square, the new distance is computed and entered into array *nmap*. Additional *if*-selections are required to take care of the maze borders. Figure 15.6 shows the stepwise generation of the distance map.

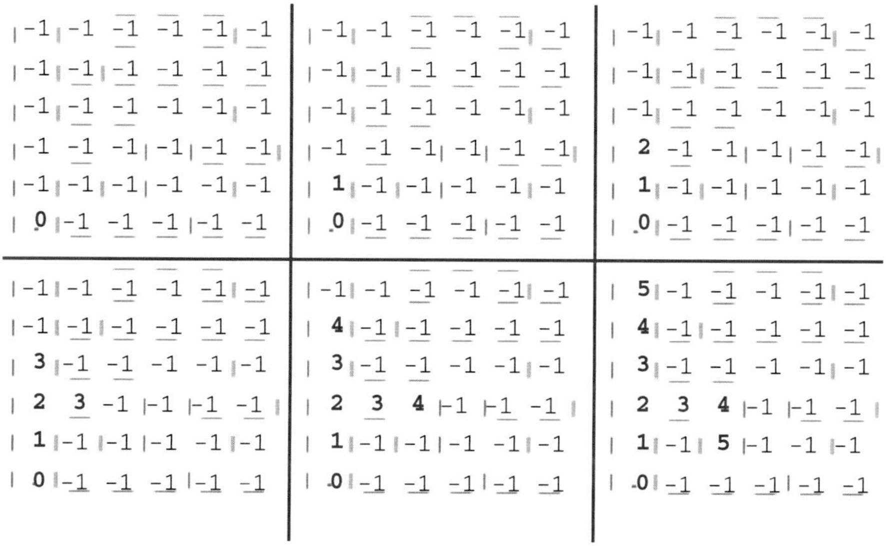

```
|-1|-1 -1 -1 -1|-1      |-1|-1 -1 -1 -1|-1      |-1|-1 -1 -1 -1|-1
|-1|-1|-1 -1 -1 -1      |-1|-1|-1 -1 -1 -1      |-1|-1|-1 -1 -1 -1
|-1|-1 -1 -1 -1|-1      |-1|-1 -1 -1 -1|-1      |-1|-1 -1 -1 -1|-1
|-1 -1 -1|-1|-1 -1|     |-1 -1 -1|-1|-1 -1|     | 2 -1 -1|-1|-1 -1|
|-1|-1|-1|-1 -1|-1      | 1|-1|-1|-1 -1|-1      | 1|-1|-1|-1 -1|-1
| 0|-1 -1 -1|-1 -1      | 0|-1 -1 -1|-1 -1      | 0|-1 -1 -1|-1 -1


|-1|-1 -1 -1 -1|-1      |-1|-1 -1 -1 -1|-1      | 5|-1 -1 -1 -1|-1
|-1|-1|-1 -1 -1 -1      | 4|-1|-1 -1 -1 -1      | 4|-1|-1 -1 -1 -1
| 3|-1 -1 -1 -1|-1      | 3|-1 -1 -1 -1|-1      | 3|-1 -1 -1 -1|-1
| 2  3 -1|-1|-1 -1|     | 2  3  4|-1|-1 -1|     | 2  3  4|-1|-1 -1|
| 1|-1|-1|-1 -1|-1      | 1|-1|-1|-1 -1|-1      | 1|-1| 5|-1 -1 -1|-1
| 0|-1 -1 -1|-1 -1      | 0|-1 -1 -1|-1 -1      | 0|-1 -1 -1|-1 -1
```

Figure 15.6: Distance map development (excerpt)

In the third and final step of our algorithm, we can now determine the shortest path to any maze square from the start square. We already have all wall information and the shortest distances for each square (see Figure 15.5). If the user wants the robot to drive to, say, maze square [1,2] (row 1, column 2, assuming the start square is at [0,0]), then we know already from our distance map (see Figure 15.6, bottom right) that this square can be reached in five steps. In order to find the shortest driving path, we can now simply trace back the path from the desired goal square [1,2] to the start square [0,0]. In each step, we select a connected neighbor square from the current square (no wall between them) that has a distance of one less than the current square. That is, if the current square has distance d, the new selected neighbor square has to have a distance of $d-1$. Figure 15.7 shows screenshots of the algorithm execution.

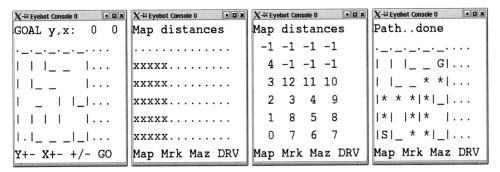

Figure 15.7: Screenshots of exploration, visited cells, distances and shortest path

Program 15.4 shows the algorithm to find the path, Figure 15.8 demonstrates this for the example square [1,2]. Since we already know the length of the path (entry in *map*), a simple *for*-loop is sufficient to construct the shortest path. In each iteration, all four sides are checked whether there is a path and the neighbor square has a distance one less than the current square. This must be the case for at least one side, or there would be an error in our data structure. There could be more than one side for which this is true. In this case, multiple shortest paths exist.

Program 15.4: Shortest Path

```
void build_path(int i, int j, int len)
{ int k;
  for (k = len-1; k>=0; k-)
   {
   if (i>0 && !wall[i][j][0] && map[i-1][j] == k)
   { i-;
     path[k] = 0; /* north */
   }
   else
   if (i<MAZESIZE-1 && !wall[i+1][j][0] && map[i+1][j] == k)
   { i++;
     path[k] = 2; /* south */
   }
   else
   if (j>0 && !wall[i][j][1] && map[i][j-1] == k)
   { j-;
     path[k] = 3; /* east */
   }
```

```
     else
     if (j<MAZESIZE-1 && !wall[i][j+1][1] && map[i][j+1] == k)
     { j++;
       path[k] = 1; /* west */
     }
     else
     { LCDPrintf("ERROR");
       KEYWait(ANYKEY);
     }
     }
     }
```

```
5 -1 -1 -1 -1 -1          5 -1 -1 -1 -1 -1          5 -1 -1 -1 -1 -1

4 -1 -1 -1 -1 -1          4 -1 -1 -1 -1 -1          4 -1 -1 -1 -1 -1

3 -1 -1 -1 -1 -1          3 -1 -1 -1 -1 -1          3 -1 -1 -1 -1 -1

2  3  4 -1 -1 -1          2  3  4 -1 -1 -1          2  3  4 -1 -1 -1

1 -1  5 -1 -1 -1          1 -1  5 -1 -1 -1          1 -1  5 -1 -1 -1

0 -1 -1 -1 -1 -1          0 -1 -1 -1 -1 -1          0 -1 -1 -1 -1 -1
```

Path: {}   Path: {S}   Path: {E,S}

```
5 -1 -1 -1 -1 -1          5 -1 -1 -1 -1 -1          5 -1 -1 -1 -1 -1

4 -1 -1 -1 -1 -1          4 -1 -1 -1 -1 -1          4 -1 -1 -1 -1 -1

3 -1 -1 -1 -1 -1          3 -1 -1 -1 -1 -1          3 -1 -1 -1 -1 -1

2  3  4 -1 -1 -1          2  3  4 -1 -1 -1          2  3  4 -1 -1 -1

1 -1  5 -1 -1 -1          1 -1  5 -1 -1 -1          1 -1  5 -1 -1 -1

0 -1 -1 -1 -1 -1          0 -1 -1 -1 -1 -1          0 -1 -1 -1 -1 -1
```

Path: {E,E,S}   Path: {N,E,E,S}   Path: {N,N,E,E,S}

Figure 15.8: Shortest path for position [y,x] = [1,2]

15.3   Simulated Versus Real Maze Program

Simulations are never enough: the real world contains real problems! We first implemented and tested the maze exploration problem in the EyeSim simulator before running the same program unchanged on a real robot (Koestler and Bräunl,[11] Figure 15.9). Using the simulator initially for the higher levels of

---

[11]A. Koestler, T. Bräunl, Mobile Robot Simulation with Realistic Error Models, International Conference on Autonomous Robots and Agents, ICARA 2004, Dec. 2004, Palmerston North, New Zealand, pp. 46–51 (6).

program design and debugging is very helpful, because it allows us to concentrate on the logic issues and frees us from most real-world robot problems. Once the logic of the exploration and path planning algorithm has been verified, one can concentrate on the lower-level problems like fault-tolerant wall detection and driving accuracy by adding sensor/actuator noise to error levels typically encountered by real robots. This basically transforms a computer science problem into a computer engineering problem.

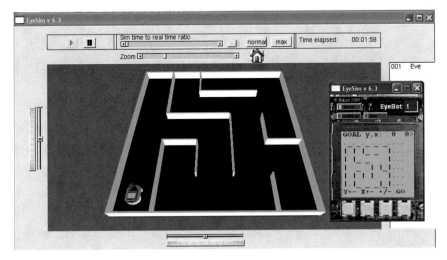

Figure 15.9: Simulated maze solving

Now we have to deal with inaccurate sensor readings and dislocations in robot positioning. We can still use the simulator to make the necessary changes to improve the application's robustness and fault tolerance, before we eventually try it on a real robot in a maze environment.

What needs to be added to the previously shown maze program can be described by the term *fault tolerance*. As mentioned before, we cannot assume that a robot has turned exactly 90° degrees or driven the exact distance after executing the corresponding command. The same holds for distance sensors and odometry readings. The logic of the program does not need to be changed, only the driving routines need to be improved.

Figure 15.10: Adaptive driving using three sensors

5.3 Simulated Versus Real Maze Program

In order to fix this problem, we need to spend some more energy for the driving straight part as shown in Program 15.5. Instead of a single line drive command, we need to program a loop where the robot constantly monitors its position and orientation and constantly updates its steering.

Program 15.5: Adaptive Straight Driving for Maze Problem

```
// Driving straight 1 square
// VWStraight(360, 360); VWWait();
VWGetPosition(&x1,&y1,&phi1);
do
{L=PSDGet(PSD_LEFT); F=PSDGet(PSD_FRONT); R=PSDGet(PSD_RIGHT);
 LCDSetPrintf(1,0, "L%d F%d R%d\n", L,F,R);
 if (100<L && L<180 && 100<R && R<180) // space check L and R
 { VWSetSpeed(SPEED, L-R); // drive left if left > right
   LCDSetPrintf(2,0, "Check L+R %d", L-R);
 }
 else
 if (100<L && L<180) // space check - LEFT
 { VWSetSpeed(SPEED, L-DIST); // drive left if left > DIST
   LCDSetPrintf(2,0, "Check Left %d", L-DIST);
 }
 else
 if (100<R && R<180) // space check - RIGHT
 { VWSetSpeed(SPEED, DIST-R); // drive left if DIST > right
   LCDSetPrintf(2,0, "Check Right %d", DIST-R);
 }
 else
 { VWSetSpeed(SPEED, 0);
   LCDSetPrintf(2,0, "Straight");
 }
 VWGetPosition(&x2,&y2,&phi2);
 drivedist = sqrt(sqr(x2-x1)+sqr(y2-y1));
 //OSWait(10);
} while (drivedist < MSIZE && F > MSIZE/2 - 50); // stop in time
```

We check the wall situation (Figure 15.10) and use if for updating our position. If there are walls to either side, we use the difference between the left and right wall distance as the angular velocity in our driving command:

```
VWSetSpeed(SPEED, L-R);
```

If we are perfectly in the middle, then L = R, so our angular speed would evaluate to 0. If we are too far toward the right wall, then L > R and their difference would be a positive value, driving the robot in a counterclockwise curve, and bringing it back toward the middle. The equivalent holds when the robot gets too close to the left wall.

In case there is only one wall (either to the left or to the right), we try to maintain the correct middle distance (*L–DIST* or *DIST–R*, depending on which wall is missing.

If there is no wall on either side, we have no other means of orientation and just drive straight (angular speed 0).

At the conclusion of the loop, we constantly monitor the distance already traveled as well as the distance toward the front (in case there is a wall) and always bring the robot to a stop in time or advance to the next drive or rotation command.

15.4   Tasks

1. Implement a simple wall-following routine for a mobile robot. The robot should follow the wall closely at a distance of about 10 cm. Optimize this program for straight walls and 90° turns.

2. Extend the wall-following program for curved walls and walls intersecting at any angle.

3. Implement the left-wall-following maze algorithm. Keep track of the robot's position in terms of maze squares and stop it when it returns to the starting square (0,0).

4. Implement the recursive maze algorithm in EyeSim and let the robot drive in a perfect simulated world. The robot needs to build a digital copy of the maze structure during its exploration.

5. After finishing the recursive search, let the user specify a goal square. First calculate the shortest path to the goal square with a flood fill algorithm as described in the chapter, then let the robot drive there and back along the shortest path.

6. Extend the recursive maze algorithms with the robust correction technique shown for the left-wall-following algorithm in this chapter. Then let the robot drive in a realistic simulation setting or in a real maze.

7. Extend the recursive maze algorithm with a Lidar sensor instead of three infrared PSDs. The better sensor equipment should be used for better robot localization and fine-tuning of the path generation (from one square to the next).

8. Implement an optimized path generation for a small enough mobile robot that can drive at a 45° angle through a sequence of zig-zag turns, saving a significant amount of time.

IMAGE PROCESSING

We want to implement an embedded vision system that can interpret an image in order to steer a robot accordingly, e.g., avoiding obstacles, following lane markings, identifying a goal by color or shape—or all of the above. Since the robot is moving and other objects in the scene may be as well, we have to be fast. Ideally, we want to achieve a frame rate of at least 10 frames per second (fps) for the whole perception–action cycle. Of course, given the limited processing power of an embedded controller, this restricts the manageable image resolution as well as the possible complexity of the image processing operations.

In the following, we will look at some basic image processing routines. These will later be reused for more complex robot applications programs. For further reading in robot vision, see Klette et al.[1] For general image processing textbooks, see Parker[2] and Gonzales and Woods.[3]

16.1 Camera Interface

The RoBIOS camera software interface for the Raspberry Pi is fairly simple. It lets you initialize the camera at the desired resolution and then read images either as color (RGB) or as grayscale byte arrays:

---

[1]R. Klette, S. Peleg, F. Sommer, (Eds.) *Robot Vision*, Proceedings of the International Workshop RobVis 2001, Auckland NZ, Lecture Notes in Computer Science, no. 1998, Springer-Verlag, Berlin Heidelberg, Feb. 2001.

[2]J. Parker, *Algorithms for Image Processing and Computer Vision*, 2nd Ed., John Wiley & Sons, New York NY, 2010.

[3]R. Gonzales, R. Woods, *Digital Image Processing*, 4th Ed., Pearson, Upper Saddle River NJ, 2017.

```
int CAMInit(int resolution) // Set camera resolution
int CAMGet(BYTE *buf)       // Read color camera image
int CAMGetGray(BYTE *buf)   // Read gray scale image
```

Remember that each pixel in a grayscale image is a byte between 0 (black) and 255 (white), while each color pixel has three bytes, one each for its red, green and blue color component. So, a *black color* is (0, 0, 0) and a *white color* is (255, 255, 255). We will talk more about color models later in this chapter. The most important image resolutions were already introduced in the sensor chapter, but here is the full list:

- QQVGA $120 \times 160 \times 3$ bytes (RGB color)—*or* $\times$ 1 byte for grayscale
- QVGA $240 \times 320 \times 3$ bytes (RGB color)—*or* $\times$ 1 byte for grayscale
- VGA $480 \times 640 \times 3$ bytes (RGB color)—*or* $\times$ 1 byte for grayscale
- CAM1MP $1296 \times 730 \times 3$ bytes (RGB color)—*or* $\times$ 1 byte for grayscale
- HDTV $1080 \times 1920 \times 3$ bytes (RGB color)—*or* $\times$ 1 byte for grayscale
- CAM5MP $2592 \times 1944 \times 3$ bytes (RGB color)—*or* $\times$ 1 byte for grayscale.

These keywords can be used for the camera initialization as well as for initializing the image processing routes:

```
int IPSetSize(int resolution);
```

Note that *CamInit* automatically calls *IPSetSize*, so the latter only needs to be called explicitly if the image data comes from a different source.

There are predefined type definitions for color and grayscale:

```
typedef QQVGAcol  BYTE [120][160][3];
typedef QQVGAgray BYTE [120][160];
typedef QVGAcol   BYTE [240][320][3];
typedef QVGAgray  BYTE [240][320];
typedef VGAcol    BYTE [480][640][3];
typedef VGAgray   BYTE [480][640];
...
```

Alternatively, the following constants can be used for color and grayscale image declarations:

```
QQVGA_SIZE, QVGA_SIZE, VGA_SIZE, CAM1MP_SIZE, CAMHD_SIZE, CAM5MP_SIZE
QQVGA_PIXELS, QVGA_PIXELS, VGA_PIXELS, CAM1MP_PIXELS, CAMHD_PIXELS,
CAM5MP_PIXELS
```

We will mostly work with QVGA, as this has a reasonable size for seeing details, still fits on our standard Raspberry Pi LCD screen and can achieve reasonable reply times on the Pi controller. The following variables will be set after the camera initialization and should be used in each application program to make it independent of the image size:

```
CAMWIDTH
CAMHEIGHT
CAMPIXELS (= CAMWIDTH*CAMHEIGHT)
CAMSIZE   (= 3*CAMPIXELS)
```

The standard Raspberry Pi camera has a fixed lens that cannot be changed or adjusted, which is unfortunate for many robotics applications. In most applications, we want a camera lens that gives us a wider viewing angle, possibly even a fisheye lens. Luckily, there are a number of third-party compatible camera modules that can be connected via USB to the Raspberry Pi. These typically have an adjustable board lens, which needs to be focused. Using the *Siemens star* in Figure 16.1 at the desired distance from the camera will make it easy to adjust the focus until the image appears sharp.

Figure 16.1: Third-party camera with board lens and Siemens star focus pattern

16.2   Image File Formats

Images formats are best understood through their file formats. We use the PNM (**p**ortable **anym**ap) format, which is arguably the simplest possible image file format, first introduced by Jef Poskanzer.[4] PNM comprises three different formats, which are distinguished through different header IDs for black and white images (portable bitmap PBM), grayscale images (portable graymap PGM) and color images (portable pixelmap PPM). For each of the three image formats, there is a binary version as well as a human-readable ASCII text version, distinguished again through different IDs:

- P1 for black&white ASCII   P4 for black&white binary
- P2 for grayscale ASCII   P5 for grayscale binary
- P3 for color ASCII   P6 for color binary.

---

[4]Wikipedia, Netbpm, online: https://en.wikipedia.org/wiki/Netbpm.

In RoBIOS, we provide functions for reading and storing images from and to files:

```
int   IPReadFile(char *filename, BYTE* img);      // Read PNM file
int   IPWriteFile(char *filename, BYTE* img);     // Write color
int   IPWriteFileGray(char *filename, BYTE* gray); // Write gray
```

IPReadFile returns the image type (color, gray or black&white), and we provide different routines for writing color and grayscale images. In the following subsections, we show examples for each of the three types, using the human-readable ASCII versions P1, P2 and P3.

16.2.1 Black and White Image Files PBM

All PNM files comprise an ASCII header that specifies type and size of the image, followed by an ASCII or binary data part. PBM files are the simplest of the three file formats and start with the ID 'P1'. An optional comment starts with symbol '#' and continues until the end of the line. This can be quite useful to encode additional data inside the image file, e.g., for classification data in learning algorithms.

In the next line follow the number of columns and the number of lines (in the example in Figure 16.2, there are five columns and four lines).

Finally, from the next line on, the image data is specified as ASCII character '1' for black or '0' for white. Spaces and line break are optional here, but adding them make the image file more readable. Each 1 and 0 represents one black or white pixel as shown in the example in Figure 16.2.

When choosing the binary format P4 instead of P1, then this data part will contain fewer bytes with 1 bit per image pixel, but the header part always remains in readable ASCII format.

```
P1
# My image file, 5 cols., 4 lines
5 4
1 0 1 0 1
0 1 0 1 0
1 1 0 1 1
1 0 0 0 1
```

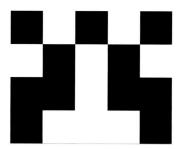

Figure 16.2: PBM example

16.2.2 Grayscale Image Files PGM

The grayscale format PGM starts with ID 'P2' and is very similar to that of black and white images. In addition to the number of columns and rows (5 and 3 in Figure 16.3), we also need to specify the maximum grayscale value, typically 255. This gives us the data range from 0 (black) to 255 (white), so each pixel value occupies 8 bits = 1 byte. Note that the PGM encoding is the opposite of the previously shown PBM format. In grayscale PGM, 0 is black, while in black&white PBM, 0 means white. In the example in Figure 16.3, we show three rows of increasing and decreasing grayscale values.

```
P2
# My image file, 5 cols., 3 lines
5 3 255
 25  50 100 150 200
200 150 100  50   0
 25  50 100 150 200
```

Figure 16.3: PGM example

16.2.3 Color Image Files PPM

The color image format PPM uses ID 'P3' and is the logic extension of the grayscale format PGM. Like in grayscale, the header has the number of columns, rows and the maximum value for each color component (in Figure 16.4, there are five columns, four rows and a maximum value of 255, so each color component occupies 1 byte).

In the data part, we now have three values (3 bytes) specifying each color pixel, 1 byte each for its red, green and blue components. In the example, we have for the first three rows decreasing color pixels for red, green and blue. In the fourth row, we have the mixed colors yellow (red + green), cyan (green + blue) and magenta (red + blue), followed by white (all components at 255) and black (all components at 0).

```
P3
# My image file, 5 cols., 4 lines
5 4 255
255 0 0    200 0 0    150 0 0    100 0 0    50 0 0
0 255 0    0 200 0    0 150 0    0 100 0    0 50 0
0 0 255    0 0 200    0 0 150    0 0 100    0 0 50
255 255    0 0 255    255 255    0 255      255 255 255  0 0 0
```

Figure 16.4: PPM example

16.3 Edge Detection

We provide a small set of useful image processing functions in RoBIOS, but we will start with processing a grayscale image directly at the pixel level. One of the most fundamental image processing operations is called *edge detection*. The idea behind edge detection is that on object in an image scene is often made up of pixels of a more or less uniform gray level, while there is a sharp contrast to the background, e.g., a white car on a gray road or a dark coffee mug on a bright table. Of course, this idea does not work in all cases, but it is a good hypothesis to start with for identifying objects.

Numerous edge detection algorithms have been introduced and used in a variety of industrial applications. Here, we will look at the Laplace and Sobel edge detectors, two very common and simple edge operators.

The Laplace operator is a local pixel-based operator that produces a derivative of a grayscale image. It achieves this by calculating for every pixel four times the current pixel value minus the four nearest neighbors left, right, top and bottom (Figure 16.5, left). This is done for every pixel in the whole image—either sequentially in a loop or in a single step with parallel hardware, e.g., on a graphics processing unit (GPU). The principle is shown in Figure 16.5, right. In areas of uniform gray levels (either bright or dark), subtracting four neighboring pixels from a similar center value times four, results in a value near zero. However, if for example, the center point and two of its neighbors are in a dark area while the other two are in a bright area, calculating this difference will result in a value of a high absolute value (in this case negative). As we do not care about the sign of the Laplace result, we will use the *abs* function to produce only positive values.

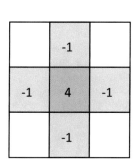
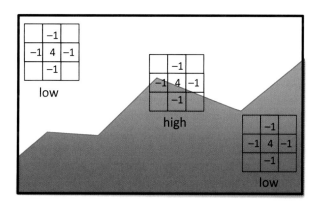

Figure 16.5: Laplace operator and principle

The coding is shown in Program 16.1 with a single loop running over all pixels. As there are no pixels beyond the outer border of an image, we need to avoid an access error by reading date elements outside the defined array bounds. Therefore, the loop starts at the second row and stops at the last but one row. Since the image is stored in a one-dimensional array, the result values for the first and last column will be incorrect, so a one-pixel-wide border should be deleted around the resulting Laplace image. The program also limits the maximum value to white (255), so that any result value remains within the byte data type.

Program 16.1: Laplace Edge Operator

```
#include "eyebot.h"
void Laplace(BYTE gray_in[], BYTE gray_out[])
{ int i, delta;
  for (i = IP_WIDTH; i < (IP_HEIGHT-1)*IP_WIDTH; i ++)
  { delta  = abs(4* gray_in[i] -gray_in[i-1] -gray_in[i + 1]
          -gray_in[i-IP_WIDTH]   -gray_in[i + IP_WIDTH]);
    if (delta > 255) delta = 255;
    gray_out[i] = (BYTE) delta;
  }
}

int main()
{ BYTE img[QVGA_PIXELS], lap[QVGA_PIXELS];

  CAMInit(QVGA);
  while (1)
  { CAMGetGray(img);
    Laplace(img, lap);
    LCDImageGray(lap);
  }
}
```

The RoBIOS library includes the Laplace function, so using it can be done much easier as shown in Program 16.2. Figure 16.6 shows a sample input image and the generated Laplace edge image.

Program 16.2: RoBIOS Laplace Operator

```
#include "eyebot.h"

int main()
{ BYTE img[QVGA_PIXELS], edge[QVGA_PIXELS];

  CAMInit(QVGA);
  while (1)
  { CAMGetGray(img);
    IPLaplace(img, edge);
    LCDImageGray(edge);
  }
}
```

Figure 16.6: Original image and Laplace edges after applying a threshold operation

The Sobel operator is often used for robotics applications and is only slightly more complex (see Bräunl[5]). It uses two separate templates, one to detect horizontal edges and one to detect vertical edges. In Figure 16.7, we see the two filter operations the Sobel filter is made of. The *dx* filter only finds discontinuities in the *x*-direction (vertical lines), while the *dy* filter only finds discontinuities in the *y*-direction (horizontal lines). Interestingly, neither filter uses its current (middle) pixel value, only neighboring values are used.

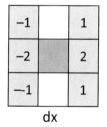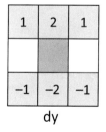

dx   dy

Figure 16.7: Sobel-x and Sobel-y masks

The two Sobel filters are then combined in two ways:

a. Using Pythagoras' formula to calculate the edge strength
$b = \sqrt{(dx^2 + dy^2)} \approx |dx| + |dy|$

b. Using the unique inverse tangent function *atan2* to calculate the edge direction
$r = atan2(dy, dx)$.

We often simplify time-consuming image processing operations. For calculating the edge strength, using the sum of the absolute values of the two Sobel operators is close enough in most cases to the square root of the sum of squares.

---

[5]T. Bräunl, *Parallel Image Processing*, Springer-Verlag, Berlin Heidelberg, 2001.

The function *atan2* used in the edge direction formula is an extremely valuable extension of the standard *atan* function. As it uses the values *dy* and *dy* as two separate parameters instead of their quotient, the *atan2* function is able to return the correct unique angle within the full [0°, 360°] range, while the standard *atan* function is limited to an output range of [−90°, +90°].

Both edge strength and edge direction results are very important. The edge strength tells us where an edge exists, and the edge direction gives us additional information about this edge (e.g., angled at 45°), which can be used for matching purposes, e.g., between stereo images. Program 16.3 shows the full code to implement the Sobel function. Only a single loop is used to run over all pixels. Again, we neglect a one-pixel-wide borderline: pixels in the first and last row and column. The program already applies a heuristic scaling (divide by three) and limits the maximum result value to constant *white* (255), so the result value remains a single byte.

Alternatively, Laplace and Sobel also exist predefined as RoBIOS library functions:

```
void  IPLaplace(BYTE* grayIn, BYTE* grayOut);
void  IPSobel(BYTE* grayIn, BYTE* grayOut);
```

Program 16.3: Sobel Edge Function

```
void Sobel(BYTE *imageIn, BYTE *imageOut)
{  int i, grad, deltaX, deltaY;

   for (i = IP_WIDTH; i < (IP_HEIGHT-1)*IP_WIDTH; i++)
   {  deltaX = 2*imageIn[i+1] + imageIn[i-IP_WIDTH+1]
                              + imageIn[i+IP_WIDTH+1]
              -2*imageIn[i-1] - imageIn[i-IP_WIDTH-1]
                              - imageIn[i + IP_WIDTH-1];

      deltaY = imageIn[i-IP_WIDTH-1]  + 2*imageIn[i-IP_WIDTH]
                                      + imageIn[i-IP_WIDTH + 1]
              -imageIn[i + IP_WIDTH-1] - 2*imageIn[i + IP_WIDTH]
                                       - imageIn[i + IP_WIDTH + 1];

      grad = (abs(deltaX) + abs(deltaY)) / 3;
      if (grad > white) grad = white;
      imageOut[i] = (BYTE) grad;
   }
}
```

Figure 16.8 shows the resulting Sobel-x, Sobel-y, edge strength and direction results for the same sample image as before.

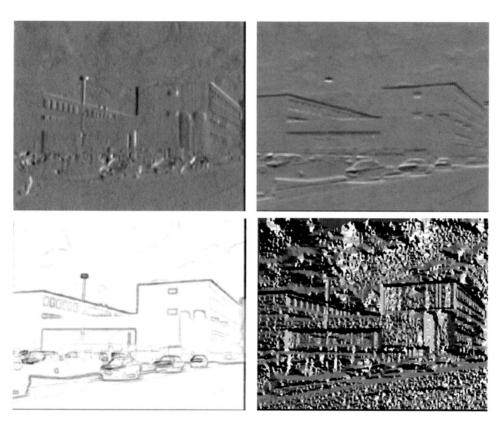

Figure 16.8: Resulting Sobel-x, Sobel-y, edge strength and edge direction

16.4   Motion Detection

The idea for a very basic motion detection algorithm is to subtract two subsequently taken images from each other (see Figure 16.9):

1. Compute the absolute grayscale difference for all pixel pairs of two images.
2. Compute the average over all pixel differences.
3. If this average value is above a certain threshold, then motion has been detected.

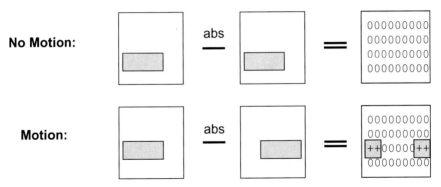

Figure 16.9: Motion detection

So, this method can detect whether or not there is any motion in an image pair, but it cannot determine the exact area or direction of the motion. Still, it would be suitable, e.g., as a burglar alarm. Program 16.4 shows the implementation of this task, aided by two subroutines. Function *image_diff* builds the pixel-wise absolute difference, and function *avg* calculates the average value of its input array.

The main program initializes the camera and reads the initial image. In an endless loop, it reads the second image (depending on odd/even iterations into array *image1* or *image2*) and then calls the auxiliary functions.

Program 16.4: Motion Detection

```
#include "eyebot.h"
#define RES  QVGA
#define SIZE QVGA_SIZE

void image_diff(BYTE i1[SIZE], BYTE i2[SIZE], BYTE d[SIZE])
{ for (int i = 0; i < SIZE; i ++)
  d[i] = abs(i1[i] - i2[i]);
}
```

```
int avg(BYTE d[SIZE])
{ int i, sum = 0;
  for (i = 0; i < SIZE; i ++)
    sum += d[i];
  return sum / SIZE;
}
```

```
int main()
{ BYTE image1[SIZE], image2[SIZE], diff[SIZE];
  int avg_diff, delay, i = 0;
```

```
CAMInit(RES);
CAMGetGray(image1);

while (1)
{ i ++;
  if (i%2 = 0) CAMGetGray(image1);
      else CAMGetGray(image2);

  image_diff(image1, image2, diff);
  LCDImageGray(diff);
  avg_diff = avg(diff);
  LCDSetPrintf(0,50, "Avg = %3d", avg_diff);

  if (avg_diff > 15) // Alarm threshold
  { LCDSetPrintf(2,50, "ALARM!!!");
    delay = 10;
  }
  if (delay) delay--;
      else LCDSetPrintf(2,50, "      "); // clear text
  }
}
```

The core of this algorithm has only three lines of code:

```
image_diff(image1, image2, diff);
avg_diff = avg(diff);
if (avg_diff > 15) ...
```

These are: calculating the difference image, then calculating the overall difference as a single (scalar) value, and finally comparing it to a fixed threshold (in this case 15).

If the difference is higher than the threshold, the program prints an alarm message and then removes it again after 10 loop iterations. If the Raspberry Pi is connected to a network, the program could equally easily send an email or text message as a warning.

Of course, there are a number of applications where a single yes/no is not enough for motion detection. Then, we would like to know more precisely where in the image the motion has occurred. The solution for this extension is surprisingly easy. Instead of running the algorithm over the whole image, we just run it three times over one-third of the image, i.e., for the left section, the middle section and the right section. Assuming there is only one motion event in the image at any time, we will know immediately in what direction the motion is and we can move the robot's camera toward this direction of interest. While moving only a small fraction during each iteration, the motion will gradually shift to the middle section in subsequent iterations, which is when the camera movement will stop.

Figure 16.10 shows the motion program for a stationary robot looking at a couple of stationary cans with a rotating robot in between. Only the moving robot appears in the motion image.

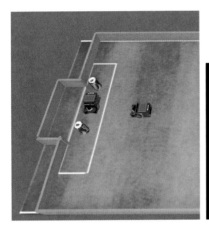

Figure 16.10: Motion example in EyeSim: robot setup (left) and motion image (right)

Nothing stops us from doing a more fine-grained analysis of the motion location. We could use 4, 10 or 100 sections or just calculate the motion value for every image column and every image row and then find the maximum. This is called generating a histogram, and we will apply this idea later when searching for colored objects (see Figure 16.11).

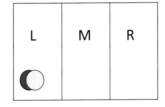

Figure 16.11: Motion localization

16.5   Color Spaces

Before looking at a color-based object detection algorithms, it makes sense to talk about different color representations or *color spaces*. So far, we have seen bitmap (black and white), grayscale and RGB color models, as well as Bayer

patterns (RGGB). There is not one superior way of representing color information, but different models have advantages for different applications.

16.5.1 Red Green Blue (RGB)

The RGB space can be viewed as a 3D cube with red, green and blue being the
three coordinate axes (Figure 16.12). The line joining black (0, 0, 0) and white
(255, 255, 255) is the main diagonal in the cube and represents all shades of
gray.

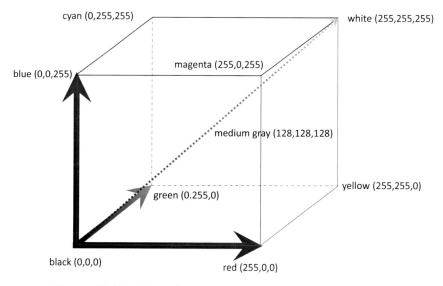

Figure 16.12: RGB color cube

In this color space, a color is determined by its red, green, and blue components in the additive color model. The main disadvantage of this color space
is that the color hue is not independent of intensity and saturation of the color.
So, for example, when trying to detect a red object, we cannot simply program
'*if (red > 200)* ....'. Assume that the lights in a room get dimmed by 50%. This
means that all pixel values will be reduced by one half. So, if a pixel had the
RGB value (220, 60, 40) before, it would then have the value (110, 30, 20).
The object and its color have not changed at all, but its RGB value has been
reduced significantly, and the simple programming method would no longer
detect it.

16.5.2 Hue Saturation Intensity (HSI)

The HSI color space (see Figure 16.13) is a cone where the middle axis represents intensity, the phase angle represents the hue of the color, and the radial distance represents the saturation. The following set of equations is adapted from Hearn et al.[6] and specifies the conversion from RGB to HSI color space:

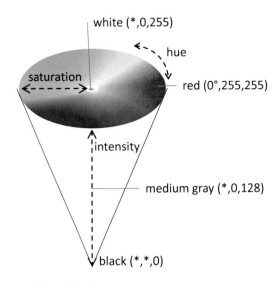

Figure 16.13: HSI color cone

$$I = \frac{1}{3}(R + G + B)$$

$$S = 1 - \frac{3}{(R + G + B)} [\min(R, G, B)]$$

$$H = \begin{cases} h & \text{if } B \leq G \\ 360° - h & \text{if } B > G \end{cases}$$

$$h = \cos^{-1} \left\{ \frac{\frac{1}{2}[(R - G) + (R - B)]}{\left[(R - G)^2 + (R - B)(G - B) \right]^{1/2}} \right\}$$

The advantage of this color space is to decorrelate the intensity information from the color information. Now, a change in ambient lighting conditions (within reason) will not significantly affect the color hue.

---

[6]D. Hearn, M. Baker, W. Carithers, *Computer Graphics with OpenGL*, 4th Ed., Pearson, Harlow, Essex, 2014.

A grayscale value is represented by its intensity, zero saturation and an arbitrary hue value. So, it can simply be differentiated between chromatic (color) and achromatic (grayscale) pixels, only by looking at the saturation value. However, this also means that the hue value by itself is not sufficient to identify pixels of a certain color. The corresponding saturation value has to be above a certain threshold.

In the HSI color model, *black* is (*, *, 0), where '*' denotes any arbitrary value, while *white* is (*, 0, 255). Grayscale values are (*, 0, gray). To find the primary color hues, we can solve the hue equation for:

- **Red** (R = 255, G = 0, B = 0)

$$H_{red} = \cos^{-1} \frac{\frac{1}{2}(R+R)}{\sqrt{R^2 + R \cdot 0}}$$

$$= \cos^{-1} \frac{R}{\sqrt{R^2}} = \cos^{-1}(1) = 0°$$

- **Green** (R = 0, G = 255, B = 0)

$$H_{green} = \cos^{-1} \frac{\frac{1}{2}(-G+0)}{\sqrt{G^2 + G \cdot 0}}$$

$$= \cos^{-1} \frac{-\frac{G}{2}}{\sqrt{G^2}} = \cos^{-1}\left(-\frac{1}{2}\right) = 120°$$

- **Blue** (R = 0, G = 0, B = 255).

$$H_{blue} = 360° - \cos^{-1} \frac{\frac{1}{2}(0-B)}{\sqrt{0^2 + B^2}}$$

$$= 360° - \cos^{-1} \frac{-\frac{B}{2}}{\sqrt{B^2}} = 360° - \cos^{-1}\left(-\frac{1}{2}\right)$$

$$= 360° - 120° = 240°$$

In order to express HSI values as three bytes, we transform the hue angle [0°, 359°] into the byte range [1, 255], reserving hue value 0 to indicate 'no hue.' We also use the integer range [0, 255] for saturation and intensity instead of the real number range [0, 1] of the original formula.

16.5.3 Normalized RGB (RGB)

Most camera image sensors deliver pixels in an RGB-like format, for example Bayer patterns (see sensor chapter). Converting all pixels from RGB to HSI might be a too intensive computing operation for an embedded controller. Therefore, we look at a faster alternative with similar properties.

One way to make the RGB color space more robust with regard to lighting conditions is to use the *normalized RGB* color space (denoted by lower case *rgb*) defined as:

$$r = \frac{R}{R+G+B} \quad g = \frac{G}{R+G+B} \quad b = \frac{B}{R+G+B}$$

This normalization of the RGB color space allows us to describe a certain color independently of the luminosity (sum of all components). This is because the luminosity in *rgb* is always equal to one:

For each pixel holds: $r + g + b = (R + G + B)/(R + G + B) = 1$

16.6 RBG-to-HSI Conversion

Program 16.5 shows the conversion of a single RGB pixel into HSI (hue, saturation, value), which is a modified version of the algorithm presented in Hearn et al.[7] We also reduced each HSI component to fit into one byte.

Program 16.5: RGB -to-Hue Conversion

```
void IPPRGB2HSI (BYTE r, BYTE g, BYTE b, BYTE* h, BYTE* s, BYTE* i)
{ BYTE delta, max, min;

  max   = MAX(r,g,b);
  min   = MIN(r,g,b);
  delta = max - min;

  *i = (r + g+b)/3;                // intensity
  if (*i > 0) *s = 255 - 255*min/(*i); // saturation
      else *s = 0;
  if (2*delta > max && (*i) > THRES) // hue
  { if   (r ==max) *h =  43 + 42*(g-b)/delta; // ±42 [ 1.. 85]
    else if (g ==max) *h = 128 + 42*(b-r)/delta; // ±42 [ 86..170]
    else if (b ==max) *h = 213 + 42*(r-g)/delta; // ±42 [171..255]
  }
  else *h = 0; // grayscale, not color
}
```

The algorithm begins with calculating the max, min and difference of the three RGB components, which is stored in variable *delta*.

For intensity, we simply take the average of the RGB components. Saturation depends on how much the minimum RGB component differs from the average intensity *i*. If color components are identical, then the quotient *min/i* becomes 1 and saturation *s* becomes 0. However, if there is a large difference, then the quotient goes toward zero and the saturation goes toward 255 (maximum value).

---

[7]D. Hearn, M. Baker, C. Carithers, *Computer Graphics with OpenGL*, 4th Ed., Pearson, Harlow, Essex, 2014.

Before calculating the hue, the saturation and intensity components are used to distinguish whether the pixel is grayscale or color. Only if there is a significant variation in RGB components values ($2*delta > max$) and the intensity is above the given threshold ($i > THRES$), we will calculate a hue value, as hue values will be arbitrary in low-light situations. Otherwise, we set the hue to 0 (*no hue*).

Depending on which RGB component is the dominant one (highest value), the corresponding *if*-statement calculates the hue value from an offset (43, 128, 213), respectively, plus the scaled positive or negative difference of the remaining two RGB components (±42). This will leave the hue result in range [1, 255], so it fits within one byte.

Note that in this simplified conversion, the HSI values for the primary colors are:

- Red   (43, 255, 85)
- Green   (128, 255, 85)
- Blue   (213, 255, 85).

Table 16.1 shows the matrix of increasing red values (rows) versus increasing green values (columns) with blue being zero. Table 16.2 shows the matrix of increasing red values (rows) versus increasing blue values (columns) with green being zero, and Table 16.3 shows the same for green and blue with red being zero. These three tables can be extended to a three-dimensional tensor, which can be used as a lookup table for a very quick RGB-to-HIS conversion.

0	0	128	128	128	128	128	128	128	128	128	128	128
0	85	107	114	118	120	121	122	123	124	124	125	125
43	64	85	100	107	112	114	116	118	119	120	121	121
43	57	71	85	97	103	107	110	113	114	116	117	118
43	53	64	74	85	95	100	104	107	110	112	113	114
43	51	59	68	76	85	93	98	102	105	107	109	111
43	50	57	64	71	78	85	92	97	100	103	106	107
43	49	55	61	67	73	79	85	92	96	99	102	104
43	48	53	58	64	69	74	79	85	91	95	98	100
43	47	52	57	61	66	71	75	80	85	91	94	97
43	47	51	55	59	64	68	72	76	80	85	90	93
43	46	50	54	58	62	65	69	73	77	81	85	90
43	46	50	53	57	60	64	67	71	74	78	81	85

Table 16.1: Hue values for red (down) and green (across) components in increments of 20

0	0	213	213	213	213	213	213	213	213	213	213	213
0	1	234	227	223	221	220	219	218	217	217	216	216
43	22	1	241	234	229	227	225	223	222	221	220	220
43	29	15	1	244	238	234	231	228	227	225	224	223
43	33	22	12	1	246	241	237	234	231	229	228	227
43	35	27	18	10	1	248	243	239	236	234	232	230
43	36	29	22	15	8	1	249	244	241	238	235	234
43	37	31	25	19	13	7	1	249	245	242	239	237
43	38	33	28	22	17	12	7	1	250	246	243	241
43	39	34	29	25	20	15	11	6	1	250	247	244
43	39	35	31	27	22	18	14	10	6	1	251	248
43	40	36	32	28	24	21	17	13	9	5	1	251
43	40	36	33	29	26	22	19	15	12	8	5	1

Table 16.2: Hue values for red (down) and blue (across) components in increments of 20

0	0	213	213	213	213	213	213	213	213	213	213	213
0	170	192	199	203	205	206	207	208	209	209	210	210
128	149	170	185	192	197	199	201	203	204	205	206	206
128	142	156	170	182	188	192	195	198	199	201	202	203
128	138	149	159	170	180	185	189	192	195	197	198	199
128	136	144	153	161	170	178	183	187	190	192	194	196
128	135	142	149	156	163	170	177	182	185	188	191	192
128	134	140	146	152	158	164	170	177	181	184	187	189
128	133	138	143	149	154	159	164	170	176	180	183	185
128	132	137	142	146	151	156	160	165	170	176	179	182
128	132	136	140	144	149	153	157	161	165	170	175	178
128	131	135	139	143	147	150	154	158	162	166	170	175
128	131	135	138	142	145	149	152	156	159	163	166	170

Table 16.3: Hue values for green (down) and blue (across) components in increments of 20

General grayscale, RGB and HSI conversion functions are also part of the RoBIOS library. The following functions work on the image level:

```
void  IPCol2Gray(BYTE* imgIn, BYTE* grayOut);          // color to gray
void  IPGray2Col(BYTE* imgIn, BYTE* colOut);           // gray to color
void  IPRGB2Col (BYTE* r, BYTE* g, BYTE* b, BYTE* imgOut);// 3*gray to col
void  IPCol2HSI (BYTE* img, BYTE* h, BYTE* s, BYTE* i); // RGB to HSI
```

The following conversion functions work on the pixel level:

```
COLOR IPPRGB2Col(BYTE r, BYTE g, BYTE b);              //RGB to color
void  IPPCol2RGB(COLOR col, BYTE* r, BYTE* g, BYTE* b); //color to RGB
void  IPPCol2HSI(COLOR c, BYTE* h, BYTE* s, BYTE* i);  //RGB to HSI
BYTE  IPPRGB2Hue(BYTE r, BYTE g, BYTE b);              //RGB to hue
void  IPPRGB2HSI(BYTE r, BYTE g, BYTE b, BYTE* h, BYTE* s, BYTE* i); //HSI
```

16.7 Color Object Detection

If it is guaranteed for a robot environment that a certain color only exists on one particular object, then we can use color detection to find this object. This assumption is widely used in mobile robot competitions, for example the AAAI robot competition to collect yellow tennis balls or the RoboCup and FIRA robot soccer competitions to kick an orange golf ball into the yellow or blue goal. See Kortenkamp et al.[8], Kaminka et al.[9]; and Cho and Lee.[10]

The following hue-histogram algorithm for detecting colored objects requires minimal computation time and is therefore very well suited for embedded vision systems (see Figure 16.14). The algorithm performs the following steps:

1. Convert the RGB color image to a hue image (HSI model).
2. Create a histogram over all image columns of pixels matching the object color.
3. Find the maximum position in the column histogram.

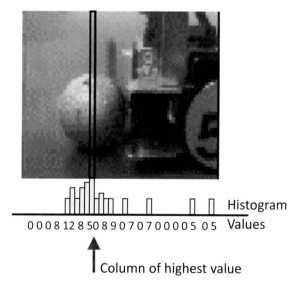

Figure 16.14: Color detection principle

---

[8]D. Kortenkamp, I. Nourbakhsh, D. Hinkle, *The 1996 AAAI Mobile Robot Competition and Exhibition*, AI Magazine, vol. 18, no. 1, 1997, pp. 25–32 (8).

[9]G. Kaminka, P. Lima, R. Rojas, (Eds.) *RoboCup 2002: Robot Soccer World Cup VI*, Proceedings, Fukuoka, Japan, Springer-Verlag, Berlin Heidelberg, 2002.

[10]H. Cho, J.-J. Lee, (Ed.) *2002 FIRA Robot World Congress, Proceedings*, Korean Robot Soccer Association, Seoul, May 2002.

The first step only simplifies the comparison whether two color pixels are similar. Instead of comparing the differences between three values (red, green and blue), only a single hue value needs to be compared. In the second step, we look at each image column separately and record how many pixels are similar to the desired object color. For a QVGA image, the histogram comprises just 320 integer values (one for each column) with values between 0 (no similar pixels in this column) and 240 (all pixels in a column are similar to the object color).

At this level, we are not concerned about continuity of the matching pixels in a column. There may be two or more separate sections of matching pixels, which may be due to either occlusions or reflections on the same object—or there might be two different objects of the same color. A more detailed analysis of the resulting histogram could distinguish between these cases.

In Figure 16.15, we show a small example demonstrating the steps involved in locating a color object in an image:

1. Converting RGB to HSI
2. HIS to binary image matching:
   1 for matching pixels and 0 for not matching pixels
3. Generating histograms (here shown for columns and rows)
4. Finding max position in histograms.

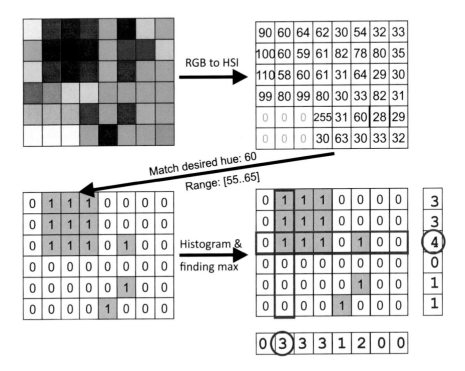

Figure 16.15: Color detection example

6.7 Color Object Detection

Step 1 we have already done, and step 2 is similar to the motion detection algorithm. For step 3, we need to generate a histogram over all *x*-positions (over all columns) of the image, as shown in Figure 16.15 and Program 16.6 (we are only doing a column histogram here). We use two nested loops going over every single pixel, but here we have the outer loop iterate over *x* (-columns) and the inner loop iterate over *y* (rows), so we process the image column by column instead of line by line. Index variable *idx* gets assigned the start position of the current pixel at image location [*y*, *x*] remembering that each pixel takes three bytes of data. We then call the RGB-to-hue conversion function, but only proceed if the detected *color* is not a grayscale (i.e., the hue value needs to be greater than 0).

The *distance* is the difference between the current pixel hue and the desired object hue. This value can be at most half of a circle (180°) in the color wheel. If this value is greater than 127 in our byte scheme, then we just take the shorter difference (*255—distance*). Finally, if the *distance* is less than the specified threshold, then the counter for this column gets incremented (*count++*).

Program 16.6: Histogram-based Color Detection

```
int ColSearch(BYTE *img, int obj_hue)
{ int x,y, pos,val, count, h, distance, idx;

  pos = -1; val = 0;  /* init */
  for (x = 0; x < CAMWIDTH; x ++)
  { count = 0;
    for (y = 0; y < CAMHEIGHT; y ++)
    { idx = 3*(y*CAMWIDTH + x);  // color image
      h = IPPRGB2Hue(img[idx],img[idx + 1],img[idx + 2]);
      if (h)  // h!= 0 which is "no hue"
      { distance = abs((int)h-obj_hue);  /* hue distance */
        if (distance > 127) distance = 255 - distance;
        if (distance < THRES) count ++;
      }
    }
    if (count > val) { val = count; pos = x; }
    LCDLine(x,CAMHEIGHT-1, x, CAMHEIGHT-1-count, RED);  // graph
  }
  return pos;
}
```

Note that we do not even need a one-dimensional array to store the histogram. As we are only interested in the position of the highest-rated column, all we need are two variables *pos* and *val* to store the position and the matching count of the best column so far. The command *LCDLine* draws a nice visualization of the histogram at the bottom of the robot's display. Figure 16.16 shows the various stages of the robot approaching a red can in EyeSim. The increasing sizes of the histogram bars are quite pronounced when approaching the can.

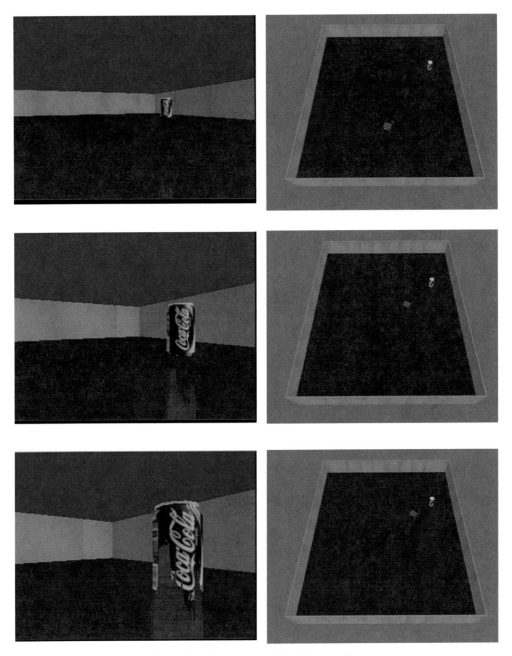

Figure 16.16: Color detection on EyeSim simulator

The remaining problem is how to drive the robot after receiving the histogram maximum, which is likely the center point of the desired object. We show a very simple, but a well-working approach in Program 16.7. Instead of calculating the exact steering angle to approach the object, we only distinguish whether the object is located in the left, middle or right section of the image—the same approach that we used for motion detection earlier in this chapter. Since this test is being executed continuously in a loop, the robot will quite smoothly drive toward the object. No further effort is required.

Program 16.7: Color Search Main Loop

```
while ((KEYRead() != KEY4))
{ CAMGet(img);
  LCDImage(img);
  pos = ColSearch(img, BALLHUE);  // search image
  if (PSDGet(PSD_FRONT) > 50)     // collision free
  { if (pos < CAMWIDTH/3) VWTurn(10, 30);             // left
      else if (pos > 2*CAMWIDTH/3) VWTurn(-10, 30); // right
          else VWStraight( 50, 100);                 // straight
  }
  OSWait(100); 0.1 sec
}
```

16.8   Image Segmentation

Detecting a single object that differs significantly either in shape or in color from the background is relatively easy. A more ambitious application is segmenting an image into disjoint regions. One way of doing this in a grayscale image is to use connectivity and edge information (see Braunl[11] and Bräunl[12] for an interactive system). The algorithm shown here, however, uses color information for faster segmentation results from Leclercq and Bräunl.[13]

This color segmentation approach transforms all images from RGB to rgb (normalized RGB) as a preprocessing step. Then, a color class lookup table is constructed that translates each rgb value to a *color class*, where different color classes ideally represent different objects. This table is a three-dimensional array with (rgb) as indices. Each entry is a reference number for a certain color class.

---

[11]T. Bräunl, Parallel Image Processing, Springer-Verlag, Berling Heidelberg, 2001.

[12]T. Bräunl, *Improve – Image Processing for Robot Vision*, http://robotics.ee.uwa.edu.au/improv, 2006.

[13]P. Leclercq, T: Bräunl, *A Color Segmentation Algorithm for Real-Time Object Localization on Small Embedded Systems*, Robot Vision 2001, International Workshop, Auckland NZ, Lecture Notes in Computer Science, no. 1998, Springer-Verlag, Berlin Heidelberg, Feb. 2001, pp. 69–76 (8).

If we know the number and characteristics of the color classes to be distinguished beforehand, we can use a static color class allocation scheme. For example, for playing robot soccer, we need to distinguish only three color classes: orange for the ball and yellow and blue for the two goals. In a case like this, the location of the color classes can be calculated to fill the table. For example, *blue goal* is defined for all points in the 3D color table for which blue dominates, or simply:

hue $\in [240° \pm \text{threshold}]$

In a similar way, we can distinguish any other color through a hue range. Since the hue conversion takes a significant amount of compute time, we can *pre-compute* hues for a limited resolution color table and then use this as a quick lookup table.

To do this for RGB values ($3 \cdot 8 = 24$ bits) directly would require a table of 2^{24} entries, which comes to 16 MB. This is a large data area for an embedded system, and it is unlikely that we would need this level of precision. Reducing each RGB component to 5 bits reduces each pixel to 15 bits, which requires a table with 2^{15} entries, so a much more manageable 32 KB.

We can now pre-compute the hue values for every table element—or we can even pre-compute the class numbers for the objects to detect, e.g., 1 for orange ball, 2 for yellow goal and 3 for blue goal.

Program 16.8 shows a small program to display the hue of a desired object in the viewing range. It reads a single pixel from the center column of a lower row of the camera image and displays its hue value. Of course, sampling a small area of, for example 3 × 3 pixels, and using a median filter would be a more robust approach. We are displaying the camera image and draw a tiny white cross over it, so we can tell which part of the image we are actually looking at. Figure 16.17 shows the scenario of the robot looking at an orange golf ball, which is the playing ball in robot soccer, while printing its hue value as a text output.

Figure 16.17: Robot looking at golf ball and camera view from robot with white marker

6.8 Image Segmentation

Program 16.8: Hue Scanning Program

```c
#include "eyebot.h"

int main()
{BYTE img[QVGA_SIZE];
 int hue, pos, x, y;

 LCDMenu(" "," "," ","END");
 CAMInit(QVGA);

 x = CAMWIDTH/2; y = CAMHEIGHT*3/4;
 pos = 3*(y*CAMWIDTH + x); // 3 bytes per pixel
 while ((KEYRead() != KEY4))
 {CAMGet(img);
   LCDImage(img);
   LCDPixel(x,y,WHITE);
   LCDPixel(x-1,y,WHITE);      LCDPixel(x + 1,y,WHITE);
   LCDPixel(x,y-1,WHITE);      LCDPixel(x,y + 1,WHITE);
   hue = IPPRGB2Hue(img[pos],img[pos + 1],img[pos + 2]);
   LCDSetPrintf(1,60, "Hue %3d", hue);
   OSWait(100); // 0.1 s
 }
}
```

The detected hue for the orange golf ball is 61, and we decide to accept a matching hue range of ±5, so [56, 66]. The yellow goal registered a hue of 79, giving us a range of [74, 84], and the blue goal has 210, so range [205, 215].

In Table 16.4 we show an example for pre-calculating these three objects as 1, 2 and 3 (ball, yellow goal and blue goal) for the red/green hue matrix of Table 16.1. Of course, this needs to be done on the full 3D tensor for all red/green/blue combinations.

–	–	–	–	–	–	–	–	–	–	–	–	–	–
–	–	–	–	–	–	–	–	–	–	–	–	–	–
–	1	–	–	–	–	–	–	–	–	–	–	–	–
–	1	–	–	–	–	–	–	–	–	–	–	–	–
–	–	1	2	–	–	–	–	–	–	–	–	–	–
–	–	1	–	2	–	–	–	–	–	–	–	–	–
–	–	1	1	–	2	–	–	–	–	–	–	–	–
–	–	–	1	–	–	2	–	–	–	–	–	–	–
–	–	–	1	1	–	2	2	–	–	–	–	–	–
–	–	–	1	1	1	–	2	2	–	–	–	–	–
–	–	–	–	1	1	–	–	2	2	–	–	–	–
–	–	–	–	1	1	1	–	–	2	2	–	–	–
–	–	–	–	1	1	1	–	–	2	2	2	–	–

Table 16.4: Matching objects for hue values for hue matrix of red (down) and green (across)

Figure 16.18 shows an example of the robot looking at a scene with different color-coded objects. We printed the classification numbers below the current image frame with the corresponding hue values to the right. Only pixels in every twentieth row and column were evaluated for clarity reasons. Those pixels are identified with a small white marker. Although to a human observer, coherent color areas and therefore objects are easy to detect, it is not trivial to extract this information from the 2D segmented output image. For smaller objects like the round golf ball, we can use the histogram method from the previous section. For larger rectangular areas, we can simply record the four corner points of matching pixels in the image.

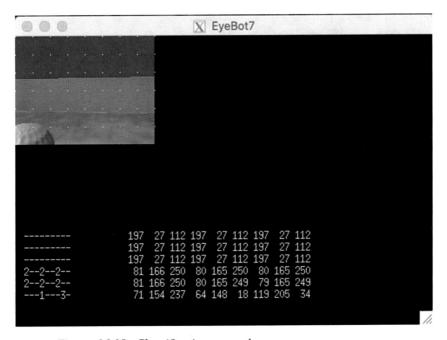

Figure 16.18: Classification example

16.9 Image Coordinates versus World Coordinates

Whenever an object is identified in an image, all we have is its image coordinates. Working in QVGA resolution (240 pixels high × 320 pixels wide), assume that our desired object is at position (210, 20) and has a size of 5 × 7 pixels, so it is located somewhere in the bottom left corner. Although this information might already be sufficient for some applications (e.g., we can steer the robot in the direction of this object), for many applications we would like to know more precisely the object's location either in the world coordinates or in the robot's local coordinate system (see Figure 16.19).

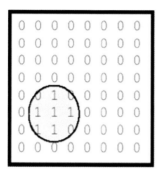
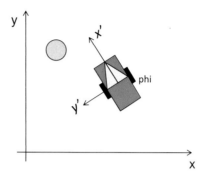

Figure 16.19: Object in front of robot and corresponding image from robot's camera

First, we only look at the object's position in the robot's local coordinate system $\{x', y'\}$, later at the global word coordinate system $\{x, y\}$. Once we have determined the object coordinates in the robot's local coordinate system and if we also know the robot's absolute pose, we can transform the object's local coordinates to global world coordinates.

For this, we will assume two simplifications:

1. We are using only objects with rotational symmetry, such as a ball or a can, because they look the same (or at least similar) from any viewing angle.
2. We assume that objects are resting on the ground, for example the level floor the robot is driving on.

Figure 16.20 demonstrates this situation with a side view and a top view from the robot's local coordinate system. We have:

- The ball's image coordinates (column and row) as pixel coordinates of the sensor
- The camera is mounted at height h and tilted down by angle α, but it is not panning sideways.
- The ball diameter is d, and the ball is lying on the ground.

From these values, we can derive the ball's local coordinates (x', y') from the robot's point of view by using the intercept theorem to calculate its local displacement. With a zero camera offset and a camera angle of zero (no tilting or panning), we have the proportionality relationships:

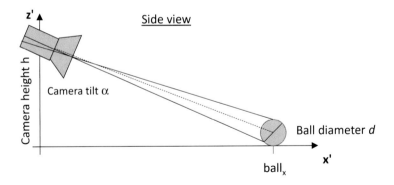

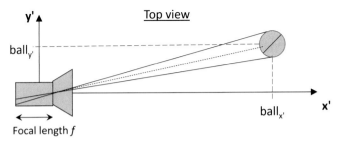

Figure 16.20: Camera position and orientation

$$\frac{ball_{x'}}{f} = \frac{d}{row}$$

$$\frac{ball_{y'}}{f} = \frac{d}{column}$$

These can be simplified when introducing a camera-specific proportionality factor $g = k \cdot f$ for converting between pixels and meters. The parameter k can either be calculated from the camera parameters or it can be determined by experiment (*camera calibration*).

$ball_{x'} = g \cdot d / row$
$ball_{y'} = g \cdot d / column$

So, with a level-mounted camera without tilt or pan angles, we can directly calculate the object coordinates. The transformation is just a constant linear factor; however, due to lens distortions and other sources of noise, these ideal conditions will not be observed in an experiment. It is therefore better to provide a lookup table for doing the transformation based on a series of distance measurements.

If the camera has an offset, the trigonometric formulas become somewhat more complex, but can be solved by trigonometric functions. An offset can either be to the side or above/below the driving plane, or a panning angle (z-axis) or a tilting angle (y-axis).

16.10   Tasks

1. Implement the color detection task using a histogram over a HSI image as described in this chapter. Mark the detected object center with a dot in the displayed image.

2. Extend the color detection algorithm to program an object collection application.
   Let the robot search for a colored object, then drive toward it and push it into a home area.

3. Let the robot find the brightest spot within a rectangular area, surrounded by walls. The robot should use its camera to search for the brightest spot and use its infrared sensors to avoid collisions with walls or obstacles. When working with real robots, you can use a flashlight to highlight a spot on the ground that the robots have to search and follow.

4. Mark a line on the ground and program the robot to follow it.

5. Implement the motion detection algorithm as shown in this chapter for a stationary robot. Calculate motion values for the three image areas left, middle and right.
   Then, in each step, slightly move the robot's camera using the pan option (or rotate the whole robot base) toward the left or right, depending in which area the motion has been detected.

6. Extend the motion detection program for a driving robot. The robot should continuously drive toward the center of the detected motion. Note that due to the robot's own movement (egomotion) all pixels in its image sequence will have an outward flow.

17

AUTOMOTIVE SYSTEMS

Modern automobiles are a real gathering place for embedded systems. Each new car has between twenty and one hundred embedded controllers, each dedicated to one particular task. There are individual controllers for engine control, dashboard displays, trip computer, keyless entry, electric seat adjustment and position memory, mirror adjustment, power windows, cruise control and airbag control. Advanced safety features such as ABS (anti-lock braking system) and ESP (electronic stability program) have their individual embedded systems, as do more advanced features such as automatic headlight switch, rain sensor, parking distance sensors and so on.

With new features being added to automobiles every day, it is cheaper to add additional embedded controllers than to develop a single monolithic automotive computer system. Also, individual embedded systems can be replaced more easily in case of a defect. However, drawbacks of having many individual controllers in a car are the need to include one or more bus systems for interfacing the controllers. Each controller has to meet the bus specification in order to not disturb the communication of others, as well as to comply with EMC (electromagnetic compatibility) restrictions.

Reliable communication in an extremely noisy environment such as a car is a challenge, which is why the automotive industry has developed their own bus standards. Most cars produced today include a CAN bus (Controller Area

Network; see ISO[1] and Zhou et al.[2]), while the newer FlexRay bus (see ISO,[3] Bretz[4] and Hönninger[5]) is gaining support.

In this chapter we want to look beyond today's automotive technology and look at research issues for autonomous automobiles and robotic cars. At The University of Western Australia's Renewable Energy Vehicle Project (REV[6]), we have worked on electric and autonomous cars for over a decade and some of our autonomous vehicle research implementations will be discussed in this chapter. While we will present advanced software libraries, such as OpenCV and ROS, all solutions presented in this chapter can also be described as *traditional engineering approaches*. In the following chapters, we will show how alternative—and arguably simpler—solutions to these problems can be found using artificial intelligence (AI) methods.

17.1   Autonomous Automobiles

The history of autonomous automobiles is still very young and has been initiated and shaped by Ernst Dickmanns from UniBw Munich (Universität der Bundeswehr München), Germany. When he first introduced his ideas on vision-guided autonomous vehicle control at regional conferences (Dickmanns et al.[7] and Dickmanns[8]), research colleagues questioned the viability of his approach and even the overall feasibility of such a project. Dickmanns proved them wrong by developing several autonomous car prototypes (VaMoRs, VaMP) and demonstrating the reliability of his autonomous driving systems on public highways in the presence of other traffic. His autonomous car trip from Bavaria to Denmark in 1995 over 1'758 km with only minimal intervention was a milestone for autonomous vehicles. Dickmanns' hardware and software

---

[1]ISO. *Road vehicles – Controller area network (CAN) – Part 1: Data link layer and physical signalling,* ISO standard 11898-1:2003, TC 22/SC 3, 2003, pp. (45).

[2]Y. Zhou, X. Wang, M. Zhou, *The Research and Realization for Passenger Car CAN Bus,* The 1st International Forum on Strategic Technology, Oct. 2006, pp. 244–247 (4).

[3]ISO. *Road vehicles – Communication on FlexRay – Part 1: General description and use case definition, ISO standard 10681-1:2007,* TC 22/SC 3, standards under development, 2007.

[4]E. Bretz, *By-wire cars turn the corner,* IEEE Spectrum, vol. 38, no. 4, Apr. 2001, pp. 68–73 (6).

[5]H. Hönninger, *Plenty of Traffic in Vehicles' Central Nervous Systems,* Bosch Research Info, News from Research and Development, no. 2, 2006, pp. (4), web: http://researchinfo.bosch.com.

[6]T. Bräunl, *The REV Project,* online: http://REVproject.com.

[7]E. Dickmanns, A. Zapp, K. Otto, *Ein Simulationskreis zur Entwicklung einer automatischen Fahrzeugführung mit bildhaften und inertialen Signalen,* 2. Symposium Simulationstechnik, Wien, Austria, 1984, pp. 554-558 (5).

[8]E. Dickmanns, *Normierte Krümmungsfunktionen zur Darstellung und Erkennung ebener Figuren,* DAGM-Symposium 1985, Erlangen, Germany, 1985, pp. 58–62 (5).

designs have been copied for other automotive research projects in industry (e.g. Daimler-Benz), as well as in academia (e.g. TU München). See Dickmanns[9] for a summary on his work (Figure 17.1).

Figure 17.1: Autonomous vehicle VaMoRs at University BW München. Photos courtesy of Prof. E. Dickmanns

By comparison, this makes the DARPA Grand Challenge in 2004 and 2005 (see DARPA[10] and Seetharaman et al.[11]) look like a simple project. For the 2005 Grand Challenge, vehicles had to navigate an empty road over 132 miles in the Nevada desert, while the exact driving path was specified by several thousand GPS waypoints. Teams were allowed to manually adjust and edit the given waypoints before the race start (e.g. with the help of satellite-based maps). Once the autonomous vehicle was on its way, no further interference was allowed. While none of the competitors was able to finish the race in 2004, five autonomous cars finished the race in 2005 (see Thrun et al.[12]) (Figure 17.2).

The most prominent sensors used for the Grand Challenge were a differential GPS receiver for navigation and a combination of several Lidar and radar sensors for fine-tuned road detection and collision avoidance. While several participating teams did use a vision subsystem to increase their road look-ahead in order to be able to drive at higher speeds, completing the Grand Challenge did not necessarily require any image processing, since the navigation path was given, and each vehicle had the road to itself.

Most team entries in the Grand Challenge and its 2007 successor competition, the Urban Challenge, were funded well in excess of one or two million U.S. Dollars, not counting staff and student labor or donations and support

---

[9]E. Dickmanns, Dynamic Vision for Perception and Control of Motion, Springer-Verlag, Heidelberg, 2007, pp. (486).

[10]DARPA, Grand Challenge, 2006, web: http://www.darpa.mil/grandchallenge/index.asp

[11]G. Seetharaman, A. Lakhotia, E. Blasch, Unmanned Vehicles Come of Age: The DARPA Grand Challenge, IEEE Computer, Dec. 2006, pp. 26–29 (4).

[12]S. Thrun et al. Stanley: The Robot that Won the DARPA Grand Challenge, Journal of Field Robotics, vol. 23, no. 9, 2006, pp. 661–692 (32).

Figure 17.2: DARPA Grand Challenge 2005 competitors. Photos courtesy of DARPA

from automotive industry partners. Of course, this makes it impossible for international universities or smaller research groups to participate.

Still, similar problem can be studied with a much cheaper setup, like the *not so Grand Challenge* student competition.[13] A GPS-based navigation task is to be solved with small robot vehicles over a track on the university campus (Figure 17.3).

The automotive industry has been reluctant to release fully autonomous driving systems, although several implementations approach market readiness. This is mainly due to liability issues and the fear of lawsuits following accidents with autonomous driving systems. Who would be liable in the case of an accident with an autonomous driving system? Since it cannot be the (non-) driver, as he or she is not in control of the vehicle, liability would default to the manufacturer.

As a consequence, the automotive industry has concentrated on developing advanced driver-assistance systems (ADAS). These systems perform exactly the same tasks as an autonomous driving system, but only interfere with the automobile's controls in certain situations, such as:

Figure 17.3: Robots and teams in the "not so Grand Challenge" at UWA

---

[13]T. Bräunl, J. Baltes, *Introducing the "not so Grand Challenge"*, 2005, online: http://robotics.ee.uwa.edu.au/nsgc/.

- Lane keeping on highways (relatively simple environment)
- Stop and go traffic (low speeds)
- Automated parking (low speed and simple environment)
- Emergency braking

Other low-level driver-assistance systems also interfere with the vehicle's driving controls. After the established ABS (anti-lock braking system) and ESP (electronic stability program), intelligent cruise control lets the driver not only set a desired speed, but also a desired minimum distance to the car in front. Whenever the actual distance to the car in front goes below this minimum distance, the car is automatically slowed down. Most of today's intelligent cruise control systems are based on radar sensors, which are considered more reliable than vision systems under all weather conditions. However, combined vision-radar systems are being used as well.

17.2 Drive-By-Wire and Safety Systems

If you have watched the popular educational TV series MythBusters on the Discovery Channel or an affiliated network, then you already know the *quick-and-dirty* way for converting a standard car to a robot car. On several episodes, Jamie Hyneman and Adam Savage have modified a standard car by adding industrial strength remote control servos to accelerator pedal, brake pedal, and steering (e.g. the *jet car* in Pilot 1 or the *pole-vaulting car* in Episode 27). This effectively makes a standard car a large version of a remote control toy car. By replacing the remote-control receiver with an embedded system and adding appropriate sensors, this would result in a low-cost autonomous car.

But if your life were at stake, such as riding as a passenger in an autonomous car, you would not choose this method. The standard way automotive manufacturers and research institutes modify cars for autonomous control is much more reliable, but unfortunately also much more expensive and requires privileged information that car manufacturers usually are not willing to share. For making a standard car autonomous, we have to interface an embedded system to drive the car *by-wire*, which means actuating the car's accelerator, brake, and steering (Figure 17.4). To be completely autonomous, the selector of the automatic gearbox (park, drive, neutral and reverse) needs to be actuated as well.

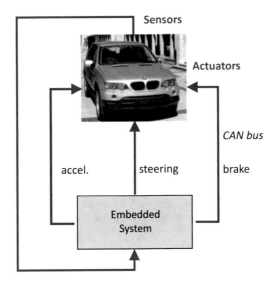

Figure 17.4: Making a car autonomous

Steps for modifying a standard car for drive-by-wire are:

1. Accelerate-By-Wire

Interfacing to the car's accelerator is fairly easy. All modern cars have an electronic accelerator pedal. Dual sensors are built into the pedal for redundancy and send data to the motor control system via a small embedded controller as a CAN bus message.

So, all that is required to connect the accelerator to an embedded system, is to interface to the car's CAN (or FlexRay) bus and send the right commands with the right timing. Unfortunately, there are no unique CAN commands for pressing the accelerator pedal between different car companies, so this needs access to privileged information.

2. Brake-By-Wire

Interfacing to the car's braking system is more difficult. Although embedded systems in modern cars already link to the braking system, such as ABS and ESP sub-systems, legislation currently still prohibits a full brake-by-wire system for safety concerns. As a consequence, modern cars still have a physical (in this case hydraulic) link between the brake pedal and the braking system, although power brakes are a standard on most new cars.

If a limited brake force is sufficient, then the same interfacing technique as for the accelerator pedal can be applied, sending CAN bus signals that mimic ABS or ESP data. Again, privileged information is required.

3. Steer-By-Wire

Modifying a car's steering for autonomous control is the most challenging task of the three actuators, as steering wheel and steering column are rigidly connected to a rack-and-pinion steering mechanism. Steer-by-wire is now implemented in many new vehicles to allow automated parking. Future steer-by-wire cars can free up space occupied by the steering column and allow a redistribution of components in the engine area as well as the driver position. However, all drive-by-wire systems do require secondary (and sometimes even tertiary) control systems to allow a fail-safe operation. And again, privileged information is required.

17.2.1 Drive-By-Wire

To build an autonomous electric Formula-SAE car (Society of Automotive Engineers), we need to implement a drive-by-wire system for accelerator, brakes and steering (see Figure 17.5). For practical reasons, we wanted the car to remain manually drivable, so all these functions need to be able to switch between manual and autonomous driving mode.

Figure 17.5: REV SAE Autonomous drive-by-wire: accelerator, brake and steering

Of these three functionalities, accelerate-by-wire is the easiest to implement. The car uses a standard electronic throttle pedal, so all we need for accelerate-by-wire is an analog multiplexer that switches between the microcontroller output (autonomous) and the accelerator pedal (manual mode).

Brake-by-wire is obviously a safety–critical function. Instead of a linear actuator pushing down the brake pedal (which would interfere with the driver's foot), we use a high-powered servo that pulls back the brake pedal from the driver and thereby actuates the hydraulic brake cylinders (see Figure 17.5). An elaborate safety system checks for all kinds of emergency situations, including the loss of a heartbeat link to the ground station. If this happens, the motor power is cut, and the brakes are automatically applied until the vehicle comes to a standstill. The latest FSAE Autonomous rules go even one step further. They require the vehicle to use stored mechanical energy for breaking (after all, this a Mechanical Engineering competition), so brakes can be automatically applied even in the event of a power failure on the vehicle. This is typically achieved through a pneumatic system with a compressed air cylinder.

Steer-by-wire has been implemented with a DC motor coupled to the steering column with a belt and an absolute encoder disk. A PID control loop running on a small embedded controller moves the steering to the desired angle. While accelerator (motor output) and braking (servo output) can be driven directly with PWM outputs, the steer-by-wire requires its own embedded controller, which receives steering commands from a high-level system via a USB-serial interface.

17.2.2 Safety Systems

Any vehicle is potentially dangerous to passengers and bystanders, so a number of independent safety measures have to be implemented to guarantee a fail-safe operation, before an autonomous vehicle can even be tested. Such a system must monitor the autonomous operation and be able to either bring the vehicle to a safe stop or to conduct a safe control hand-over to the driver, if a problem is being detected.

On our autonomous SAE car, we implemented the safety system as a separate sub-system on its own embedded controller, which can directly shut down the autonomous drive system and engage the car's brakes. On the car itself, this includes the following safety features:

- Independent system status monitoring
  Automated shutdown of autonomous operation and engaging of brakes in case of any fault detection.

- Manual Override
  Immediate shutdown of autonomous operation if the driver applies the brake or tries to steer.

- Shut-off button for autonomous driving

- Emergency stop button for electric drive system

- Electronic heartbeat between the vehicle and a base station
  If the signal from the base station is lost, the vehicle will come to a controlled halt. This also includes a remote emergency stop feature from the base station.

- Geofencing
  If the vehicle leaves the assigned GPS area, it will be brought to a controlled halt.

17.3   Computer Vision for Autonomous Driving

As we have seen so far, computer vision is not necessarily the first project to work on when preparing a car for autonomous operation or even when designing a non-interfering driver-assistance system. However, computer vision may well be the most important research topic for future intelligent automotive assistance and driving systems. Already Dickmanns' first autonomous vehicle system was based on real-time computer vision and many industrial and academic research groups work on driver-assistance systems that use vision either in combination with other sensors or as the sole sensor.

The first decision to make is on the camera equipment. The options are:

- Monocular or stereo camera
- Fix-focus, adjustable focus or use of multiple cameras with different focal lengths (e.g. for near and far sight)
- Grayscale or color cameras
- Fixed-mounted camera, actuated camera or actuated mirror in front of camera

The use of a stereo camera system gives valuable depth information for all points of interest in an image and with current computer performance, the depth map can be derived in video real-time. While stereo systems are heavily used in some research centers, others use dual cameras with different focal lengths for near and far sight. The obvious advantage here is that such a system can look further ahead (with a sufficiently high resolution) than a single or stereo camera system. While a near-sight camera system can remain stationary, a far-sight camera always requires actuation, in order to stay focused on an object such as another car driving in front. Actuating a small and light mirror has a number of advantages over moving the whole camera. However, a system without any moving parts is always preferred for an eventual market introduction because of reliability and durability issues.

Grayscale cameras are sufficient for interpreting most driving scenarios. However, color cameras are required for detecting and interpreting traffic signs and traffic lights, as well as brake lights and turn lights from other vehicles.

Finally, the preferred position of the camera is behind the rear-view mirror, so it does not block the driver's view through the windshield, but still has a viewing field similar to the that of the driver. Other possibilities include positioning multiple cameras near the headlights or integrating them into the left and right mirrors. Additional cameras can be installed viewing to the left and right (important for automated overtaking, driving in inner-city traffic or automatic parking) and the rear (for automated parking or rear collision warning).

In the following sections we will present some driver-assistance projects, following the historic developments, from lane detection to car recognition and tracking. The first driver-assistance system by Dickmanns et al.[14] in 1984 had the goal to drive autonomous on a highway by detecting lane markings. Although it may seem paradoxical to start with a high-speed environment instead of inner-city traffic, it turns out that highway driving is by far the simpler problem when compared to inner-city traffic. On a highway there is no oncoming traffic, there are usually well-marked and clearly identifiable lanes,

---

[14]E. Dickmanns, A. Zapp, K. Otto, *Ein Simulationskreis zur Entwicklung einer automatischen Fahrzeugführung mit bildhaften und inertialen Signalen*, 2. Symposium Simulationstechnik, Wien, Austria, 1984, pp. 554–558 (5).

and there is limited road curvature. For inner-city traffic, there are much harder problems to solve: there is oncoming and intersecting traffic, there are bicycles, pedestrians, and there are plenty of difficult to read road markings and signs.

17.4 OpenCV and KITTI

Robotics research advances through the availability of better tools, and OpenCV (*Open Computer Vision*, open source by Intel[15]) is certainly one of them for advanced computer vision applications. Not having to program each image processing function at the lowest level frees up time for more advanced tasks and lifts robotics and autonomous vehicle applications to another level.

Our image processing framework *Improv* (Image Processing for Robot Vision, Bräunl,[16] uses the OpenCV library in an interactive tool that allows the creation of complex image processing applications of combining modules. Each module's parameters can be adjusted via slide rulers without re-compilation and results can be tested on either live camera data or pre-recorded video sequences. Figure 17.6 demonstrates an *ImprovCV* application implemented by Hawe[17] using the Hough transform of line and lane detection.

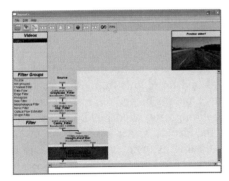

Figure 17.6: ImprovCV, implemented by Simon Hawe, UWA and TU Munich

KITTI[18] (Karlsruhe Institute of Technology and Toyota Technological Institute) is a joint project between Karlsruhe's KIT and the Toyota Tech. Institute in Chicago. It comprises a large computer vision benchmark suite for

[15]Intel, *Open Source Computer Vision Library*, 2008, online: http://www.intel.com/technology/computing/opencv/.

[16]T. Bräunl, *Improv – Image Processing for Robot Vision*, online: http://robotics.ee.uwa.edu.au/improv, 2006.

[17]S. Hawe, *A Component-Based Image Processing Framework for Automotive Vision*, Diploma/Master Thesis, Technical University München TUM, supervised by T. Bräunl and G. Färber, 2008, pp. (87).

[18]A. Geiger, P. Lenz, C. Stiller, R. Urtasun, *The KITTI Vision Benchmark Suite*, Karlsruhe Institute of Technology, 2020, online: http://www.cvlibs.net/datasets/kitti/.

autonomous driving, including ground truth from additional sensors, such as Lidar and GPS. KITTI—and other data collections like it—allow a direct comparison of the performance of different autonomous driving routines, as they can execute exactly the same data sequence in real time.

17.5   ROS

Another great package for advanced robotics and autonomous driving applications is ROS,[19] which provides a vast number of advanced robot driving, navigation and mapping functions. ROS stands for *Robot Operating System* and was originally developed by Willow Garage, the manufacturers of the PR2 robot,[20] and later taken over by the Open Robotics[21] consortium (previously called Open Source Robotics Foundation, OSRF). ROS is open source, so everyone can examine and extend the current code base.

ROS contains a large number of open-source AI components, such as SLAM (simultaneous localization and mapping) and intelligent path planning, which greatly simplify building an intelligent robot program.

The learning curve for ROS is quite steep, so you may not want to use it for small or introductory level projects. However, the rewards of using ROS for any larger system are great, especially when teams are working on various aspects of a complex autonomous driving system.

17.5.1 ROS Concepts and Core Functions

ROS is based on a number of principles and comprises a growing number of open-source libraries. The basic concepts and tools in ROS are:

- Process Nodes
  Robot program parts are implemented as *nodes* running in parallel.
- Communication System
  Data is exchanged between *ROS Nodes* with the help of a publish–subscribe message passing mechanism
- Recording file format *rosbag*
  The concept of a *rosbag* allows the recording of a real or simulated robot experiment and its playback, e.g. in order to re-test a previous experiment with improved program code.
- Visualizer *rviz*
  rviz can graphically display the position and orientation of a robot in its environment, together with a visual representation of its sensors data (e.g. Lidar).

---

[19]ROS, *Robot Operating System*, online: https://www.ros.org.

[20]Willow Garage, *PR2 Overview*, online: http://www.willowgarage.com/pages/pr2/overview.

[21]Open Robotics, *Powering the world's robots*, online: https://www.openrobotics.org.

- User Interface Tool *rqt*
  Based on the Qt framework, *rqt* can be used to develop ROS-specific user interfaces.
- Simulator *Gazebo*
  Gazebo is a 3D robot simulator that is now heavily used with ROS. Originally developed by the University of Southern California, today the Open Source Robotics Foundation (OSRF) coordinates the further development of Gazebo (see Figure 17.7).
- Robot Description *URDF*
  The Unified Robot Description Format (URDF) allows the definition for arbitrary mobile robots and robot manipulators. Many commercial robot manufacturers provide URDF files plus ROS device drivers for their robot systems.

ROS requires Ubuntu as its operating system and its language bindings are either C++ or Python. There are no versions for Windows or MacOS. Even an introduction to ROS can easily fill a whole book, so here we refer to the many available online tutorials, such as:

```
https://wiki.ros.org/ROS/Tutorials
```

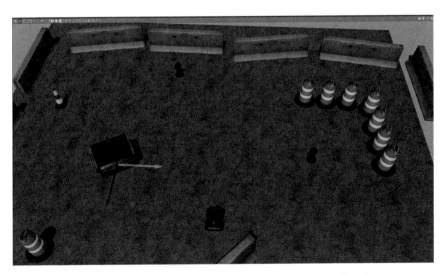

Figure 17.7: ROS example scene with Gazebo visualization tool

Sensor data processing, especially image processing is often handled via separate libraries outside of ROS. Especially for image processing, the open source library OpenCV (see chapter on image processing) is in widespread use:

```
https://opencv.org
```

17.5.2 ROS Packages

In addition to its core functions, ROS comprises a growing number of open-source libraries, including advanced implementations of SLAM (Simultaneous Localization And Mapping). Some of these useful ROS packages are:

- Gmapping[22]
  An early Lidar-based SLAM implementation.

- Hector_slam[23]
  An alternative SLAM implementation.

- Cartographer[24]
  A SLAM implementation in 2D and 3D that can achieve real-time loop closure.

- ORB-SLAM2[25]
  Vision-based SLAM algorithm for monocular, stereo and RGB-D image sensors.

- move_base[26]
  Motion planning algorithm for local robots.
- sbpl[27] (search-based planning)
  A global motion planning package for either robot manipulator arms (sbpl_arm_planner) or mobile robots (sbpl_lattice_planner).

- teb_local_planner[28]
  A local planning algorithm to navigate from waypoint to waypoint using elastic bands.

- amcl[29]
  A probabilistic localization software for a 2D robot using adaptive particle filters for sampling the robot's pose, based on Lidar data.

---

[22]Gmapping, online: http://wiki.ros.org/gmapping.

[23]Hector_slam, online: http://wiki.ros.org/hector_slam.

[24]Cartographer, inline: http://wiki.ros.org/cartographer.

[25]ORB-SLAM2, online: http://wiki.ros.org/orb_slam2_ros.

[26]Move_base, online: http://wiki.ros.org/move_base.

[27]sbpl, online: http://wiki.ros.org/sbpl.

[28]teb, online: http://wiki.ros.org/teb_local_planner.

[29]Amcl, online: http://wiki.ros.org/amcl#amcl-1.

17.6 Carla Simulator

Carla[30] is an open-source simulation system for autonomous driving that runs under Windows and Linux. Carla is built on top of the game platform *Unreal Engine*[31], including its physics simulation package, and uses the OpenD-RIVE[32] standard for constructing roads and scenes. This allows the construction of realistic looking street scenarios and provides a set of nicely rendered, cars, houses and vegetation. Carla's API provides bindings for Python and C++ and there is a *ROS Bridge* to link up Carla with ROS-based driving applications.

Figure 17.8, left, shows a Carla scene of the UWA campus with Winthrop Hall in the foreground. Figure 17.8, right, shows a more detailed scene with a model of the *n*UWA*y* autonomous electric shuttle bus. We have modeled the full university campus plus adjoining streets with all relevant buildings, roads and vegetation and use this for algorithm development and testing of our autonomous vehicle software before we deploy it in the real vehicle.

Figure 17.8: Carla scene of UWA campus (left) and rendering of nUWAy shuttle bus (right) by Y. Chen, G. Chen, G. Mendoza, Y. Shi, S. Arowosafe and T. Bräunl at UWA

A big plus of Carla is its ability to simulate Lidar sensors as shown in Figure 17.9. Since many autonomous vehicle algorithms are still based on Lidar sensors, this is an essential feature for any autonomous driving simulation package. Carla further allows multiple vehicles driving in a scene and provides simple movement rules for passive, i.e. not sensor-driven, vehicles. It further provides graphics for pedestrians and rule-based software for their movement.

---

[30]Carla, *Open-source simulator for autonomous driving research*, online: https://carla.org.

[31]Unreal Engine, online: https://www.unrealengine.com/.

[32]OpenDRIVE standalone mode, online: https://carla.readthedocs.io/en/latest/adv_opendrive/.

7.6 Carla Simulator

Figure 17.9: Lidar simulation in Carla by Y. Chen, G. Chen, G. Mendoza, Y. Shi, S. Arowosafe and T. Bräunl at UWA

In our simulation setup, we use two different computer systems for the Carla simulation: a standard PC with a high-end Nvidia graphics card and a copy of the actual drive computer of the autonomous vehicles (Nvidia Xavier embedded GPU). With this, we implement a HIL-simulation (hardware-in-the loop), so the embedded controller receives its sensor data from Carla and sends its drive commands back to Carla. The embedded controller does not know whether it is driving the real or the simulated vehicle. The hardware setup is shown in Figure 17.10, details can be seen at Brogle et al.[33]

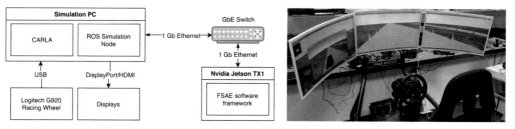

Figure 17.10: Hardware setup for Carla simulation as schematics and actual system by C. Brogle, UWA

---

[33]C. Brogle, C. Zheng, K. Lim, T. Bräunl, *Hardware-in-the-Loop Autonomous Driving Simulation without Real-Time Constraints*, IEEE Transactions on Intelligent Vehicles, vol. 4, no. 3, 2019, pp. 375–384 (10).

17.7 Lane Detection

The first driver-assistance systems developed, both academically by UBM/Dickmanns in 1984 and commercially by DaimlerChrysler in 2001, were lane detection or lane keeping systems. A possible method for finding lane information from automotive image sequences using straight line segments is the following:

1. Edge detection (and possibly thinning)
2. Line detection (e.g. using the Hough transform)
3. Deleting short and stray lines
4. Matching lines to lanes

In the following, we describe a method that uses straight line segments, implemented by Zeisl[34] at TU Munich. This is simpler than using more advanced curve models, such as splines[35] or clothoides, but also has some limitations, especially regarding the maximum detectable lane curvature.

For a straight road section, lane markings are parallel lines on the ground. In the image, however, with perspective distortion from the driver's point of view, all lane markings intersect in one point, the vanishing point. This property can be used to find position and orientation of lane markings in an image frame (Figure 17.11).

Figure 17.11: Lane detection

[34]B. Zeisl, *Robot Control and Lane Detection with Mobile Phones*, Bachelor thesis, Technical University München TUM, supervised by T. Bräunl and G. Färber, 2007, pp. (93).

[35]Y. Wang, D. Shen, E. Teoh, *Lane detection using spline model*, Pattern Recognition Letters, vol. 21, no. 8, 2000, pp. 677–689 (13).

Road and lane markings show a huge variety of shapes, which makes it difficult to use a single feature extraction technique. Edge-based techniques work well with solid and segmented lines, see Kasprzak and Niemann.[36] However, this method will fail if an image contains many lines not representing lane markings, so splitting the image into a foreground and a background range is helpful.

An advanced method is to take the expected direction of lane markings into account in the filtering procedure. Steerable filters offer such a tool to tune the edge filter in the direction of the expected lane orientation, as shown by Freeman and Adelson.[37] Adaptive road templates build upon a matching of current road scenes with predefined textures. The method will fail if the assumption of a constant road surface texture does not hold. However, it is usable for the far field of a road scene, where lane markings are difficult to identify by other methods, as shown by Kaske et al.[38] Statistical criteria such as energy, homogeneity and contrast can be used as well to distinguish between road and non-road areas. This approach of lane boundary detection especially addresses the peculiarities of country roads, where other methods might fail because of fuzzy road boundaries.

17.7.1 Edge Detection

There is a large number of different edge detection methods that could be used for this pre-processing step. Here, we compare a modified (mirrored) Sobel filter with first- and second-order *steerable filters*.

All edge filters search for grayscale discontinuities in images, therefore they will detect dark-to-bright transitions as well as bright-to-dark transitions. In a single horizontal scan line, this will result in an inner and an outer edge for each lane marking (see Figure 17.12, bottom).

When applying the simple Sobel filter, we want to avoid this problem at the lowest possible level, rather than having to post-process all edge images later. So we want the edge filter to return only the narrower inner edges for each lane marking. We modified the Sobel operator to only return bright-to-dark edges for the left half of the image, and only dark-to-bright edges (mirrored filter) for the right half of the image. This will effectively only return the inner edges for lane markings.

An alternative method to achieve is are steerable filters. When applying a steerable filter set to an image frame, the image has to be split into several segments (see Figure 17.12, top). For each of these segments, one dedicated

---

[36]W. Kasprzak, H. Niemann, *Adaptive Road Recognition and Ego-state Tracking in the Presence of Obstacles*, International Journal of Computer Vision 28(1), 526 (1998) Kluwer Academic Publishers, vol. 28, no1, 1998, pp. 5–26 (22).

[37]W. Freeman, E. Adelson, *The design and use of steerable filters*, IEEE Transactions on Pattern Analysis and Machine Intelligence, vol. 13 no. 9, 1991, pp. 891–906 (16).

[38]A. Kaske, D. Wolf, R. Husson, *Lane boundary detection using statistical criteria*, International Conference on Quality by Artificial Vision, QCAV'97, Le Creusot France, 1997, pp. 28–30 (3).

orientation is being defined, matching the typical expected lane angles for that particular image part (see Figure 17.12, middle, for second-order steerable filters). A comparison of the image preprocessing methods, mirrored Sobel and first-order steerable filter, is presented in Figure 17.13. For the examples shown, the first-order steerable filter gives the best results. All lane markings are detected well and only few other edges occur in the filtered image.

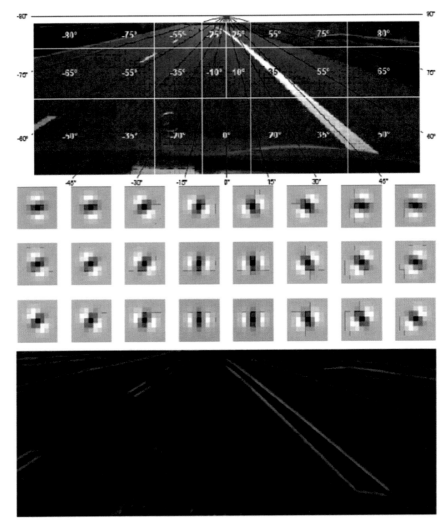

Figure 17.12: Steerable filters applied to lane markings by Zeisl, TU Munich

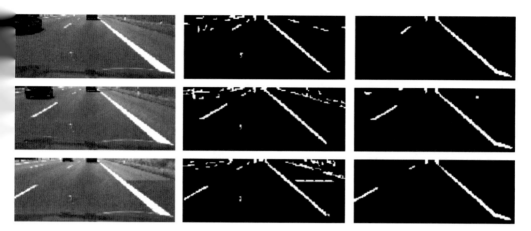

Figure 17.13: Comparison of original, mirrored Sobel, steerable filter by Zeisl, TU Munich

17.7.2 Image Tiling

The next step is to find and extract lines from the filtered binary images. Our goal is to find a scalable line-detection algorithm, which is suitable for implementing on embedded systems. Ideally, this should avoid using the compute-intense Hough transform.

Our novel approach to this task is to divide the image into several square tiles, which are then being processed independent of each other. This means the algorithm can easily cope with images of different sizes or resolutions and individual tiles can be processed in parallel, either by multiple processors or by reconfigurable hardware. With the appropriate tile size chosen, most tiles will contain only a single image line.

For each tile, the center point of all line pixels is calculated, and the line orientation is determined through the variance of the image tile. After discarding outliers, detected lines are clustered to find a set of lines representing the lane markings. If a tile contains pixels belonging to just one line, then its local centroid matches exactly with the global line. The centroid (marked as a gray dot in Figure 17.14) is given by the first moment of the tile and its coordinates can be calculated as:

$$x_M = \frac{\sum_x \sum_y x \cdot b(x,y)}{\sum_x \sum_y b(x,y)} \qquad y_M = \frac{\sum_x \sum_y y \cdot b(x,y)}{\sum_x \sum_y b(x,y)}$$

Due to interferences of two or more discontinuous features, the calculated centroid might not match with a point on a line as can be seen in Figure 17.14, top, row 2, column 5. A subsequent cleaning process is required to detect and remove such outliers.

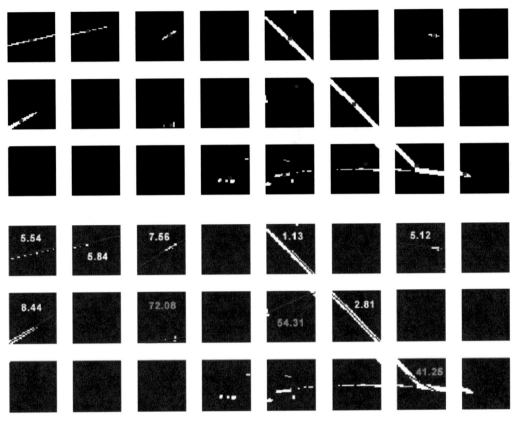

Figure 17.14: Image tile centroids (top) and vanishing point distances (bottom), B. Zeisl, TU Munich

To be able to find the direction vector of the line, we decided to perform a Principal Component Analysis for each remaining tile. The computation of the principal axes of a binary object can be easily done, implementing an Eigenvalue decomposition of the covariance matrix for every tile. Because the covariance matrix is a symmetric matrix, both Eigenvalues are positive. They describe the variances of the binary tile along the major and minor principal axes. The Eigenvector belonging to the greater Eigenvalue is pointing in the direction of the greatest variance:

$$\sigma_{xx}^2 = \sum_{x'} \sum_{y'} x'^2 \cdot b(x', y')$$

$$\sigma_{xy}^2 = \sum_{x'} \sum_{y'} x' \cdot y' \cdot b(x', y')$$

$$\sigma_{yy}^2 = \sum_{x'} \sum_{y'} x'^2 \cdot b(x', y')$$

$$\lambda_{1,2} = \frac{\sigma_{xx}^2 + \sigma_{yy}^2}{2} \pm \sqrt{\left(\frac{\sigma_{xx}^2 + \sigma_{yy}^2}{2}\right)^2 - \left(\sigma_{xx}^2 \cdot \sigma_{yy}^2 - \left(\sigma_{xy}^2\right)^2\right)}$$

$$q_1 = \begin{bmatrix} \lambda_1 - \sigma_{yy}^2 \\ \sigma_{xy}^2 \end{bmatrix} = k_1 \cdot \begin{bmatrix} \sigma_{xy}^2 \\ \lambda_1 - \sigma_{xx}^2 \end{bmatrix} \qquad q_2 = \begin{bmatrix} \lambda_2 - \sigma_{yy}^2 \\ \sigma_{xy}^2 \end{bmatrix} = k_2 \cdot \begin{bmatrix} \sigma_{xy}^2 \\ \lambda_2 - \sigma_{xx}^2 \end{bmatrix}$$

As a result, a tile like the top left in Figure 17.14, bottom, containing only a single line, has a high variance in the direction of the line and a low variance in the direction normal to the line. Therefore, the ratio of its Eigenvalues will be high. We use this property to eliminate all tiles with a ratio below a certain threshold. We also delete tiles without any lines in it, but we are still considering tiles with a thick line that results in a lower ratio.

As an image frame of a road scenery shows a perspective distortion, all lines representing lane markings intersect in one point, called the vanishing point. For further elimination of incorrectly detected line segments, we use their minimum line distance to the vanishing point. Lines with a large distance to the vanishing point are unlikely to represent lane markings and are therefore discarded.

As in most images of a road scenery the vanishing point lies in the middle of the top border, we assume this position to initialize the first frame of an image sequence. For all subsequent image frames we calculate the vanishing point dynamically by using a least square optimization, intersecting all qualifying line segments from the previous image frame. This requires detection of at least two lane marking lines in the previous image, otherwise the previous vanishing point is retained.

17.7.3 Line Segment Clustering

So far lines segments are only described locally by the center of mass in reference to the tile in which they appear and by an Eigenvector in direction of the line. We are not using a global Hough transform in order to significantly reduce the computational effort, so we need a different algorithm to merge local line segments to global lines.

Using the Moore–Penrose pseudo-inverse, we transform each tile's line equation to the form:

$$a \cdot x + b \cdot y + c = 0$$

As each lane marking, whether continuous or dashed, is likely to span over several image tiles, we will have multiple parameter sets (one from each tile) representing the same line. For clustering of line segments in the image, we match their respective parameter triplets $(a,\ b,\ c)$ in a three-dimensional parameter space. The distance function of parameter triplets is calculated as:

$$d = (I_1 - I_n)^T \cdot (I_1 - I_n) = (a_1 - a_n)^2 + (b_1 - b_n)^2 + (c_1 - c_n)^2$$

Parameters a and b lie in the range $[-1, +1]$, while parameter c has a value range about 100 times greater. To equally weight the parameters, their values would have to be scaled considering their statistical distribution. The optimal approach would require choosing the weights so that the variance of the different parameters is the same. However, this implies more computational effort without significantly improving the clustering. For our application it is sufficient to scale parameters a and b by a constant, matching their value range to c.

Figure 17.15 shows each line segment as a point in 3D space, represented by its scaled line parameters a, b, c. Line segments from different tiles representing the same lane marking are close together, while different lane markings are clusters significantly further apart.

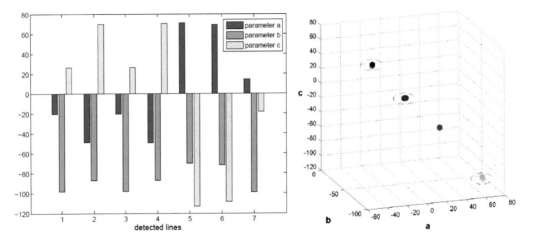

Figure 17.15: Parameter triplet matching by Zeisl

If the distance is below a certain threshold, the lines will be clustered. After each iteration, all lines in the new cluster are removed from the set. The cluster itself is represented by calculating the mean parameter values of the lines included. If a cluster contains only a single line, it is discarded. Hence, it is necessary to find at least two similar line segments in different tiles to detect a lane marking. With this additional restriction, incorrect lines and outliers are rejected, but we might also miss correct lines, especially for poorly painted or dashed lane markings.

To improve algorithm performance, we use temporal coherence by including lanes found in the previous image frame in the clustering process. The rationale for this is that lane markings do not change abruptly between successive image frames but move gradually. This modification allows us to detect lane markings that are supported by only a single tile but have also appeared in the previous image frame.

Figure 17.16 shows the lane detection results. The algorithm is efficient enough to run on an embedded vision system or even on a mobile phone.

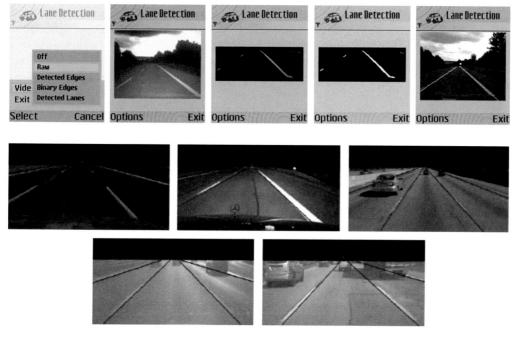

Figure 17.16: Lane detection results and mobile phone implementation

Further methods for lane detection can be found in Dickmanns, Mysliwetz[39]; Kreucher and Lakshmanan[40]; Yim and Oh[41]; and McCall and Trivedi.[42]

[39]E. Dickmanns. B. Mysliwetz, *Recursive 3-D road and relative ego-state recognition*, IEEE Transaction on Pattern Analysis and Machine Intelligence, vol. 14, 1992, pp. 199–213 (15).

[40]C. Kreucher, S. Lakshmanan, *LANA: a lane extraction algorithm that uses frequency domain features*, IEEE Transactions on Robotics and Automation, vol. 15, no. 2, 1999, pp. 343–350 (8).

[41]Y. Yim, S. Oh, *Three-feature based automatic lane detection algorithm (TFALDA) for autonomous driving*, IEEE Transactions on Intelligent Transportation Systems, vol. 4, no. 4, 2003, pp. 219–225 (7).

[42]J. McCall, M. Trivedi, *Video Based Lane Estimation and Tracking for Driver Assistance: Survey, Systems and Evaluation*, IEEE Transactions on Intelligent Transportation Systems, vol. 7, no. 1, 2006, pp. 20–37 (18).

17.8   Vehicle Recognition and Tracking

After lane keeping, the next logical step is the recognition and tracking of other vehicles in traffic, in order to detect and possibly avoid hazardous situations. Daimler, BMW, and Jaguar were the first to introduce radar-based *adaptive cruise control* systems in 2001. These systems will override the set speed of the cruise control if the vehicle in front comes closer than a preset minimum distance. Consequently, most driver-assistance systems developed after 2001 make dual use of the radar information to initialize, identify or track other vehicles. While these systems rely on a working radar system to function, we will present a vehicle recognition and tracking system that is solely based on image processing. Our system can work without radar information, however, if radar information is present, it can be used to improve results through sensor fusion.

Our algorithm for vehicle detection is based on symmetry properties of cars' rear views and uses the following steps:

- Horizon detection
- Spatial feature clustering using optic flow
- Lane detection for reduction of search area
- Elimination of lane marking features
- Temporal feature clustering
- Determining of vehicle center point via symmetry properties
  - Compact Symmetry Operator
  - Generalized Symmetry Transform
- Vehicle extraction and car fitting for fine adjustment

Figure 17.17 shows the first three steps of the algorithm. The original image is clipped at the horizon line (top) in order to restrict image information to areas that potentially have cars in them. Next, we perform a lane detection (middle) and a feature-based tracking using optic flow (bottom). This already gives us moving features, but it remains uncertain whether these features actually belong to a car or to some other object (e.g. a traffic sign) and there are many separate feature tags instead of a single one per car.

17.8.1 Symmetry Operators

The rear view of a car is usually highly symmetrical about a vertical axis. The license plate is in the middle and we have lights on either side of the car. Even if we see the car at a slight angle on a curved road, this skewed symmetry is

7.8 Vehicle Recognition and Tracking

Figure 17.17: Lane detection and feature tracking in optic flow

usually sufficient for a vehicle detection. We tried a number of symmetry operators and found that a combination of the Compact Symmetry Operator[43] and the General Symmetry Transform[44,45] gives best results (see Bourgou[46]).

The compact symmetry operator (Figure 17.18, left) works directly on a scan line, so it is very easy to implement, but will give inferior results if the camera's and the car's horizontal axes are not lined up. The formula expressing the symmetry property at point pi with search window size m is:

[43]K. Huebner, *A 1-Dimensional Symmetry Operator for Image Feature Extraction in Robot Applications*, 16th International Conference on Vision Interface (VI'03), 2003, pp. 286–291 (6).

[44]D. Reisfeld, H. Wolfson, Y. Yeshurun, *Context-Free Attentional Operators: The Generalized Symmetry Transform*, International Journal of Computer Vision, vol. 14, 1995, pp. 119–130 (12).

[45]I. Choi, S. Chien, *A Generalized Symmetry Transform with Selective Attention Capability for Specific Corner Angles*, IEEE Signal Processing Letters, vol. 11, no. 2, Feb. 2004, pp. 255–257 (3).

[46]S. Bourgou, *Objekterkennung und Tracking für autonome Fahrzeuge*, Bachelor thesis, Technical University Munich TUM, supervised by T. Bräunl and G. Färber, 2007, pp. (41).

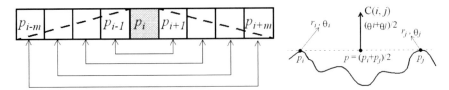

Figure 17.18: Compact symmetry (left) and general symmetry (right)

$$ComSym(p_i, m) = 1 - \frac{1}{m \cdot Maxgray} \sum_{j=1}^{m} \left(1 - \frac{|j|}{m+1}\right) \cdot \left\| P_{i-j} - P_{i+j} \right\|^2$$

This means we are only looking at image gray values directly and sum up weighted differences of pixel pairs equidistant from the center p_i. The further away a pixel pair is from p_i, the less its differences subtract from the symmetry score. The total number of pixel pairs considered equals the window size m.

In contrast to this, the more complex general symmetry transform can work with any symmetry axis, but we have simplified it to detect symmetries about a vertical axis only (Figure 17.18, right). This symmetry operator works on edges instead of on raw grayscale image data and it takes into account edge strength and edge direction. The symmetry score at point p is a combination of edge distance, phase difference (edge direction difference) and edge strengths. The following formula is used for points p_i and p_j to the left and right of the symmetry center p with their respective edge strengths r_i, r_j and edge directions θ_i, θ_j:

$$GenSym(p) = \sum_{i,j} f \cdot e^{-\frac{\left\| P_1 - P_j \right\|}{2\sigma}} \cdot (1 - \cos(\theta_i + \theta_j)) \cdot (1 - \cos(\theta_i - \theta_j)) \cdot r_i \cdot r_j$$

Figure 17.19, bottom, shows the results of the car extraction algorithm, which uses the edge image of the area surrounding the symmetry points. The car fitter algorithm by Betke et al.[47] scans the edge histogram around a symmetry point until a threshold is reached. This cut-out rectangular area is then the approximation of the vehicle area. The following formulas are used for this:

---

[47]M. Betke, E. Haritaoglu, L. Davis, *Multiple Vehicle Detection and Tracking in Hard Real Time*, Computer Vision Laboratory, Center for Automation Research, Institute for Advanced Computer Studies, Technical Report CS-TR-3667, University of Maryland, College Park, July 1996.

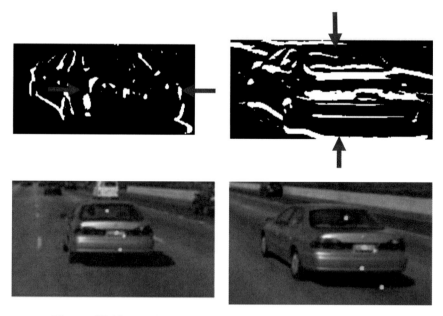

Figure 17.19: Finding car symmetry centers and car extraction

$$V = (v_1, \ldots, v_{height}) = \left(\sum_{i=1}^{width} Sobel_H(x_i, y_1), \ldots, \sum_{i=1}^{width} Sobel_H(x_i, y_{height}) \right)$$

$$W = (w_1, \ldots, w_{weight}) = \left(\sum_{j=1}^{height} Sobel_V(x_1, y_j), \ldots, \sum_{j=1}^{height} Sobel_V(x_{width}, y_j) \right)$$

$$\Theta_v = \frac{1}{2} \cdot \max\{v_i | 1 \leq i \leq width\} \quad \Theta_w = \frac{1}{2} \cdot \max\{w_j | 1 \leq j \leq height\}$$

17.8.2 Vehicle Tracking

The final step is tracking the detected vehicle over subsequent image frames. This will make use of temporal coherance as cars tend to be at very similar image coordinates in subsequent image frames. A full image scan is conducted at a much lower frequency in order to reduce the computational effort, but is still necessary to detect new cars coming into the field of view.

A template matching is conducted with the original car region found by the car tracker from the previous step. For the matching, a correlation coefficient method from the OpenCV library is used.

The rear view of a car changes from image frame to image frame because of a number of factors, e.g. the distance to it gets shorter or larger or the car turns in a curved section of the road or the general lighting conditions change. Because of this, we have to dynamically change the template in every step of the correlation-based matching function, in order to ensure the most recent model for the car is used for the matching process.

Figure 17.20 shows an image frame from the tracking process and its corresponding ImprovCV implementation.

Figure 17.20: Vehicle tracking: image frame (Bourgou, left) and ImprovCV (Hawe, right)

17.9   Automatic Parking

Parking aids have been introduced in luxury vehicles a number of years ago and have since made their way into smaller cars and after-market systems. While simpler systems only measure the distance between the front and rear bumper of a vehicle to an obstacle, more complex systems can automatically park a car on the press of a button.

Sonar and radar sensors have been the choice for most commercial parking aid systems, while Lidar sensors have been used mainly in research applications. In the patent by Bräunl and Franke[48] a camera-based approach is

---

[48]T. Bräunl, U. Franke, *Method and device for the video-based monitoring and measurement of the lateral environment of a vehicle – Verfahren und Vorrichtung zur videobasierten Beobachtung und Vermessung der seitlichen Umgebung eines Fahrzeugs*, Patent application – Schutzrechtsanmeldung 102 44 148.0-32, submitted 23. Sep. 2002, DC Akte P1 12799/DE/1, in cooperation with Daimler Research Esslingen/Ulm, March 2003, Submitted as international patent in Europe, Japan, and USA, 23. Sep. 2003, online: http://v3. espacenet.com/textdoc?DB=EPODOC&IDX=WO2004029877&F=0&QPN=WO2004029877, Daimler internal report, June 2001.

described that can serve as an alternative low-cost sensor, providing significantly more accurate and detailed information than sonar sensors (Figure 17.21).

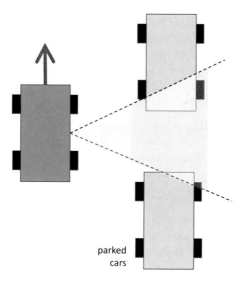

parked
cars

Figure 17.21: Camera-based automatic parking by Bräunl and Franke

The principle used in this patent application is to apply stereo processing to subsequent images of a moving monocular camera (*motion stereo*). The camera is mounted perpendicular to the vehicle's driving direction, monitoring the right curb for potential parking space (or the left curb in countries driving on the left side of the road). While the vehicle (and therefore the camera) is in motion, subsequent image frames have a similar but variable stereo baseline compared to image pairs captured with a stereo camera pair. The baseline width depends on the speed of the vehicle and the frame rate of the camera. An online stereo-matching process can reproduce the car surroundings in 3D. It can then either advise the driver on whether a potential parking space is large enough, issue a warning signal when encountering obstacles during the parking process, suggest an optimal parallel parking procedure, or automatically park the vehicle if it is fitted with the necessary drive-by-wire actuators.

Quoting from the patent by Bräunl and Franke: *"The first commercially-available driver-assist systems required a compromise between the resolution of the scanning and the extent of the scanned region, according to application. Conventional video-based systems have a good compromise between resolution and recording range, however, do not generally provide direct distance information. According to the new arrangement of the object of the invention, it is possible to achieve a system, which, on installation in a road vehicle, can record complex dynamic scenes, for example the lateral 3D geometry to the road edge from the point of view of the dynamically operating vehicle and use the same to advantage on parking. According to the invention, the monitoring and measuring of the lateral environment of a vehicle is displayed on the one hand by means of a camera with the digital images and on the other hand with a computer unit which serves to provide the images with a time stamp and to buffer the same. The movements of the vehicle are further recorded, in order to select image pairs from the buffered images, based on the above data. A local 3D depth image of the lateral environment of a vehicle can thus be generated by an algorithm for stereo image processing."*

Figure 17.22 shows a sample image sequence and the pair-wise reconstructed 3D model. The individual algorithm steps are outlined as follows:

- Each input image from the camera stream needs to be marked with the corresponding vehicle odometry data and a time stamp.
- Not always are subsequent images on the camera stream selected for stereo matching, as the time stamp difference between images in a pair needs to be translated to a stereo base distance of at least 30 cm.
- A stereo matching algorithm is applied to all image pairs selected in the previous step. This produces a 3D depth map for the lateral view at a particular point in time, which translates to a particular point in space, provided the vehicle's trajectory is known.
- The local 3D views (scatter plots) generated in each step of the stereo matching process are then combined in a single geometry data structure. This generates an accumulated 3D view to the side of a vehicle over a certain length.

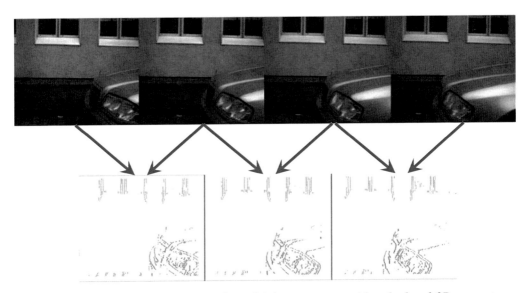

Figure 17.22: Frames from driving sequence with calculated 3D geometry

As shown in Figure 17.23 the 3D data is accumulated in a global geometry data structure using discretization or spatial sub-sampling. Due to a number of inaccuracies in the overall process, data points will not always line up creating some amount of noise (Figure 17.23, top). After applying some noise filtering techniques (Figure 17.23, middle), we subdivide the volume of interest into

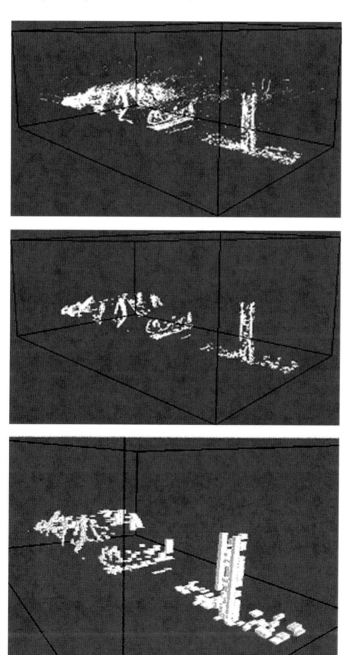

Figure 17.23: Raw and filtered voxel structures

small voxel cells and accumulate all generated geometry data into an octree of larger cubic voxel cells (matching the desired spatial resolution). Each voxel stores its number of accumulated points as its weight. A simple thresholding will then delete voxels with a low weight (Figure 17.23, bottom).

17.10 Autonomous Formula-SAE

The Renewable Energy Vehicle Project (REV[49] and Drage et al.[50]) at UWA was one of the first teams in the world to develop an autonomous race car based on a Formula-SAE[51] vehicle. SAE (Society of Automotive Engineers) holds annual competitions in several countries, where student teams present and race their self-designed and self-built one-seater race cars. Starting as a pure petrol-based competition, Formula-SAE now also comprises competitions for electric and autonomous race cars, the Formula-Student Autonomous Competition (FSAC). For this event the race track is lined with cones and the car has to autonomously drive around the track, not being given a map (see Figure 17.24).

Figure 17.24: REV-SAE-Autonomous vehicle completing a driverless lap around the race track

The vehicle is equipped with a SICK single-layer Lidar in front, an ibeo LUX multi-layer Lidar on top, a FLIR stereo camera system, an Xsens GPS-IMU combination, and student-built odometry sensors (Figure 17.25).

[49]T. Bräunl, *The REV Project*, online: http://revproject.com.

[50]T. Drage, J. Kalinowski, T. Bräunl, T. Drage, J. Kalinowski, T. Bräunl, *Integration of Drive-by-Wire with Navigation Control for a Driverless Electric Race Car*, IEEE Intelligent Transportation Systems Magazine, pp. 23-33 (11), Oct. 2014.

[51]SAE, *Formula-SAE Student Events*, online: https://www.sae.org/attend/student-events/.

Figure 17.25: REV-SAE-Autonomous vehicle with sensors for autonomous driving

The REV team's autonomous driving hardware and software was perfected over the years and included the development of a realistic hardware-in-the-loop simulation system based on the open source package Carla (see Lim et al.[52] and Brogle et al.[53]). The car's simulation model in Figure 17.26 has been directly derived from its CAD model, which was designed by Ian Hooper[54].

Figure 17.26: Simulation model of Formula SAE race car

Figure 17.27 shows a simulation run of the autonomous SAE vehicle with simulated camera view and Lidar scans.

[52]K. Lim, T. Drage, C. Zheng, C, Brogle, W, Lai, T. Kelliher, M. Adina-Zada, T. Bräunl, *Evolution of a Reliable and Extensible High-Level Control System for an Autonomous Car*, IEEE Transactions on Intelligent Vehicles, vol. 4, no. 3, 2019, pp. 396–405 (10).

[53]C, Brogle, C. Zheng, K. Lim, T. Bräunl, *Hardware-in-the-Loop Autonomous Driving Simulation without Real-Time Constraints*, IEEE Transactions on Intelligent Vehicles, vol. 4, no. 3, 2019, pp. 375–384 (10).

[54]I. Hooper, *Development of in-wheel motor systems for Formula SAE electric vehicles*, Master by Research thesis, supervised by T. Bräunl, 2011, pp. (149).

Figure 17.27: Carla simulation of autonomous SAE car by Craig Brogle UWA/REV

A much simpler simulation is possible using the EyeSim[55,56] simulator. This does not require ROS and Ubuntu and thanks to the Unity 3D engine runs on almost any platform, including MacOS, Windows and Linux. Figure 17.28 shows the simulated vehicle on the track lined with cones, while the inserted image in this Figure shows the controller's display with the simulated camera image, left, and the simulated Lidar data, right. The positions of the four next cones (two left and two right of the center line) can be clearly seen.

Figure 17.28: EyeSim simulation of autonomous SAE car

A simple algorithm based on cone positions can now be used as shown in Program 17.1. Functions *getleftcone* and *getrightcone* find the nearest cone

[55]T. Bräunl et al., *EyeSim VR –EyeBot Mobile Robot Simulator*, 2021, online: https://robotics. ee.uwa.edu.au/eyesim/.
[56]T. Bräunl, *Robot Adventures in Python and C*, Springer Nature, Switzerland, 2020, pp. (183).

form the middle Lidar position. The *main* program then calculates the desired steering angle based at the middle between the next cones. Driving is always done at a constant speed.

Program 17.1: Autonomously Driving Around a Cone Track

```
int getleftcone(int a[], int mid)
{ int i;
  for (i=mid-1; i>20; i-)
    if (a[i] < 9000) return i; // cone detected !
  return -1; // no cone
}

int getrightcone(int a[], int mid)
{ int i;
  for (i=mid+1; i<160; i++)
    if (a[i] < 9000) return i; // cone detected !
  return -1; // no cone
}

int main ()
{ int  i, m, l,r,middle=90, drive=0, dir=0;
  int  scan[POINTS];
  float scale;
  BYTE img[QQVGA_SIZE];
  CAMInit(QQVGA);
  LIDARSet(180, 0, POINTS); // range, tilt, points
  MOTORDrive(1, SPEED);
  do
  { LIDARGet(scan);
    m = getmax(scan);
    scale = m/150.0;
    l = getleftcone (scan, middle); // left-most cone
    r = getrightcone(scan, middle); // right-most cone
    if (l>0 && r>0 && l<r)
    { middle = (l+r)/2;        // middle position of [0..POINTS]
      dir = (POINTS/2 - middle);  // range +/-POINTS/2 * factor
      SERVOSet(1, 128+dir);     // 0=right, 128=middle, 255=left
    }
    // plot distances
    for (i=0; i<PLOT; i++)
    { LCDLine(180+i,150-scan[SCL*i]/scale, 180+i,150, BLUE);
      LCDLine(180+i,150-scan[SCL*i]/scale, 180+i,  0, BLACK);
    }
    LCDLine(180+middle/SCL,0, 180+middle/SCL,150, RED); // center
  } while (k!=KEY4);
}
```

While this works in the simplified simulation environment, a lot more work needs to be done to drive the real car around a cone-lined track. Cone positions are not that easy to identify, there are not always matching cones on the left and right side of the car, the vehicle's suspension lets it pitch up and down, constantly changing the Lidar's scanning angle, and so on.

17.11   Autonomous Shuttle Bus

Autonomous electric shuttle buses have become quite popular in recent years, with a number of companies providing almost identical products. Most prominent are Navya and EasyMile in France, and Local Motors in the U.S. with combined over 100 installations worldwide in 2021. These buses are quite small with only six seats and standing space for a few more; they are also quite slow, typically driving at speeds of 10–20 km/h, so not much faster than walking speed; and they require a safety officer on board, who needs to take over manual control when the vehicle's self-driving system gives up. So clearly, these vehicles cannot solve any traffic problem, they are more of a tourist attraction or a technology exhibition. But at prices around a million dollars for a vehicle with a five-year service plan, they are not exactly cheap, and they provide no insight or technology transfer to their customers or operators, as they represent a closed system. The manufacturer explores and programs the fixed path that the shuttle will drive. Any changes (e.g. driving to a different building) can only be done through the manufacturer repeating this process. The vehicles rely heavily on RTK-GPS (real-time kinematic), which allows centimeter accuracy through the help of a 4G correction signal broadcast by a local RTK server. At this stage all commercial shuttle buses follow a fixed preprogrammed route exactly and will not deviate even a few centimeters to pass a parked bicycle or just a piece of cardboard. Instead, they will stop and ask the safety operator to take over. Their movement mode resembles a train on invisible rails (see also *trackless trams*[57]).

In 2020, REV acquired a second-hand electric shuttle bus, which we called *nUWAy*. The manufacturer had erased the hard disk before handover, so we added our own hardware and software and developed our own version of a flexible, call-on-demand autonomous shuttle bus with dynamic route planning and collision avoidance (Figure 17.29).

---

[57]P. Newman, *Why trackless trams are ready to replace light rail*, The Conversation, 26 Sep. 2018, online: https://theconversation.com/why-trackless-trams-are-ready-to-replace-light-rail-103690.

Figure 17.29: nUWAy autonomous shuttle bus

The shuttle is equipped with a number of sensors:

- Four single-layer SICK lidars, one on each corner. These sensors are linked to a safety PLC which will stop driving if a possible collision is detected.
- Two Velodyne puck lidars facing front and back; these were not connected to the shuttles control PC when we received the vehicle.
- Two ibeo LUX multi-layer lidars on top of the shuttle for localization and mapping.
- Two FLIR grayscale cameras; these were originally not connected to the shuttle's control PC.
- Odometry from wheel encoders; unfortunately not accessible to us, as we do not have manufacturer support.
- Combined RTK-GPS and IMU from NovAtel.

The shuttle's drive system has the following controllers and actuators:

- Curtis motor controller for driving forward/backward,
- Curtis steering controllers for steering front and rear wheels,
- Curtis master controller,
- With the exception of the passenger door on only one side, the shuttle has a symmetric front and rear, so it can drive back without having to reverse.

Figure 17.30 shows the shuttle's outside controls and charging port (left top) and one of the two 48 V battery packs opened (right). We added some modifications to the control panel and charge port (left bottom), in order to make it automotive compliant (IEC Type-2 socket) and to have instrumentation for charge status, state of charge (SoC), current and voltage for the 48 V system as well as voltage for the 12 V system.

Figure 17.30: Shuttle bus charging controls and battery pack

The motor controllers are usually driven through CAN bus signals from the onboard PC. However, since the manufacturer did not disclose the CAN codes to us and we could not listen to drive commands, as all software had been removed from the vehicle before delivery, we had to develop an interface board to link to the safety-operator's manual joystick controller, in order to drive the vehicle from the on-board PC (see Figure 17.31).

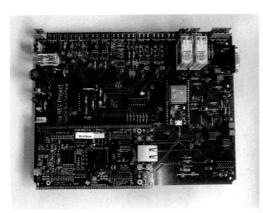

Figure 17.31: Drive control interface board developed by Thomas Drage, UWA 2021

This system uses a separate embedded controller to generate analog outputs for steering and driving. Matching the original analog joystick controller, the acceptable output ranges for steering and driving are set to 0.5–4.5 V. For the steering motors, the middle value of 2.5 V with a deadband of about ±0.5 V sets the steering to straight. For the drive system, the middle value of 2.5 V with a deadband of about ±0.5 V means the vehicle is at rest. Higher voltages will drive it forward; lower voltages will drive it backward. Note that this is a potentially dangerous setup, as a drop in the analog control voltage could let the bus drive backward at maximum speed. We compensated this with additional hardware guards on the generated voltages and an automatic shutdown in case of voltages being out of range or any error states encountered. Our interface board also includes a Bluetooth controller, which lets us interface to a wireless gamepad that we can use as our version of a manual joystick to override the autonomous control. The gamepad's joystick commands are being transmitted via Bluetooth to the interface board, which converts each signal to the corresponding analog joystick voltages.

For autonomous control, the interface board connects to the main onboard PC via a serial link. A ROS node has been created with a simple API for steering and driving, so autonomous drive commands can be issued from the main PC.

We further added a GPU-based Nvidia Xavier embedded processor plus a gigabyte switch, in order to provide the processing power for image and Lidar processing from the cameras and Velodyne Lidar sensors, which were never connected in the original shuttle.

The software stack for autonomously driving the shuttle bus was built in ROS with our own software development, making use of its many robotics packages, as listed in the previous section.

Figure 17.32 shows the main electronics panel inside the bus with the side panel removed. The green box on the right is the shuttle's central PC, the yellow controllers on the top are a quite sophisticated PLC (programmable logic controller) that controls auxiliary functions such as opening and closing of the doors (via additional controllers on top of the doors), as well as a firmware collision avoidance system. For this, the four safety Lidars from each corner of the shuttle send minimum distance values to the PLC. If any object (pedestrian, car, bicycle, wall, etc.) comes closer than a set distance, the PLC will immediately stop the shuttle.

The four single-layer SICK Lidars at the shuttle's corner also report their full distance range data to the central PC, as do the two ibeo Lux ranging Lidar sensors and the two (originally not connected) Velodyne Puck Lidars. This allows us to generate a full Lidar point cloud around the shuttle bus, merging the data from all Lidar sensors, as shown in Figure 17.33.

For mapping, we are using SLAM algorithm Cartographer[58] within ROS, which creates a map of the shuttle's driving environment as shown in Figure 17.34 top. The REV lab is located in the narrow vertical lane, where localization Lidar beams are superimposed onto the map. The larger horizontal

---

[58]ROS Cartographer, online: https://google-cartographer-ros.readthedocs.io/en/latest/.

Figure 17.32: Shuttle bus central electronics panel

Figure 17.33 Lidar point cloud from nUWAy's eight Lidar sensors by Craig Brogle UWA/REV

area in Figure 17.34, bottom, contains parking bays on the top row and internal road space to access the nearby buildings as well as the access to the public road.

Figure 17.34: SLAM map of driving area around the REV lab by Yuchen Du UWA/REV

For localization we are merging results from ROS Cartographer and the combined GPS/IMU output of a NovAtel sensor package. Before the generated maps can be used, we label drivable areas manually, as not all obstacles can be identified by Lidar or camera, e.g. empty parking bays or the mandate to stay on the left side of the road (we are in Australia). Finally, path planning is done with the ROS package *sbpl_lattice_planner*.[59] The image sequence in Figure 17.35 shows the initially planned path and then the progression of the continually updated path while the shuttle bus is in motion. Once the vehicle has reached its destination within an accuracy of about one meter, the goal has been reached. If this location is defined as a bus stop, the vehicle will remain stationary for a few seconds, to see whether passengers want to alight or enter the shuttle bus by pressing the inside or outside door buttons.

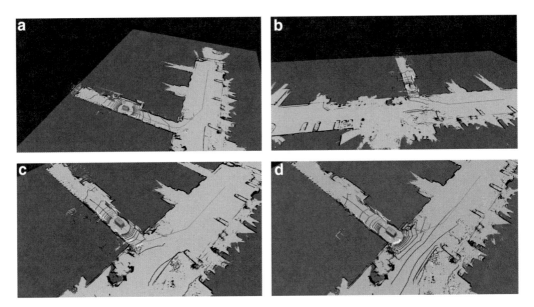

Figure 17.35: Path planning via ROS sbpl by Yuchen Du, Farhad Ahmed and Kyle Crescencio UWA/REV

The resulting navigation system is good enough for autonomously driving the *n*UWA*y* shuttle bus in a regular service on the university campus. We added a user interface that allows passengers to select desired bus stops from a map on the onboard touch screen. Via our mobile phone app, users can call the bus to their current location as an on-demand service (see Figure 17.36).

---

[59]ROS sbpl_lattice_planner, online: http://wiki.ros.org/sbpl_lattice_planner.

Figure 17.36: nUWAy user interface

17.12   Tasks

1. Implement traffic cone detection based on image processing

2. Implement traffic cone detection based on Lidar scanning.

3. Combine camera and Lidar methods for a more robust detection algorithm.

4. Detect all cones in front of an autonomous vehicle and calculate their respective 2D coordinates. Then plan a vehicle path that stays within the track lined by the cones.

5. While driving the track for the first time, feed the data into a SLAM algorithm to construct a map. Once the first lap has been completed, calculate optimized paths and speeds for faster lap times based on the generated map.

6. Explore OpenCV algorithms and the KITTI vision benchmark suite to create a lane detection algorithm.

7. Explore the ROS packages for localization and navigation, especially Cartographer, and use a combination of them for autonomous vehicle control.

PART IV
ARTIFICIAL INTELLIGENCE

AI CONCEPTS

18

With this chapter, we start a sequence of four chapters on intelligent systems. We begin with an overview of artificial intelligence (AI) and behavior-based systems, then continue with neural networks (NN), genetic algorithms (GA) and finally advanced deep learning systems.

AI technologies have achieved remarkable results that would not have been possible with traditional engineering-based software design. Maybe one of the key insights was that software design is hard and labor-intensive and therefore expensive. The idea to generate reliable software without the need to program is therefore very appealing. The two required steps of collecting training data and executing a relatively simple repetitive learning process can both be mostly automated and are therefore cheap.

18.1 Software Architecture

Most traditional robotics control systems require a large effort in software development. Any possible detail has to be foreseen and programmed in the appropriate way. Extending a working robot program, e.g., through adding a new sensor or adding a new functionality, is also very time-consuming.

This classical robot software architecture is known by many names: hierarchical model, functional model, engineering model, or three-layered model (Figure 18.1, left). It is a predictable software structure with top-down implementation. Sensor data from the vehicle is preprocessed over two levels until the highest *intelligent* level makes the action decisions. Execution of the actual driving (e.g., navigation and lower-level driving function) is left to the layers below. The lowest layer is the vehicle interface, transmitting driving commands to the robot's actuators. The dominant paradigm of this approach has been the sense–plan–act[1] (SPA) organization: a mapping from perception, through construction of an internal world model, planning a course of action based upon this model and finally execution of the plan in the real-world environment.

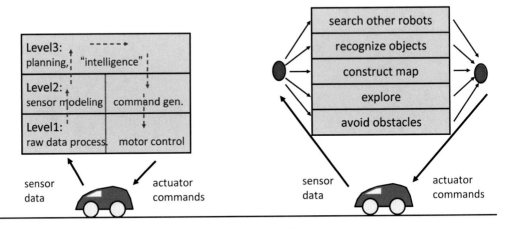

Figure 18.1: Software architecture models

The hierarchical model places emphasize on construction of a world model and planning actions based on this model. The computation time required to construct a symbolic model has a significant impact on the performance of the robot, while the disparity between the internal model and the actual environment may result in robot actions not achieving the intended effect. Behavior-based systems as described by Arkin[2] try a completely different approach, getting away from a monolithic coded robot program, toward a collection of independent robot behaviors that work together concurrently. The behavior-based model in Figure 18.1, right, is a bottom-up design that creates not easily predictable robot actions. Instead of designing large blocks of code, each robot functionality has been encapsulated in a small self-contained module, here called a *behavior*. All behaviors are executed in parallel, while explicit synchronization is not required. One of the goals of this design is to simplify extendibility, e.g., for adding a new sensor or a new behavioral feature to the robot program. While all behaviors can access all vehicle sensors, a problem occurs at the reduction of the behaviors to produce a single output for the vehicle's actuators. The original *subsumption architecture* introduced by Brooks[3] uses fixed priorities among behaviors, which results in mostly predictable reactive systems, while modern implementations use more flexible selection schemes.

---

[1]Carnegie Mellon Robotics Academy, *Sense Plan Act (SPA)*, online: http://cmra.rec.ri.cmu. edu/products/cortex_video_trainer/lesson/media_files/hp_spa.pdf.

[2]R. Arkin, *Behavior Based Robotics*, MIT Press, Cambridge MA, 1998.

[3]R. Brooks, *A Robust Layered Control System for a Mobile Robot*, IEEE Journal of Robotics and Automation, vol. 2, no.1, 1986, pp. 14–23 (7).

18.2   Behavior-Based Systems

The term *behavior-based robotics* is broadly applicable to a range of control approaches. Concepts taken from the original subsumption design have been adapted and modified by commercial and academic research groups, to the point that the nomenclature has become generic. Some of the most frequently identified traits of behavior-based architectures are:

- Tight coupling of sensing and action
  At some level, all behavioral robots are reactive to stimuli with actions that do not rely upon deliberative planning. Deliberative planning is avoided in favor of computationally simple modules that perform a simple mapping from input to action, facilitating a rapid response. Brooks[4] expressed this philosophy with the observation that *"Planning is just a way of avoiding figuring out what to do next."*
- Avoiding symbolic representation of knowledge
  Rather than construct an internal model of the environment to perform planning tasks, the world is used as "its own best model," as Brooks described it. The robot derives its future behavior directly from observations of its environment, instead of trying to produce an abstract representation of the world that can be internally manipulated and used as a basis for planning future actions.
- Decomposition into contextually meaningful units
  Behaviors act as situation–action pairs, being designed to respond to certain situations with a definite action.
- Time-varying activation of concurrent relevant behaviors
  A control scheme is utilized to change the activation level of behaviors during run-time to accommodate the task that the robot is trying to achieve.

In a behavior-based system, a certain number of behaviors run as parallel processes. While each behavior can access all sensors (read), only one behavior can have control over the robot's actuators or driving mechanism (write). Therefore, an overall controller is required to coordinate *behavior selection* or some mechanism of *behavior output merging* at appropriate times to determine the robot's actions.

Early behavior-based systems, such as the subsumption model, used a fixed priority ordering of behaviors. For example, the wall avoidance behavior always has priority over the foraging behavior. Obviously, such a rigid system is very restricted in its capabilities and becomes difficult to manage with increasing system complexity. Therefore, the goal must be to design a behavior-based system that uses an adaptive control mechanism. Such a controller would use machine learning techniques to develop the correct selection response from the specification of desired outcomes. The controller is the

---

[4]R. Brooks, *Planning is just a way of avoiding figuring out what to do next*, MIT AI Lab, working paper 303, Sep. 1987, online: https://dspace.mit.edu/bitstream/handle/1721.1/41202/AI_WP_303.pdf.

intelligence behind the system, deciding from sensory and state input which behaviors to activate at any particular time. The combination of a reactive and planning component (adaptive controller) produces a hybrid system.

The terms *emergent functionality*, *emergent intelligence* or *swarm intelligence* (if many robots are involved) are used to describe the manifestation of an overall behavior from a combination of smaller behaviors that may not have been designed for the original task; see Moravec[5] and Steels and Brooks[6]. The appearance of this behavior can be attributed to the complexity of interactions between simple tasks instead of the tasks' coding themselves. Behavior is generally classified as emergent if the response produced was outside the analysis of the system design but proves to be beneficial to system operation.

Arkin[7] argues that the coordination between simpler subunits does not explain emergence completely. If coordination of a robot is achieved by a deterministic algorithm, a sufficiently sophisticated analysis should be able to perfectly predict the behavior of a robot. In contrast, the emergent intelligence phenomenon is attributed to the non-deterministic nature of real-world environments. These can never be modeled completely accurately, and so there is always a margin of uncertainty in the system that could cause an unexpected behavior to occur.

18.3 Behavior Framework

The objective of a behavior framework is to simplify the design and implementation of behavior-based programs for a robot platform such as the EyeBot. Its foundation is a programming interface for consistently specified behaviors.

We adapt the convention of referring to simple behaviors as schemas and extend the term to encompass any processing element of a control system. The specification of these schemas is made at an abstract level so that they can be generically manipulated by higher-level logic or other schemas without specific knowledge of their implementation details.

Schemas may be recursively combined either by programming or by generation from a user interface. Aggregating different schemas together enables more sophisticated behaviors to be produced. The mechanism of arbitration between grouped schemas is up to the system designer. When combined with coordination schemas to select between available behaviors, the outputs of the contributing modules can be directed to actuator schemas to produce actual robot actions. A commonly used technique is to use a weighted sum of all schemas that drive an actuator as the final control signal.

[5]H. Moravec, *Mind Children: The Future of Robot and Human Intelligence*, Harvard University Press, Cambridge MA, 1988.

[6]L. Steels, R. Brooks, *Building Agents Out of Autonomous Behavior Systems*, in L. Steels, R. Brooks (Eds.), The Artificial Life Route to AI: Building Embodied, Situated Agents, Erlbaum Associates, Hillsdale NJ, 1995.

[7]R. Arkin, *Behavior-Based Robotics*, MIT Press, Cambridge MA, 1998.

A flexible method for behavior selection is to use a neural network controller (Figure 18.2), as was implemented by Daniel Venkitachalam[8] at UWA. The neural network receives input from all sensors (direct or preprocessed), a clock and status lines from each behavior. It generates output to select the currently active behavior and thereby determines the robot's action. The structure of the network is itself developed with a genetic algorithm designed to optimize a fitness function describing the task criteria. This framework architecture was inspired by the reactive system AuRA[9] and takes implementation cues from the TeamBots[10] environment realization.

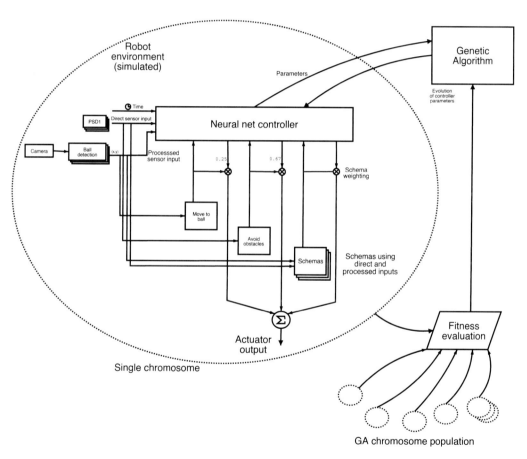

Figure 18.2: Behaviors and selection mechanism in robot environment by D. Venkitachalam, UWA

[8]Venkitachalam, D. *Implementation of a Behavior-Based System for the Control of Mobile Robots*, B.E. Honours Thesis, The Univ. of Western Australia, Electrical and Computer Eng., supervised by T. Bräunl, 2002.

[9]R. Arkin, T. Balch, *AuRA: Principles and Practice in Review*, Journal of Experimental and Theoretical Artificial Intelligence, vol. 9, no. 2–3, 1997, pp. 175–189 (15).

[10]T. Balch, *TeamBots simulation environment*, available from http://www.teambots.org, 2006.

The basic unit of this framework is a schema, which may be perceptual (e.g., a sensor reading) or behavioral (e.g., command *move to a location*). A schema is defined as a unit that produces an output of a pre-defined type. In our implementation, the simplest types emitted by schemas are integer, floating-point and Boolean scalar values. More complex types that have been implemented are two-dimensional floating-point vector and image types. A floating-point vector may be used to encode any two-dimensional quantity commonly used by robot schemas, such as velocities and positions. The image type corresponds to the image structure used by the RoBIOS image processing library.

Schemas may optionally embed other schemas for use as inputs. Data of the pre-defined primitive types is exchanged between schemas. In this way, behaviors may be recursively combined to produce more complex behaviors.

In a robot control program, schema organization is represented by a processing tree. Sensors form the leaf nodes, implemented as embedded schemas. The complexity of the behaviors that embed sensors varies from simple movement in a fixed direction to object detection using an image processing algorithm. The output of the tree's root node is used every processing cycle to determine the robot's next action. Usually, the root node corresponds to an actuator output value. In this case, output from the root node directly produces robot action.

The behavioral framework has been implemented in C++, using the RoBIOS API to interface with the EyeBot. These same functions are simulated and available in EyeSim, enabling programs created with the framework to be used on both the real and simulated platforms.

The hierarchy of schema connections forms a tree, with actuators and sensors mediated by various schemas and schema aggregations. Time has been discretized into units, and schemas in the tree are evaluated from the lowest level (sensors) to the highest from a master clock value generated by the running program.

A working set of schemas using the framework was created for use in a neural network controller design task. The set of schemas with a short description of each is listed in Table 18.1. The schemas shown are either perceptual (e.g., *Camera*, *PSD*), behavioral (e.g., *Avoid*) or generic (e.g., *Fixed* vector). Perceptual schemas only emit a value of some type that is used by the behavioral schemas. Behavioral schemas transform their input into an egocentric output vector that would fulfill its goal.

A visual editor allows point-and-click assemblage of a new schema from preprogrammed modules. The representation of the control program as a tree of schemas maps directly to the interface presented to the user (Figure 18.3). From the interconnection of the visual modules, the user interface generates code for the application program.

Schema	Description	Output
Camera	Camera perceptual schema	Image
PSD	PSD sensor perceptual schema	Integer
Avoid	Avoid obstacles based on PSD reading	2D vector
Detect ball	Detects ball position by hue analysis	2D vector
Fixed vector	Fixed vector representation	2D vector
Linear movement	Moves linearly from current position to another point	2D vector
Random	Randomly directed vector of specified size	2D vector

Table 18.1 Behavior schemas

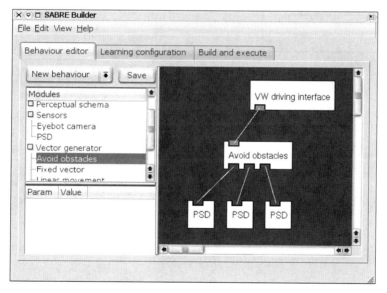

Figure 18.3: Graphical user interface for assembling schemas by D. Venki-tachalam, UWA

A similar approach has been developed by Rahim, Yusof and Bräunl[11] in their *Genetically Evolved Action Selection Mechanism* (GEASM; see Figure 18.4). A dynamic *behavior coordinator* fuses sensor input and evolved behavior from a genetic algorithm (GA) engine through a neural network (NN) for action selection. The weights of the neural network are being determined by the genetic algorithm.

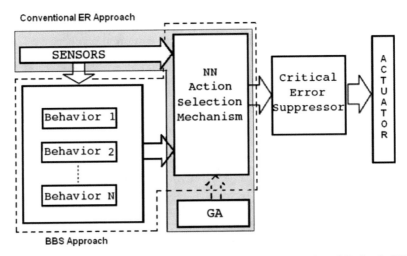

Figure 18.4: GEASM architecture of Rahim, Yusof and Bräunl, UWA

Saufiah Abdul Rahim[12] implemented the GEASM system and applied it to a mobile robot exploration task. She compared the GEASM performance to the more conventional action selection methods priority-based (PB), vector summation (VS) and conventional evolutionary robotics (ER). GEASM shows a superior performance in Figure 18.5 when compared to the other methods

[11]S. Rahim, A. Yusof, T. Bräunl, *Genetically evolved action selection mechanism in a behavior-based system for target tracking*, Neurocomputing, vol. 133, June 2014, pp. 84–94 (11), online: https://www.sciencedirect.com/science/article/abs/pii/S0925231214000575.

[12]S. Rahim, *A Genetically Evolved Neural Network for an Action Selection Mechanism in Behavior-based Systems*, PhD thesis, supervised by T. Bräunl, UWA, May 2016, pp. (195).

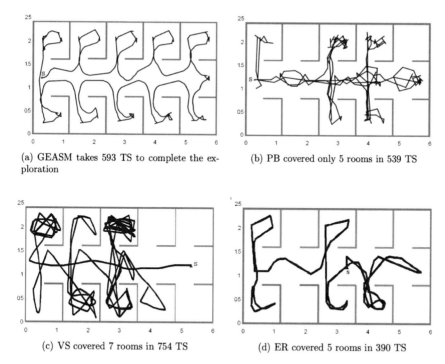

(a) GEASM takes 593 TS to complete the exploration

(b) PB covered only 5 rooms in 539 TS

(c) VS covered 7 rooms in 754 TS

(d) ER covered 5 rooms in 390 TS

Figure 18.5: Comparison of evolutionary action selection methods by Saufiah Rahim, UWA

18.4 Behavior-Based Applications

Typical behavior-based applications involve a group of interacting robots mimicking some animal behavior pattern and thereby exhibiting a form of swarm intelligence. Communication between the robots can be either direct (e.g., wireless) or indirect via changes in the shared environment (*stigmergy*). Depending on the application, communication between individuals can range from essential to not required. Some prominent applications are:

- Foraging
  One or more robots search an area for "food" items (usually easy to detect objects, such as colored blocks), collect them and bring them to a home area. Note that this is a very broad definition and also applies to tasks such as collecting "rubbish."

- Predator–Prey
  Two or more robots interact, with at least one robot in the role of the predator, trying to catch prey robots. Prey robots are in turn trying to avoid the predator robots.

- Clustering
  Mimicking the social behavior of termites, which individually follow very simple rules when cooperating on building a mound.

Some of our implementations of these behavior-based systems use neural networks, genetic algorithms or both. We will explain neural networks and genetic algorithms in detail in the following two chapters.

18.4.1 Clustering

As an example, we look at the clustering behavior. When termites are building a mound, each termite applies new building material to the largest mound within its vicinity—or at a random position when starting. The complex mound structure that results from the interaction of each of the colony's termites is a typical example of the *emergence* phenomenon.

This phenomenon can be exhibited by either computer simulations or real robots interacting in a shared environment; see Iske, Rückert[13] and Du, Bräunl[14]. In our implementation (see Figure 18.6), we let a single robot or multiple robots search an area for red cubes. Once a robot has found a cube, it pushes it to the position of the largest collection of red cubes it has seen previously—or, if this is the first cube the robot has encountered, it uses this cube's coordinates for the start of a new cluster.

Over time, several smaller clusters will appear, which will eventually be merged into a single large cluster, containing all cubes from the robots' driving area. No communication between the robots is required to accomplish this task, but of course the process can be sped up by using communication. The number of robots used for this task also affects the average completion time. Depending on the size of the environment, using more and more robots will result in a faster completion time, up to the point where too many robots encumber each other (by stopping to avoid collisions or by accidentally destroying each other's cluster), resulting in an increasing completion time.

---

[13]B. Iske, U. Rückert, *Cooperative Cube Clustering using Local Communication*, Autonomous Robots for Research and Edutainment - AMiRE 2001, Proceedings of the 5th International Heinz Nixdorf Symposium, Paderborn, Germany, 2001, pp. 333–334 (2).

[14]J. Du, T. Bräunl, *Collaborative Cube Clustering with Local Image Processing*, Proc. of the 2nd Intl. Symposium on Autonomous Minirobots for Research and Edutainment, AMiRE 2003, Brisbane, Feb. 2003, pp. 247–248 (2).

Figure 18.6: Cube clustering with real robots and in simulation by J. Du, UWA

Figure 18.7 shows a somewhat harder variation of the clustering problem with the addition of an internal barrier, solved by Rahim's GEASM system.

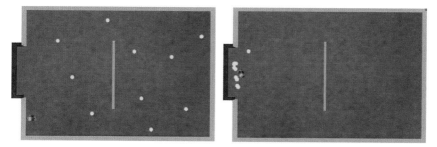

Figure 18.7: Cube clustering using the GEASM behavior-based system by S. Rahim, UWA

18.4.2 Tracking

The task in this application is to continually follow a certain object, for which we use a colored ball. First, we identify the primitive schemas that can be combined to perform this task. These are selected by the evolved controller during program execution to perform the overall task. A suitable initial fitness function for the task was constructed and then an initial random population generated for refinement by the genetic algorithm.

We identified the low-level motor schemas that could conceivably perform this task when combined together. Each schema produces a single normalized 2D vector output, as described in Table 18.2.

Behavior	Normalized vector output
Move straight ahead	In the direction, the robot is facing
Turn left	Directed left of the current direction
Turn right	Directed right of the current direction
Avoid obstacles	Directed away from detected obstacles

Table 18.2 Primitive schemas

The *avoid detected obstacles* schema embeds PSD sensor schemas as inputs, mounted on the front, left and right of the robot. These readings are used to determine a vector away from any close obstacle (see Figure 18.8). Activation of the *avoid detected obstacles* schema prevents collisions with walls or other objects and getting stuck in areas with a clear exit.

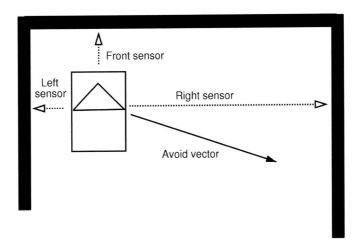

Figure 18.8: Avoidance schema

Ball detection is achieved by a hue recognition algorithm that processes images captured from the EyeBot camera (see Figure 18.9) and returns ball position in the x-direction and ball height as "high-level sensor signals." The system should learn to activate the *turn left* behavior whenever the ball drifts toward the left image border and the *turn right* behavior whenever the balls drift to the right. If the sensors detect the ball roughly in the middle, the system should learn to activate the *drive straight* behavior.

Figure 18.9: EyeBot robot seeing ball in simulation

We constructed a very simple NN with only one input node, two hidden nodes and two output nodes (see Figure 18.10). As mentioned, the ball position is provided as a direct value to the NN's input node. This gives a total of six connection weights, which had to be optimized through the GA component. The fitness function for each step was simply the advancement toward the ball (new distance minus old distance), capped at zero as lower bound.

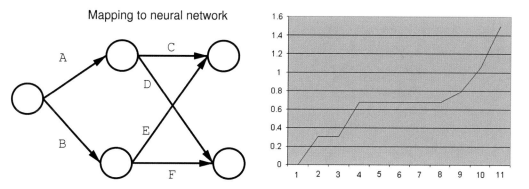

Figure 18.10: Simple NN (left) and GA evolution (right) for action selection

The resulting robot movement of the robot following the ball—and inadvertently kicking the ball and then following it again—is shown in Figure 18.11. There is no stopping behavior when the robot is close to the ball, so this *kick and rush* continues indefinitely.

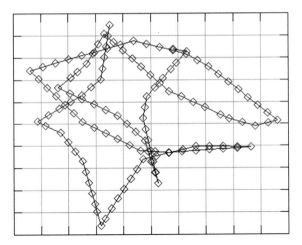

Figure 18.11: Resulting robot driving behavior

18.4.3 Predator–Prey

In the predator–prey scenario, one robot is chasing after another. Below is a simple example adapted from *Robot Adventures*.[15] We set up two robots in the same borderless driving environment with the help of the *SIM* file in Program 18.1. The environment file is *Field.wld*, for which we chose an open square area.

Program 18.1: Configuration (.sim) File

```
# Environment
world Field.wld
# robots
Labbot 2000 500 0 prey.py
S4 500 500 0 predator.x
```

Next, we send the prey robot (robot model LabBot) in a continuous circle path (program *prey.py*). In Python, this only takes a single line of code for the drive command *VWSetSpeed*, which specifies linear and angular speed, plus one line of code for the importing the *eye* library (see Program 18.2).

---

[15]T. Bräunl, *Robot Adventures in Python and C*, Springer Verlag, Heidelberg, 2020.

Program 18.2: Prey Program in Python for Driving a Continuous Circle

```
from eye import *
VWSetSpeed(300, 15)
```

For the predator program in C (*predator.x*) in Program 18.3, we use an S4 SoccerBot robot with differential drive steering. Its main loop is being executed until *KEY4* is pressed to terminate the program. In every iteration, a full Lidar scan is generated (variable *scan*) and a simple loop will find the lowest entry. As there are no other obstacles or walls in this simple environment, the lowest Lidar distance value must be caused by the prey robot we are trying to chase. A sample Lidar scan can be seen in Figure 18.12. The prey can be identified through the gap that is left of the center. Remember that each scan value represents an angular distance measurement from the predator robot.

Once we have found the angular position of the prey in variable *min_pos*, we can use this value for updating the robot's angular speed using the driving functions *VWSetSpeed*. We already used the same function for the prey robot.

Finally, we print the position and distance information on the robot's LCD and check the overall distance of the prey robot. If it is more than 5'000 mm away, we assume it has disappeared and we stop the predator by setting its speed values to (0,0).

Program 18.3: Predator Program in C

```
#include "eyebot.h"

int main ()
{ int i, min_pos, scan[360];

  while (KEYRead()!=KEY4)
  { LCDClear();
    LCDMenu(" ", " ", " ", "END");
    LIDARGet(scan);
    min_pos = 0;
    for (i=0; i<360; i++)
    { if (scan[i] < scan[min_pos]) min_pos = i;
      LCDLine(i,250-scan[i]/50, i,250, BLUE);
    }
    VWSetSpeed(300, 180-min_pos);
    LCDSetPrintf(20,0, "min %d (%d)\n", min_pos, scan[min_pos]);
    if (scan[min_pos] > 5000) VWSetSpeed(0,0); // stop
    OSWait(100); // 0.1 sec
  }
}
```

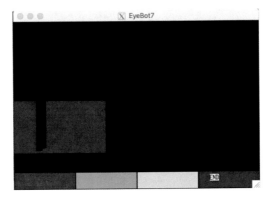

Fig. 18.12 Sample Lidar scan on predator robot; the "gap" indicates the prey location

In Figure 18.13, we are using EyeSim's tracing option for prey and predator to show their paths for the developing chase between the two robots. Any more complex evasive action from the prey would result in an equally more complex and *seemingly intelligent* chasing action of the predator.

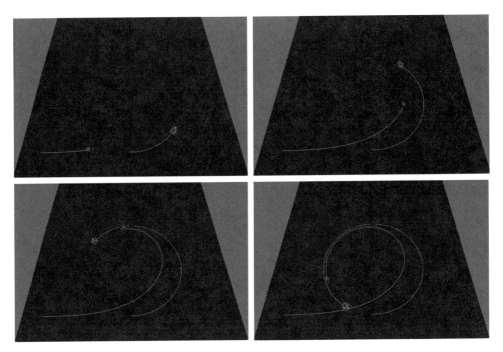

Fig. 18.13 Predator–prey program execution in EyeSim

18.5   Tasks

1. Implement a behavior-based system with dynamic priorities. Each behavior gets to transmit its desired action plus its own priority to an action selection instance. This instance also receives data from all robot sensors plus a real-time clock to make its decision.

2. Implement the foraging scenario described in the chapter as a behavior-based system.

3. Implement the cube clustering task described in the chapter as a behavior-based system.

4. Implement the ball tracking task described in the chapter as a behavior-based system.

5. Implement the predator–prey scenario described in this chapter as a behavior-based system.

NEURAL NETWORKS

19

An artificial neural network (ANN), often simply called neural network (NN), is a processing model loosely derived from biological neurons; see Gurney.[1] Neural networks are often used for classification problems or decision-making problems that do not have a simple or straightforward algorithmic solution. The beauty of a neural network is its ability to learn an input-to-output mapping from a set of training cases without explicit programming and then being able to generalize this mapping to cases not seen previously. There is a large research community as well as numerous industrial users working on neural network principles and applications; see Rumelhart, McClelland[2] and Zaknich.[3]

In this chapter, we will first look at the basic NN principles and then concentrate on NN applications relevant to mobile robots. Here, we will only look at the *traditional* three-layer NN, while we will look at *deep learning* networks in a separate chapter later on.

19.1 Neural Network Principles

A neural network is constructed from a number of individual units called neurons that are linked with each other via connections. Each individual neuron has a number of inputs, a processing node and a single output. Each connection

---

[1]K. Gurney, *Neural Nets*, UCL Press, London, 2002.

[2]D. Rumelhart, J. McClelland (Eds.), *Parallel Distributed Processing*, 2 vols., MIT Press, Cambridge MA, 1986.

[3]A. Zaknich, *Neural Networks for Intelligent Signal Processing*, World Scientific, Singapore, 2003.

from one neuron to another is associated with a weight. Processing in a neural network takes place in parallel for all neurons. Each neuron constantly (in an endless loop) evaluates (reads) its inputs, calculates its local activation value according to a formula shown below and produces (writes) an output value.

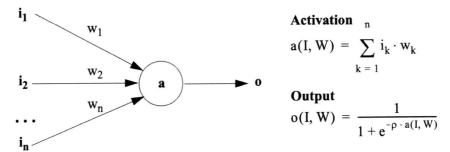

Activation

$$a(I, W) = \sum_{k=1}^{n} i_k \cdot w_k$$

Output

$$o(I, W) = \frac{1}{1 + e^{-\rho \cdot a(I, W)}}$$

Figure 19.1: Individual artificial neuron

The activation function of a neuron $a(I, W)$ is the weighted sum of its inputs; i.e., each input is multiplied by the associated weight and all these terms are added. The neuron's output is determined by the output function $o(I, W)$, for which numerous different models exist.

In the simplest case, just thresholding is used for the output function. For our purposes, however, we use the nonlinear *sigmoid* output function defined in Figure 19.1 and shown in Figure 19.2, which has superior characteristics for learning. This sigmoid function approximates the Heaviside step function, with parameter ρ controlling the slope of the graph (set to 1 in the simplest case).

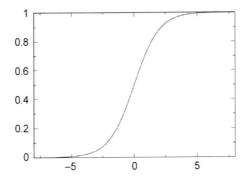

Figure 19.2: Sigmoidal output function

19.2   Feed-Forward Networks

A neural net is constructed from a number of interconnected neurons, which are usually arranged in layers. The outputs of one layer of neurons are connected to the inputs of the following layer. The first layer of neurons is called the *input layer*, since its inputs are connected to external data, for example, sensors to the outside world. The last layer of neurons is called the *output layer*, accordingly, since its outputs are the result of the total neural network and are made available to the outside. These could be connected, for example, to robot actuators or external decision units. All neuron layers between the input layer and the output layer are called *hidden layers*, since their actions cannot be observed directly from the outside.

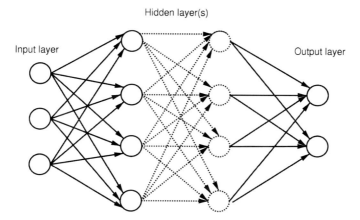

Figure 19.3: Fully connected feed-forward network

If all connections go from the outputs of one layer to the inputs of the next layer, and there are no connections within the same layer or connections from one later layer back to an earlier layer, then this type of network is called a *feed-forward network*. Feed-forward networks (Figure 19.3) are used for the simplest types of ANNs and differ significantly from feedback networks, which we will not look further into here.

For most practical applications, a single hidden layer is sufficient, so the typical NN for our purposes has exactly three layers:

- Input layer (e.g., input from robot sensors)
- Hidden layer (connected to input and output layers)
- Output layer (e.g., output to robot actuators).

Incidentally, the first feed-forward network proposed by Rosenblatt[4] had only two layers, one input layer and one output layer. However, these so-called

---

[4]F. Rosenblatt, F. *Principles of Neurodynamics*, Spartan Books, Washington DC, 1962.

perceptrons were severely limited in their computational power because of this restriction, as was soon after discovered by Minsky and Papert.[5] Unfortunately, this publication almost brought neural network research to a halt for several years, although the principal restriction applies only to two-layer networks, not for networks with three layers or more.

In the standard three-layer network, the input layer is usually simplified in a way that the input values are directly taken as neuron activation. No activation function is called for input neurons. The remaining questions for our standard three-layer NN type are:

- How many neurons should be used in each layer?
- Which connections should be made between layer i and layer $i + 1$?
- How are the weights determined?

Some answers to these questions straightforward, others not so much:

- *How many neurons to use in each layer?*
  The number of neurons in the input and output layers is determined by the application. For example, if we want to have an NN drive a robot around a maze with three PSD sensors as input and two motors as output, then the network should have three input neurons and two output neurons.
  Unfortunately, there is no rule for the *right* number of hidden neurons. Too few hidden neurons will prevent the network from learning, since the net has insufficient storage capacity. Too many hidden neurons will slow down the learning process because of extra overhead. The right number of hidden neurons depends on the complexity of the given problem and has to be determined through experimenting. In this example we are using six hidden neurons.

- *Which connections should be made between layer i and layer $i + 1$?*
  We simply connect every output from layer i to every input at layer $i + 1$.
  This is called a *fully connected* neural network. There is no need to leave out individual connections, since the same effect can be achieved by giving such a connection a weight of zero. That way we can use a much more general and uniform network structure.

- *How are the weights determined?*
  This is the really tricky question. Apparently, the whole intelligence of an NN is somehow encoded in the set of weights being used. What used to be a program (e.g., driving a robot in a straight line while avoiding obstacles) is now reduced to a set of floating-point numbers. With sufficient insight, we could just *program* an NN by specifying the correct (or let us say working) weights. However, since this would be virtually impossible, even for networks of a small size, we need another technique.

---

[5]M. Minsky, S. Papert, *Perceptrons*, MIT Press, Cambridge MA, 1969.

A standard method for generating NN weights is supervised learning through error *backpropagation*. The same task is repeatedly run through the NN, and the output layer values are compared to the known results. Errors made by the network are fed back into the network from the output layer via the hidden layer to the input layer, amending the weights of each connection along the way (backpropagation principle).

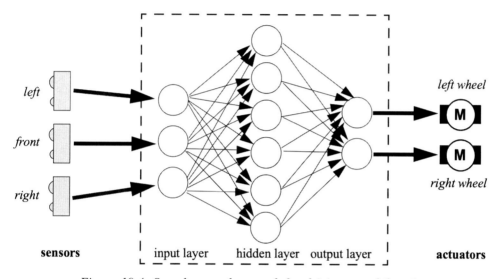

Figure 19.4: Sample neural network for driving a mobile robot

Figure 19.4 shows the experimental setup for an NN that should drive a mobile robot collision-free with constant speed through a maze (e.g., left-wall following). Since we are using three sensor inputs plus two motor outputs and we chose six hidden neurons, our network has $3 + 6 + 2 = 11$ neurons in total. The input layer receives the sensor data from the infrared PSD distance sensors, and the output layer produces driving commands for the left and right motors of a robot with differential drive steering.

For simplicity, let us calculate the output of an NN for a smaller case with $2 + 4 + 1 = 7$ neurons. Figure 19.5, top, shows the labeling of the neurons and connections between the three layers, and Figure 19.5, bottom, shows the network with sample input values and weights. For a network with three layers, only two sets of connection weights are required.

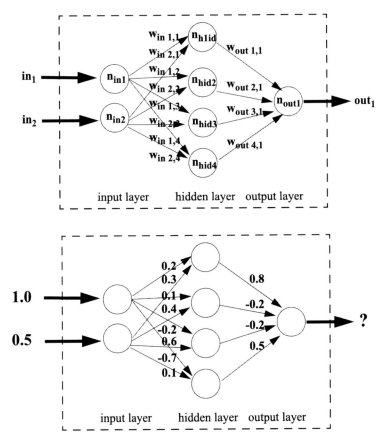

Figure 19.5: Example neural network

- Weights from the input layer to the hidden layer, summarized as matrix $w_{\text{in}\,i,j}$ (weight of connection from input neuron i to hidden neuron j).
- Weights from the hidden layer to the output layer, summarized as matrix $w_{\text{out}\,i,j}$ (weight of connection from hidden neuron i to output neuron j).
- No weights are required from sensors to the first layer or from the output layer to actuators. These weights are just assumed to be always 1. All other weights are normalized to the range $[-1, +1]$.

Calculation of the output function starts with the input layer on the left and propagates through the network. For the input layer, there is one input value (sensor value) per input neuron. Each input data value is used directly as the neuron activation value:

$$a(n_{in1}) = o(n_{in1}) = 1.00$$
$$a(n_{in2}) = o(n_{in2}) = 0.50$$

For all subsequent layers, we first calculate the activation function of each neuron as a weighted sum of its inputs and then apply the sigmoid output function. The first neuron of the hidden layer has the following activation and output values:

$$a(n_{hid1}) = 1.00 \cdot 0.2 + 0.50 \cdot 0.3 = 0.35$$
$$o(n_{hid1}) = 1/(1 + e^{-0.35}) = 0.59$$

The subsequent steps for the remaining two layers are shown in Figure 19.6 with the activation values printed in each neuron symbol and the output values below it, always rounded to two decimal places.

Once the values have percolated through the feed-forward network, they will not change until the input values change. Obviously, this only holds for networks without feedback connections. Program 19.1 shows the implementation of the feed-forward process. This program already takes care of two additional so-called *bias neurons* with fixed activation level of 1, which are required later for backpropagation learning.

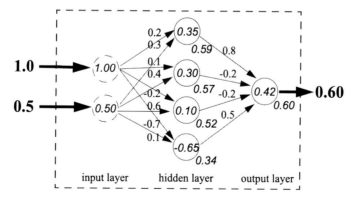

Figure 19.6: Feed-forward evaluation

Program 19.1: Feed-Forward Execution

```c
#include <math.h>
#define NIN (16*16+1)  // input neurons + bias neuron
#define NHID (256+1)            // hidden neurons + bias neuron
#define NOUT (10)      // output neurons:

#define NUM_TRAINING 200  // training images MAX images 1500
#define NUM_TEST 100       // test images MAX images 1500

//global neural network weights
float w_in [NIN][NHID], w_out[NHID][NOUT];

float sigmoid(float x)
{ return 1.0 / (1.0 + exp(-x));
}
```

```c
void  feedforward(float  N_in[NIN],float  N_hid[NHID],float  N_out
[NOUT])
{ int i,j;
  N_in[NIN-1] = 1.0; // set bias input neuron
  for (i=0; i<NHID-1; i++) // calculate activation
  { N_hid[i] = 0.0;
    for (j=0; j<NIN; j++) N_hid[i] += N_in[j] * w_in[j][i];
    N_hid[i] = sigmoid(N_hid[i]);
  }
  N_hid[NHID-1] = 1.0; // set bias hidden neuron
  for (i=0; i<NOUT; i++) // calculate output neurons
  { N_out[i] = 0.0;
    for (j=0; j<NHID; j++) N_out[i] += N_hid[j] * w_out[j][i];
    N_out[i] = sigmoid(N_out[i]);
  }
}
```

19.3 Backpropagation

A large number of different techniques exist for learning in neural networks. These include supervised and unsupervised techniques, depending on whether a *teacher* presents the correct answer to a training case or not, as well as online or offline learning, depending on whether the system evolves inside or outside the execution environment. Classification networks with the popular back-propagation learning method, a supervised offline technique, can be used to identify a certain situation from the network input and produce a corresponding output signal. The drawback of this method is that a complete set of all relevant input cases together with their solutions have to be presented to the NN. Another popular method requiring only incremental feedback for input/output pairs is *reinforcement learning* (see Sutton and Barto[6]). This online technique can be seen as either supervised or unsupervised, since the feedback signal only refers to the network's current performance and does not provide the desired network output. In the following, the backpropagation method is presented.

A feed-forward neural network starts with random weights and is presented a number of test cases called the *training set*. The network's outputs are compared to the known correct results for the particular set of input values and any deviations (*error function*) are propagated back through the net.

Having done this for a number of iterations, the NN has hopefully learned the complete training set and can now produce the correct output for each input pattern in the training set. The real benefit, however, is that the network should be able to generalize, which means it will be able to produce similar output values corresponding to similar input patterns it has not seen before. Without the capability of generalization, no useful learning can take place, since we would simply store and reproduce the training set.

The backpropagation algorithm works as follows:

1. Initialize network with random weights.
2. For all training cases:
   a. Present training inputs to network and calculate outputs.
   b. For all layers (starting with output layer, back to input layer):
      i. Compare network output with correct output (error function).
      ii. Adapt weights in current layer.

When implementing this learning algorithm, we do know for the output layer what the correct results should be, because they are supplied together with the training inputs. However, it is not clear for the other layers, so let us do this step by step.

Firstly, we look at the error function. For each output neuron, we compute the difference between the actual output value out_i and the desired output $d_{out\,i} d_{out\,i}$. For the total network error, we calculate the sum of square difference:

$$E_{out\,i} = d_{out\,i} - out_i$$

$$E_{total} = \sum_{i=0}^{num(n_{out})} E_{out\,i}^2$$

---

[6]R. Sutton, A. Barto, *Reinforcement Learning: An Introduction*, MIT Press, Cambridge MA, 1998.

The next step is to adapt the weights, which is done by a gradient descent approach:

$$\Delta w = -\eta \cdot \frac{\partial E}{\partial w}$$

So the adjustment of the weight will be proportional to the contribution of the weight to the error, with the magnitude of change determined by constant η. This can be achieved by the following formulas (from Rumelhart, McClelland):

$$\text{diff}_{out\ i} = (o(n_{out\ i}) - d_{out\ i}) \cdot (1 - o(n_{out\ i})) \cdot o(n_{out\ i})$$
$$\Delta w_{out\ k,i} = -2 \cdot \eta \cdot \text{diff}_{out\ i} \cdot \text{input}_k(n_{out\ i})$$
$$= -2 \cdot \eta \cdot \text{diff}_{out\ i} \cdot o(n_{hid\ k})$$

Assuming the desired output $d_{out\ 1}$ of the NN in Figure 19.5 to be 1.0, and choosing $\eta = 0.5$ to simplify the calculation, we can now update the four weights between the hidden layer and the output layer. Note that all calculations have been performed with full floating-point accuracy, while only two or three decimals are printed.

$$\text{diff}_{out1} = (o(n_{out1}) - d_{out1}) \cdot (1 - o(n_{out1})) \cdot o(n_{out1})$$
$$= (0.60 - 1.00) \cdot (1 - 0.60) \cdot 0.60 = -0.096$$
$$\Delta w_{out1,1} = -\text{diff}_{out1} \cdot \text{input}_1(n_{out1})$$
$$= -\text{diff}_{out1} \cdot o(n_{hid1})$$
$$= -(-0.096) \cdot 0.59 = +0.057$$
$$\Delta w_{out2,1} = 0.096 \cdot 0.57 = +0.055$$
$$\Delta w_{out3,1} = 0.096 \cdot 0.52 = +0.050$$
$$\Delta w_{out4,1} = 0.096 \cdot 0.34 = +0.033$$

The new weights will be:

$$w'_{out1,1} = w_{out1,1} + \Delta w_{out1,1} = 0.8 + 0.057 = 0.86$$
$$w'_{out2,1} = w_{out2,1} + \Delta w_{out2,1} = -0.2 + 0.055 = -0.15$$
$$w'_{out3,1} = w_{out3,1} + \Delta w_{out3,1} = -0.2 + 0.050 = -0.15$$
$$w'_{out4,1} = w_{out4,1} + \Delta w_{out4,1} = 0.5 + 0.033 = 0.53$$

The only remaining step is to adapt the w_{in} weights. Using the same formula, we need to know what the desired outputs $d_{hid\ k}$ are for the hidden layer. We get these values by backpropagating the error values from the output layer multiplied by the activation value of the corresponding neuron in the hidden layer and adding up all these terms for each neuron in the hidden layer. Here, we use the old (unchanged) value of the connection weight, which improves convergence. The error formula for the hidden layer (difference between desired and actual hidden value) is:

$$E_{hid\ i} = \sum_{k=1}^{num(n_{out})} E_{out\ k} \cdot w_{out\ i,k}$$

$$diff_{hid\ i} = E_{hid\ i} \cdot (1 - o(n_{hid\ i})) \cdot o(n_{hid\ i})$$

In the example in Figure 19.5, there is only one output neuron, so each hidden neuron has only a single term for its desired value. The value and difference values for the first hidden neuron are therefore:

$$E_{hid1} = E_{out1} \cdot w_{out1,1}$$
$$= 0.4 \cdot 0.8 = 0.32$$
$$diff_{hid1} = E_{hid1} \cdot (1 - o(n_{hid1})) \cdot o(n_{hid1})$$
$$= 0.32 \cdot (1 - 0.59) \cdot 0.59 = 0.077$$

Program 19.2 Backpropagation Execution

```
float backprop(float train_in[NIN], float train_out[NOUT])
{ int i,j;
    float err_total;
    float N_out[NOUT], err_out[NOUT], diff_out[NOUT];
    float N_hid[NHID], diff_hid[NHID];

    //run network, calculate difference to desired output
    feedforward(train_in, N_hid, N_out);
    err_total = 0.0;

    for (i=0; i<NOUT; i++)
    { err_out[i] = train_out[i]-N_out[i];
      diff_out[i]= err_out[i] * (1.0-N_out[i]) * N_out[i];
      err_total += err_out[i]*err_out[i];
    }
    // update w_out and calculate hidden difference values
    for (i=0; i<NHID; i++)
    { diff_hid[i] = 0.0;
      for (j=0; j<NOUT; j++)
      { diff_hid[i] += diff_out[j] * w_out[i][j]
                    * (1.0-N_hid[i]) * N_hid[i];
```

```
        w_out[i][j] += diff_out[j] * N_hid[i];
      }
    }
    // update w_in
    for (i=0; i<NIN; i++)
      for (j=0; j<NHID; j++)
        w_in[i][j] += diff_hid[j] * train_in[i];
    return err_total;
  }
```

The weight changes for the two connections from the input layer to the first hidden neuron are as follows. Remember that the input of the hidden layer is the output of the input layer:

$$\Delta w_{\text{in } k,i} = 2 \cdot \eta \cdot \text{diff}_{\text{hid } i} \cdot \text{input}_k(n_{\text{hid } i}) \quad (\textit{for } \eta = 0.5)$$
$$= \text{diff}_{\text{hid } i} \cdot o(n_{\text{in } k})$$
$$\Delta w_{\text{in } 1,1} = \text{diff}_{\text{hid } 1} \cdot o(n_{\text{in } 1})$$
$$= 0.077 \cdot 1.0 = 0.077$$
$$\Delta w_{\text{in } 2,1} = \text{diff}_{\text{hid } 1} \cdot o(n_{\text{in } 2})$$
$$= 0.077 \cdot 0.5 = 0.039$$

and so on for the remaining weights. The first two updated weights will therefore be:

$$w'_{\text{in } 1,1} = w_{\text{in } 1,1} + \Delta w_{\text{in } 1,1} = 0.2 + 0.077 = 0.28$$
$$w'_{\text{in } 2,1} = w_{\text{in } 2,1} + \Delta w_{\text{in } 2,1} = 0.3 + 0.039 = 0.34$$

The backpropagation procedure iterates until a certain termination criterion has been fulfilled. This could be a fixed number of iterations over all training patterns, or an iteration until a sufficient convergence has been achieved, for example, if the total output error over all training patterns falls below a certain threshold.

Program 19.2 demonstrates the implementation of the backpropagation process. Note that in order for backpropagation to converge, we need one additional input neuron and one additional hidden neuron, called *bias neurons*. The activation levels of these two neurons are always fixed to 1. The weights of the connections to the bias neurons are required for the backpropagation procedure to converge (see Figure 19.7).

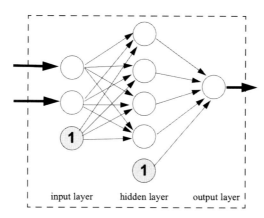

Figure 19.7: Bias neurons and connections for backpropagation

19.4   Neural Network Examples

A simple example for testing a neural network implementation is trying to learn the digits 0...9 from a seven-segment display representation. Figure 19.8 shows the arrangement of the segments and the numerical input and training output for the neural network, which could be read from a data file. Note that there are ten output neurons, one for each digit, 0, ..., 9. This will be much easier to learn than a four-digit binary encoded output (outputs 0000 to 1001).

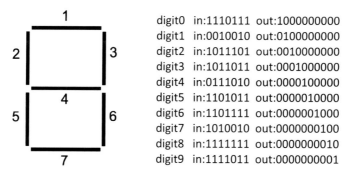

Figure 19.8: Seven-segment digit representation

Figure 19.8 shows the decrease of total error values by applying the backpropagation procedure on the complete input data set for some 700 iterations (Figure 19.9). Eventually, the goal of an error value below 0.1 is reached and the algorithm terminates. The weights stored in the neural net are now ready to take on previously unseen real data. In this example, the trained network could be tested against 7-segment inputs with a single defective segment (e.g., one segment always on or always off).

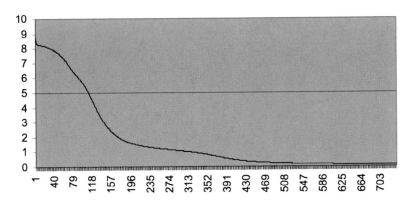

Figure 19.9: Learning curve for 7-segment example

Of course, using a 7-segment display as an example was only for demonstration purposes and does not show the real classification capabilities of a neural network. A much more useful but also more complex problem is the classification of handwritten digits, such as for an automated postal mail sorter that scans the ZIP code. This problem is also known as optical character recognition (OCR). A good source for handwritten digits (unless you want to write and scan your own) is the MNIST database.[7] This freely accessible database contains a total of 70'000 handwritten digits from approximately 250 writers.

Figure 19.10 shows some sample digits together with a visualization tool we developed, which uses the same program code for backpropagation training as the 7-segment display example above.

---

[7]Y. LeCun, C. Cortes, *The MNIST Database of Handwritten Digits*, online: http://yann. lecun.com/exdb/mnist/, 2008.

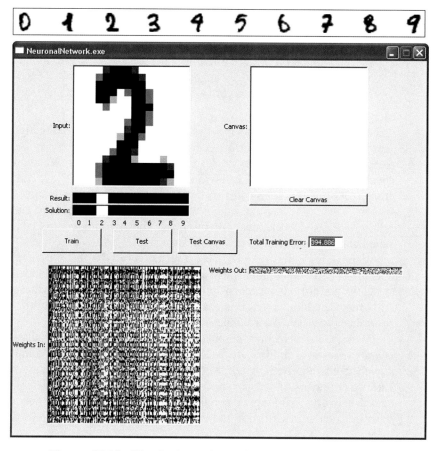

Figure 19.10: Handwritten digits from MNIST (top) and visualization tool (bottom) by Torsten Sommer UWA and TU Munich

19.5 Neural Robot Control

Control of mobile robots requires tangible actions from sensor inputs. A controller for a robot receives input from its sensors, processes the data using relevant logic and sends appropriate signals to the actuators. For most large tasks, the ideal mapping from input to action is not clearly specified nor readily apparent. Such tasks require a control program that must be carefully designed and tested in the robot's operational environment. The creation of these control programs is an ongoing concern in robotics as the range of application domains is expanding, increasing the complexity of tasks for autonomous robots.

A number of questions need to be answered before a feed-forward NN such as the one in Figure 19.4 can be implemented. Among them are:

- How can the success of the network be measured?
  The robot should perform a collision-free left-wall following.
- How can the training be performed?
  In simulation or on the real robot.
- What is the desired motor output for each situation?
  Differential drive motor output, keep constant distance to left wall, avoid collisions.

Neural networks have been successfully used to mediate directly between sensors and actuators to perform certain tasks. Past research has focused on using neural net controllers to learn individual behaviors. Vershure et al.[8] developed a working set of behaviors by employing a neural net controller to drive a set of motors from collision detection, range finding and target detection sensors. The online learning rule of the neural net was designed to emulate the action of Pavlovian classical conditioning. The resulting controller has associated actions beneficial to task performance with positive feedback.

Adaptive logic networks (ALNs), a variation of NNs that only use Boolean operations for computation, were successfully employed in simulation by Kube et al.[9] to perform simple cooperative group behaviors. The advantage of the ALN representation is that it is easily mappable directly to hardware once the controller has reached a suitable working state.

Trying to implement a complex robotics task with only a three-layered neural network is unlikely to succeed because of the limited computational power of this approach. More successful will either be deep neural networks with tens of layers of neurons, as described in the chapter on *deep learning*, or neural networks working in conjunction with behaviors or action selection schemas such as described in the chapter on *AI concepts*, as well as discussed below.

We implemented a neural controller to perform action selection in a behavior-based system (see Figure 19.11 by Rahim et al.[10] and description in previous chapter). During each processing cycle, the NN selects the currently active behavior. The active behavior will then take control over the robot's actuators and drives the robot in the direction it desires. In principle, the neural network receives information from all sensor inputs plus status inputs from all behaviors and a clock value to determine the priority level of each behavior. Inputs may be in the form of raw sensor readings or processed sensor results

[8]P. Vershure, J. Wray, O. Sprons, G. Tononi, G. Edelman, *Multilevel Analysis of Classical Conditioning in a Behaving Real World Artifact*, Robotics and Autonomous Systems, vol. 16, 1995, pp. 247–265 (19).

[9]C. Kube, H. Zhang, X. Wang, *Controlling Collective Tasks with an ALN*, IEEE/RSJ IROS, 1993, pp. 289–293 (5).

[10]S. Rahim, A. Yusof, T. Bräunl, *Genetically evolved action selection mechanism in a behavior-based system for target tracking*, Neurocomputing, vol. 133, June 2014, pp. 84–94 (11).

such as distances, positions and preprocessed image data. Information is processed through a number of hidden layers and fed into the output layer. The controller's output neurons are then responsible for selecting the active behavior.

The tasks implemented with this controller were still relatively simple and ranged from object detection to area exploration and collision avoidance tasks.

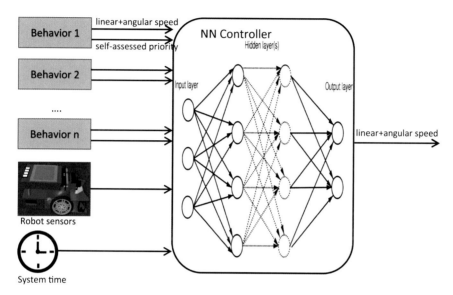

Figure 19.11: Action selection NN for neural robot controller

19.6 Tasks

1. Implement the 7-segment detection task with a hand-coded neural network.

2. Implement the MNIST handwritten digit recognition task with a hand-coded neural network.

3. Explore options of NN libraries and use them for solving the MNIST handwritten digit recognition task.

4. Use a NN library to detect a small number of different objects from a mobile robot's point of view, e.g., cans, cubes, model cars and other robots. When driving around and recognizing one of these objects, let the robot announce each detected object on its display and via speech output.

GENETIC ALGORITHMS

Evolutionary algorithms (EA) are a family of search and optimization techniques that make use of principles from Darwin's theory of evolution[1] from 1859 to find approximate solutions for hard problems. Genetic algorithms (GA) are a prominent part of EAs. They operate by iteratively evolving a solution from a pool of potential solutions, which are manipulated by a number of biologically inspired operations. They have been proven to provide a powerful and robust means of solving problems.

The advantage of genetic algorithms is that they can be applied to problems for which no obvious algorithmic solution exists. A number of complex problems in robot control fall into this category, as, for example, developing an optimal walking gait for a biped robot. For this, a large variety of robot gaits could be generated from a set of parameters, which can then be optimized by a GA.

The effectiveness of using a genetic algorithm to find a solution depends on the problem domain, the existence of an optimal solution to the problem at hand and a suitable fitness function. Applying genetic algorithms to problems that can be solved algorithmically would be a waste of CPU time. GAs are best used for solving tasks that are difficult to solve, such as NP-hard problems, which have a large search space, but can be easily verified once a candidate solution has been found. For further reading see Goldberg[2] and Langton.[3]

---

[1] C. Darwin, *On the Origin of Species by Means of Natural Selection, or Preservation of Favoured Races in the Struggle for Life*, John Murray, London, 1859.

[2] D. Goldberg, *Genetic Algorithms in Search, Optimization and Machine Learning*, Addison-Wesley, Reading MA, 1989.

[3] C. Langton, (Ed.) *Artificial Life – An Overview*, MIT Press, Cambridge MA, 1995.

© The Author(s), under exclusive license to Springer Nature Singapore Pte Ltd. 2022
T. Bräunl, *Embedded Robotics*,
https://doi.org/10.1007/978-981-16-0804-9_20

20.1 Genetic Algorithm Principles

In this section, we describe the terminology used and outline the operation of a genetic algorithm. We then examine the components of the algorithm in detail, describing different ways of implementing them.

20.1.1 Genotype and Phenotype

Genetic algorithms borrow terminology from biology to describe their interacting components. We are dealing with phenotypes, which are possible solutions to a given problem (e.g., in our case a simulated robot with a particular control structure), and genotypes, which are encoded representations of phenotypes, typically represented by bit strings. Genotypes are sometimes also called chromosomes and can be split into smaller chunks of information, called genes (Figure 20.1).

The genetic operators work only on genotypes (chromosomes), while it is necessary to construct phenotypes (individuals) in order to determine their fitness.

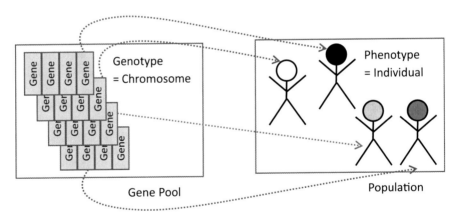

Figure 20.1: GA terminology

20.1.2 GA Execution

The basic operation of a genetic algorithm can be summarized as follows:

1. Randomly initialize a population of chromosomes (bit strings).
2. While the terminating criteria have not been satisfied:

    a. Evaluate the fitness of each chromosome:

        i. Construct the individual (e.g., simulated robot) from the chromosome.

        ii. Evaluate the individual's fitness (e.g., its simulated walking abilities).

    b. Generate a new generation of chromosomes:

        i. Use a selection scheme that selects pairs of chromosomes from the current generation, proportional to their relative fitness.

        ii. Apply *crossover* to each selected pairs of chromosomes to produce a new pair of chromosomes.

        iii. Apply random *mutation* (with low probability) to some new chromosomes.

        iv. Optionally: Copy one or more of the top performing chromosomes unchanged to the new generation (*elitism*).

The algorithm can start with either a set of random chromosomes or ones that already represent approximate solutions for the given problem. The gene pool is evolved from generation to generation through using the genetic operators and a selection scheme that depends on the fitness of each chromosome. The selection scheme determines which chromosomes should reproduce. Each chromosome's selection probability should be proportional to its relative fitness value. Therefore, chromosomes with a high fitness value will have a higher probability to be selected for reproduction (even multiple times) than chromosomes with a low fitness value. The genetic operators *crossover* and *mutation* are constantly modifying chromosomes, thereby effectively covering a large search space.

Each iteration of the overall procedure creates a new population of chromosomes. The total set of chromosomes is known as a generation and changes with every iteration of the algorithm. As the algorithm continues, it searches through the solution space, refining the chromosomes, until either it finds one with a sufficiently high fitness value (matching the desired criteria of the original problem), or the evolutionary progress slows down to such a degree that finding a matching chromosome is unlikely.

20.1.3 Fitness Function

Each problem to be solved requires the definition of a fitness function describing the characteristics of an appropriate solution. The purpose of the fitness function is to rate the suitability of a solution with respect to the overall goal. Given a particular chromosome, the fitness function returns a numerical value corresponding to the chromosome's quality. For many applications, the selection of a fitness function will be straightforward; however, in some applications there are no obvious performance measurements of the goal. In these cases, a suitable fitness function has to be constructed from a combination of desired factors, characteristic of the problem.

20.1.4 Selection Mechanisms

In nature, organisms that reproduce the most have the greatest influence on the composition of the next generation. This effect is employed in the genetic algorithm selection scheme that determines which individuals of a given population will contribute to form the new individuals for the next generation. *Tournament selection*, *random selection* and *roulette wheel selection* are three commonly used selection schemes.

Tournament selection operates by selecting two chromosomes from the available pool and by comparing their fitness values by evaluating them against each other. The better of the two is then permitted to reproduce. Thus, the fitness function chosen for this scheme only needs to discriminate between the two entities.

Random selection randomly selects the parents of a new chromosome from the existing pool. Any returned fitness value below a set operating point is instantly removed from the population. Although it would appear that this would not produce beneficial results, this selection mechanism can be employed to introduce randomness into a population that has begun to converge to a suboptimal solution.

In roulette wheel selection (sometimes referred to as fitness proportionate selection), the chance for a chromosome to reproduce is proportional to the fitness of the entity. Thus, if the fitness value returned for one chromosome is twice as high as the fitness value for another, then it is twice as likely to reproduce. However, its reproduction is not guaranteed as in tournament selection.

It has been shown that the convergence rate can be significantly increased by duplicating unchanged copies of the fittest chromosomes for the next generation, a method called *elitism*.

20.2 Genetic Operators

Genetic operators comprise the methods by which one or more chromosomes are combined to produce a new chromosome. Traditional schemes utilize only two operators: crossover and mutation (see Beasley, Bull, Martin[4]).

---

[4]D. Beasley, D. Bull, R. Martin, *An Overview of Genetic Algorithms: Part 1, Fundamentals*, University Computing, vol. 15, no. 2, 1993a, pp. 58–69 (12).

20.2.1 Crossover

Crossover takes two individuals and divides the string into two portions at a randomly selected point inside the encoded bit string. This produces two *head* segments and two *tail* segments. The two tail segments for the chromosomes are then interchanged, resulting in two new chromosomes, where the bit string preceding the selected bit position belongs to one parent, and the remaining portion belongs to the other parent. This process is shown in Figure 20.2.

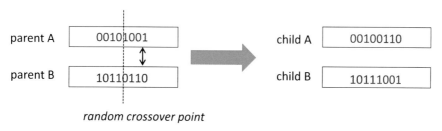

Figure 20.2: Crossover operator

20.2.2 Mutation

The mutate operator (Figure 20.3) randomly selects one bit in the chromosome string and inverts its value with a defined probability. Historically, the crossover operator has been viewed as the more important of the two techniques for exploring the solution space; however, without the mutate operator, portions of the solution space may not be searched, as the initial chromosomes may not contain all possible bit combinations (see Beasley, Bull, Martin[5]).

Figure 20.3: Mutate operator

There are a number of possible extensions to the set of traditional operators. The two-point crossover operates similarly to the single-point crossover described, except that the chromosomes are now cut in two places rather than one.

Further extensions rely on modifying the bit string under the assumption that portions of the bit string represent non-binary values (such as 8-bit integer values or 32-bit floating-point values). Two commonly used operators that rely

[5]D. Beasley, D. Bull, R. Martin, *An Overview of Genetic Algorithms: Part 2, Research Topics,* University Computing, vol. 15, no. 4, 1993b, pp. 170–181 (12).

on this interpretation of the chromosome are the *Non-binary Average* and *Non-binary Creep* operators. Non-binary Average interprets the chromosome as a string of higher cardinality symbols and calculates the arithmetic average of the two chromosomes to produce the new individual. Similarly, Non-binary Creep treats the chromosomes as strings of higher cardinality symbols and increments or decrements a randomly selected value in these strings by a small randomly generated amount.

Figure 20.4: Non-binary Average operator

The operation of the Non-binary Average operator is shown in Figure 20.4. In the example shown, the bit string is interpreted as a sequence of two-bit symbols and is averaged using rounding. So for the first number pair, zero and two average to one, i.e.,

$$(00_2 + 10_2)/2 = 01_2$$

For the second pair, three and three average to three, i.e.,

$$(11_2 + 11_2)/2 = 11_2$$

and so on.

Figure 20.5: Non-binary Creep operator

The Non-binary Creep example shown in Figure 20.5 also represents the bit string as a sequence of two-bit symbols. Here the operator decrements the second symbol by a value of one.

20.2.3 Encoding

The encoding method chosen to transform the parameters to a chromosome can have a large effect on the performance of the genetic algorithm. A compact encoding allows the genetic algorithm to perform efficiently. There are two common encoding techniques applied to the generation of a chromosome. Direct encoding explicitly specifies every parameter within the chromosome, whereas indirect encoding uses a set of rules to reconstruct the complete parameter space. Direct encoding has the advantage that it is a simple and powerful representation, but the resulting chromosome can be quite large. Indirect encoding is far more compact, but it often represents a more restrictive set compared to the original structure.

20.3 Evolution Example

In this section, we demonstrate a simple GA example that we will solve manually. The fitness function we wish to optimize is a simple quadratic formula (Figure 20.6):

$$f(x) = 625 - (x - 6)^2 \quad \text{for } 0 \le x \le 31 \ (\text{so } x \text{ can be represented by 5 bits})$$

The fitness value range is designed so all values are positive, which helps when implementing a proportional selection function.

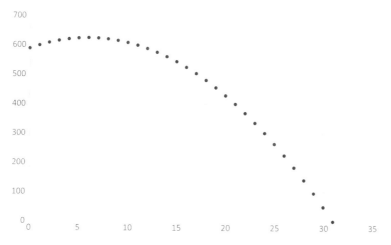

Figure 20.6: Graph of fitness function f(x)

Using a genetic algorithm to search a solvable equation with a small search space like this is inefficient and not advised for reasons stated earlier in this chapter. In this particular example, the genetic algorithm may be even less efficient than an exhaustive search. Still, this problem has been chosen because its small size allows us to examine the workings of the genetic algorithm.

For our GA, we will use a simple binary encoding, one-point crossover and elitism. In this particular problem, each chromosome (of just one gene) consists of a single integer number between 0 and 31. This restriction of the search space allows us to use a 5-bit binary encoding for our genes. Hence, all chromosomes are represented by bit strings of length 5. We begin by producing a random population and evaluating each chromosome through the fitness function. If the terminating criterium of a high enough fitness value is met, we stop the algorithm. In cases where the optimal fitness value is not known, we can express the termination criteria in relation to the rate of convergence of the top performing members of the population.

Bit string	Decimal	Fitness
00000	0	589
11110	30	49
00010	2	609
10100	20	429
01011	11	600

Table 20.1 Initial random population

The initial random population, encodings and fitness levels are shown in Table 20.1. Table 20.2 shows the same table sorted by fitness levels. Chromosome $x = 2$ has the highest fitness value and will therefore be copied unchanged to the next generation (elitism). In exchange, the lowest performing chromosome (30) will be removed from the gene pool.

Bit string	Decimal	Fitness	%
00010	**2**	**609**	**27.3%**
01011	11	600	26.9%
00000	0	589	26.4%
10100	20	429	19.3%
~~11110~~	~~30~~	49	
Gen. 0	Total:	2'227	100%

Table 20.2 Population sorted by fitness level

This leaves four chromosomes for selection: 2, 11, 0 and 20. The next step is to perform crossover between the chromosomes. We select two pairs of chromosomes randomly, however, with a selection probability proportional to their contribution to the overall fitness.

First, the pair (2, 11) gets selected; second, the pair (2, 20). So, chromosome 2 gets selected twice, while chromosome 0 does not get selected at all.

Now, each pair of chromosomes undergoes a crossover operation to produce two new chromosomes. A random cut point is selected for the crossover operation (see Figure 20.7).

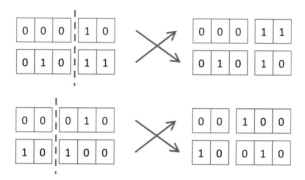

Figure 20.7: Crossover step 1

After crossover, the new chromosomes are: 2 (elitism), 3, 10, 4, 18. Table 20.3 lists the next generation.

Bit string	Decimal	Fitness
00010	2	609
00011	3	616
01010	10	609
00100	4	621
10010	10	481

Table 20.3 Population after first crossover stage

Next, we apply the mutation operation, however, with a low probability. For this example, only a single chromosome gets selected by random: 10. Figure 20.8 shows the mutation of a random bit position, which changes 10 to 26.

Figure 20.8: Mutation step 1

Table 20.4 shows the updated and sorted population after the completed first iteration. The gene pool has improved, both in the fitness of the top chromosome (621 vs. 609) and in the total (or average) chromosome fitness (2'327 vs. 2'227). For the next iteration, we will again copy over the best chromosome (4) and delete the worst (26). As before, two pairs will be selected randomly depending on their relative fitness, and then crossover and mutation will be applied.

Bit string	Decimal	Fitness	%
00100	**4**	**621**	**26.7%**
00011	3	616	26.5%
00020	2	609	26.2%
10010	18	481	20.7%
~~11010~~	~~26~~	~~225~~	
Gen. 1	Total:	2'327	100%

Table 20.4 Population sorted after first iteration

This iterative process will stop when we either get a sufficiently good chromosome (e.g., a predetermined fitness value), when the rate of fitness improvement has fallen below a certain threshold, or when we have reached the set maximum number of iterations.

In our sample, if we had set a fitness threshold of 620, then the iteration would stop with approximate solution 4 → 621. This is not the optimal solution (6 → 625), but it may be sufficient for the given application.

20.4   Implementing Genetic Algorithms

In the following, we describe an elementary genetic algorithm framework in C, which can be used for simple robotics projects. The base data type is a bitstring of a certain length. The population size *POP* determines how many individuals (chromosomes) we are handling.

There are also free fully featured genetic algorithm libraries available, such as GA Lib[6] and OpenBeagle.[7] These allow to design a working genetic algorithm without having to implement the low-level functions.

Program 20.1: GA Basic Functions

```
BYTE pool[POP][SIZE], // 0 (false) or 1 (true) only
next[POP][SIZE];

// globals
float fitlist[POP]; // list of all fitness values
float fitsum;       // sum of all fitness values
```

```
// init all gene pool and goal values
void init()
{ // set all bits in pool to random
  for (int i=0; i<POP; i++)
    for (int j=0; j<SIZE; j++) pool[i][j] = rand()%2;
}
```

```
// select gene according to rel. fitness
int selectgene()
{ int i, wheel;
  float count;

  wheel = rand() % lroundf(fitsum); // range [0..fitsum-1]
  i=0;
  count = fitlist[0];
  while (count < wheel)
  { i++;
    count += fitlist[i];
  }
  return i;
}
```

```
// crossover 2 selected genes
void crossover(int g1, int g2, int pos)
{ int cut = rand()%(SIZE-1) +1; // range [1.. SIZE-1]
  memcpy(next[pos ], pool[g1], cut);
  memcpy(next[pos ]+cut, pool[g2]+cut, SIZE-cut);
  memcpy(next[pos+1], pool[g2], cut);
```

---

[6]GALib Galib – *A C++ Genetic Algorithms Library*, online: http://web.mit.edu/galib/www/GAlib.html.

[7]J. Beaulieu, C. Gagné, *Open BEAGLE – A Versatile Evolutionary Computation Framework*, Dèpartement de Génie Électrique et de Génie Informatique, Université Laval, Québec, Canada, online: https://github.com/chgagne/beagle.

```
  memcpy(next[pos+1]+cut, pool[g1]+cut, SIZE-cut);
}
```

```
// mutation: flip a single bit position
void mutation()
{ int pos = rand() % (POP-1) + 1; // 1.. POP-1 (keep best at [0])
  int bit = rand() % SIZE;
  next[pos][bit] = ! next[pos][bit]; // flip bit
  printf("flip pos %d bit %d\n", pos, bit);
}
```

The basic GA functions in Program 20.1 show the fundamental routines of the generic algorithm, which are random initialization, crossover and mutation. We define *pool* as an array of bit strings (chromosomes), using data type *BYTE* instead of *Boolean* for convenience. Population size (*POP*) and bit string length (*SIZE*) can be adjusted by constants. We use a second array (*next*) for copying over the whole population for the next generation.

Program 20.2: GA Selection and Evaluation

```
// select gene according to rel. fitness
int selectgene()
{ int i, wheel;
  float count;
  wheel = rand() % lroundf(fitsum); // range [0..fitsum-1]
  i=0;
  count = fitlist[0];
  while (count < wheel)
  { i++;
    count += fitlist[i];
  }
  return i;
}
```

```
// fitness function for gene j
float fitness(int gene)
{ float fit=0.0;
  for (int k=0; k<ELEM; k++)
    fit += 1.0 - fabs(goal[k] - slice(gene,k));
  return fit;
}
```

```
// evaluate all genes in pool
void evaluate()
{ float fit;
```

```
fitsum = 0.0;
maxfit = 0.0;
for (int i=0; i<POP; i++)
{ fit = ELEM;
  for (int j=0; j<ELEM; j++)
    fit -= fabs(goal[j] - slice(i,j));
  fitlist[i] = fit;
  fitsum += fit;
  if (fit>maxfit) // record max fitness to terminate
  { maxfit=fit; maxpos=i;
  }
}
}
```

The *selectgene* function in Program 20.2 uses a method we originally devised for a parallel programming implementation and that we named *wheel of fitness* in a reference to the *Wheel of Fortune* game show. If we visualize the list of genes with their relative fitness values as circle segments, then the complete gene pool resembles a fortune wheel (or a pie chart). Giving it a spin with random force will select a gene with a probability relative to its fitness (see Figure 20.9). So, a gene with a high fitness is more likely to be chosen (even multiple times) than a gene with a low fitness. But even a gene with a low fitness has a nonzero chance of being selected.

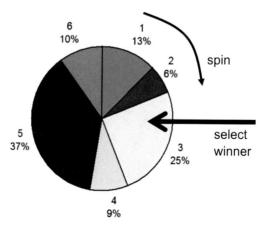

Figure 20.9: Sample pie chart with six genes and their relative fitness (wheel of fitness)

The random value used in the *selectgene* function resembles the spinning force of a fortune wheel, but it is limited to one full revolution through using the modulo operation. The remainder of this function is finding out which gene corresponds to the random number. This implementation has the significant advantage over other methods that it does not require any sorting of fitness values or gene indices—a large computational saving of $O(n \cdot \log(n))$ steps.

The *evaluate* function in Program 20.2 loops through all chromosomes to determine their fitness and also calculate the sum of all fitness values, which is require for the next round of the *wheel of fitness*. The *fitness* function itself needs to be adjusted for each problem to be solved. In our example, we compare the chromosome to a predefined number sequence by cutting it into 5-bit number slices. The difference of each slice to the goal number is then added to the overall fitness value.

The *main* routine in Program 20.3 now gets quite straightforward. After calling the *init* function, the central *for*-loop calls *evaluate*, which calculates the fitness function for the given problem. Function *memcopy* is used to preserve the best performing individual. For the remaining $n - 1$ places in the next generation, pairwise selection and crossover are executed. Mutations are handled in a separate *for*-loop. Finally, *memcopy* copies back the whole array for the next generation.[8]

Program 20.3: GA Main Program

```
int main()
{ int iter, pos, s1, s2;

  init();

  for (iter=0; iter<MAX_ITER; iter++)
  { evaluate();
    // SELECT + CROSSOVER
    memcpy(next[0], pool[maxpos], SIZE); // preserve best gene
    for (pos=1; pos<POP; pos+=2)
    { s1 = selectgene();
      s2 = selectgene();
      crossover(s1,s2, pos);
    }
    for (int m=0; m<MUT; m++) mutation(); // perform mutations
    memcpy(pool, next, POP*SIZE); // copy back generation
  }
}
```

[8]Note: This functionality could have been achieved much quicker and in half the memory space by using pointers.

Figure 20.10, left, shows the approximation of a sine curve with 25 chromosomes over 50 generations. Displayed in Figure 20.10, right, are the curves for best and average fitness values over all generations.

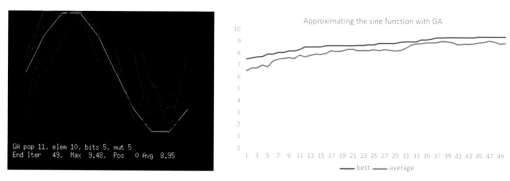

Figure 20.10: GA approximation of sine curve (left) and fitness value improvements over generations (right)

20.5   Genetic Robot Control

Genetic algorithms have been applied to the evolution of neural controllers for robot locomotion by a number of researcher groups. A genetic algorithm evolves the weights between interconnected neurons to construct a controller that achieves the desired movement. Neuron inputs are taken from various sensors on the robot, and the outputs of certain neurons are directly connected to the robot's actuators. Lewis, Fagg, Bekey[9] successfully generated gaits for a Hexapod robot using a traditional genetic algorithm with one-point crossover and mutate. A simple neural network controller was used to control the robot, and the fitness of the generated individuals was evaluated by human operators.

Ijspeert[10] evolved a controller for a simulated salamander using an enhanced genetic algorithm. The neural model employed was biologically based and quite complex. The developed system was capable of operating without human intervention.

At UWA, we are using genetic algorithms for finding spline parameters for biped robot locomotion,[11,12] as outlined in detail in the following section.

---

[9]M. Lewis, A. Fagg, G. Bekey, *Genetic Algorithms for Gait Synthesis in a Hexapod Robot, in Recent Trends in Mobile Robots*, World Scientific, New Jersey, 1994, pp. 317–331 (15).

[10]A. Ijspeert, *Evolution of neural controllers for salamander-like locomotion*, Proceedings of Sensor Fusion and Decentralised Control in Robotics Systems II, 1999, pp. 168–179 (12).

[11]A. Boeing, T. Bräunl, *Evolving a Controller for Bipedal Locomotion*, Proceedings of the Second International Symposium on Autonomous Minirobots for Research and Edutainment, AMiRE 2003, Brisbane, Feb. 2003, pp. 43–52 (10).

[12]A. Boeing, T. Bräunl, *Evolving Splines: Evolution of Locomotion Controllers for Legged Robots*, in: Tzyh-Jong Tarn, Shan-Ben Chen and Changjiu Zhou (Eds.), Robotic Welding, Intelligence and Automation, Lecture Notes in Control and Information Sciences, vol. 299, Springer-Verlag, 2004, pp. (14).

Genetic algorithms have been used in a variety of different ways to newly produce or optimize existing behavioral controllers. Ram et al.[13] used a genetic algorithm to control the weights and internal parameters of a simple reactive schema controller. In schema-based control, primitive motor and perceptual schemas do simple distributed processing of inputs (taken from sensors or other schemas) to produce outputs. Motor schemas asynchronously receive input from perceptual schemas to produce response outputs intended to drive an actuator. A schema arbitration controller produces output by summing contributions from independent schema units, each contributing to the final output signal sent to the actuators according to a weight (Figure 20.11).

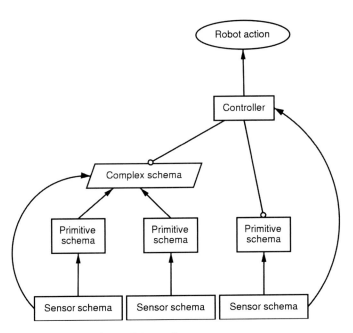

Figure 20.11: Schema hierarchy

The approach taken by Ram et al. was to use a genetic algorithm to determine an optimal set of schema weights for a given fitness function. By tuning the parameters of the fitness function, robots optimized for the qualities of safety, speed and path efficiency were produced. The behavior of each of these robots was different from any of the others. This demonstrates how behavioral outcomes may be easily altered by simple changes in a fitness function.

---

[13]A. Ram, R. Arkin, G. Boone, M. Pearce, *Using Genetic Algorithms to Learn Reactive Control Parameters for Autonomous Robotic Navigation*, Journal of Adaptive Behaviour, vol. 2, no. 3, 1994, pp. 277–305 (29).

Harvey, Husbands and Cliff[14] used a genetic algorithm to evolve a robot neural net controller to perform the tasks of wandering and maximizing the enclosed polygonal area of a path within a closed space. The controller used sensors as its inputs and was directly coupled to the driving mechanism of the robot. A similar approach was taken in Venkitachalam[15]; however, the outputs of the neural network were used to control schema weights. The neural network then produces dynamic schema weights in response to input from perceptual schemas.

20.6 Starman

Mr. Star-Man is a simulated 2D mobile robot introduced by Ngo and Marks.[16] It has five identical actuated legs (one hinge joint per leg) on a cylindrical body, as shown in Figure 20.12. It is an ideal application for genetic algorithms since the problem is sufficiently hard, so there is no obvious procedural solution and a genetic algorithm might reveal surprising locomotion techniques (reminiscent of *The Ministry of Silly Walks*[17]).

Figure 20.12: Starman model

The Starman model is based on the following assumptions:

- The robot lives in 2D.
  So the robot can never fall over sideways.

---

[14]I. Harvey, P. Husbands, D. Cliff, *Issues in Evolutionary Robotics*, in J. Meyer, S. Wilson (Eds.), From Animals to Animats 2, Proceedings of the Second International Conference on Simulation of Adaptive Behavior, MIT Press, Cambridge MA, 1993.

[15]D. Venkitachalam, *Implementation of a Behavior-Based System for the Control of Mobile Robots*, B.E. Honours Thesis, The University of Western Australia, Electrical and Computer Engineering, supervised by T. Bräunl, 2002.

[16]A. Fukunaga, L. Hsu, P. Reiss, A. Shuman, J. Christensen, J. Marks, J. Ngo, *Motion-synthesis techniques for 2D articulated figures*. Technical Report TR-05-94, Center for Research in Computing Technology, Harvard University, March 1994, MERL-TR-94-11 August 12 1994, pp. (10).

[17]Monty Python's Flying Circus, episode 14, 1970.

- Fitness can simply be measured by the distance the robot has travelled within a fixed simulated timeframe.
- The robot can rotate each of its five legs about its body center by applying a force (or impulse in some physics engines).
- The maximum rotation angle of each leg is $\pm 36° = (360°/5)/2$. That way legs can never collide with each other.

For encoding Starman's motions in a way suitable for genetic algorithms, we are using a simplified variation of the algorithm presented:

- All leg motions are repetitive with a fixed cycle time (e.g. 1 s).
- Each of the five leg motions is represented by a spline function with a fixed number of control points (e.g., 10 points for a 1 s iteration).
- Starman's complete motion behavior can be described by $L \cdot T$ Bytes with L = number of legs, and T = number of time steps, e.g., $5 \cdot 10 = 50$ Bytes.
  After 1 s of simulation is up, the spline functions will be repeated from the start.
- An open-source physics engine (such as Unity[18] or Bullet[19]) is required for carrying out the physics simulation. Its handling of friction between the legs and the ground is crucial.

Figure 20.13 shows some Starman simulation results using the Bullet physics engine and the above-mentioned GA method. Note that the body has been extruded to a full 3D model.

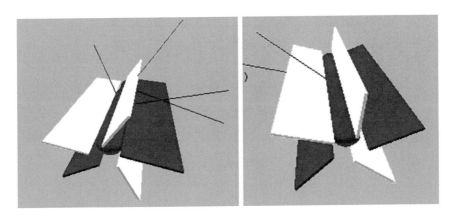

Figure 20.13: Starman simulation

[18]Unity, *Create with Unity*, online: https://unity.com/products.
[19]Pybullet, *Pybullet/Bullet Physics*, online: https://pybullet.org/wordpress/.

20.7 Evolving Walking Gaits

Designing or optimizing control systems for legged locomotion is a complex and time-consuming process. Human engineers can only produce and evaluate a limited number of configurations, although there may be numerous competing designs that should be investigated. Automation of the controller design process allows the evaluation of thousands of competing designs, without requiring prior knowledge of the robot's walking mechanisms (see Ledger[20]). Development of an automated approach requires the implementation of a control system, a test platform and an adaptive method for automated design of the controller. Thus, the implemented control system must be capable of expressing control signals that can sufficiently describe the desired walking pattern. Furthermore, the selected control system should be simple to integrate with the adaptive method.

One method for automated controller design is to utilize a spline controller and evolve its control parameters with a genetic algorithm (see Boeing, Bräunl[21,22]). To decrease the evolution time and remove the risk of damaging robot hardware during the evolution, a dynamic simulation system can be employed as a first step. After this, the simulation results have to be adapted to the real walking robot.

20.7.1 Splines

Splines are a set of special parametric curves with certain desirable properties. They are piecewise polynomial functions, expressed by a set of control points. There are many different forms of splines, each with their own attributes (Bartels, Beatty, Barsky[23]); however, there are two desirable properties:

- Continuity, so the generated curve smoothly connects its parts.
- Locality of the control points, so the influence of a control point is limited to a neighborhood region.

The Hermite spline is a special spline with the unique property that the curve generated from the spline passes through the control points that define the spline. Thus, a set of predetermined points can be smoothly interpolated by

---

[20]C. Ledger, *Automated Synthesis and Optimization of Robot Configurations*, Ph.D. Thesis, Carnegie Mellon University, 1999.

[21]A. Boeing, T. Bräunl, *Evolving Splines: An alternative locomotion controller for a bipedal robot*, Proceedings of the Seventh International Conference on Control, Automation, Robotics and Vision (ICARV 2002), CD-ROM, Nanyang Technological University, Singapore, Dec. 2002, pp. 1–5 (5).

[22]A. Boeing, T. Bräunl, *Dynamic Balancing of Mobile Robots in Simulation and Real Environments*, in Dynamic Balancing of Mechanisms and Synthesizing of Parallel Robots, in Dan. Zhang, Bin Wei (Eds.), Springer International, Cham Switzerland, Dec. 2015, pp. 457–474 (18).

[23]R. Bartels, J. Beatty, B. Barsky, *An Introduction to Splines for Use in Computer Graphics and Geometric Models*, Morgan Kaufmann, San Francisco CA, 1987.

simply setting these points as control points for the Hermite spline. Each segment of the curve is dependent on only a limited number of neighboring control points. Thus, a change in the position of a distant control point will not alter the shape of the entire spline. The Hermite spline can also be constrained to achieve C^{K-2} continuity.

The function used to interpolate the control points, given starting point p_1, ending point p_2, tangent values t_1 and t_2, and interpolation parameter s, is shown below:

$$f(s) = h_1 p_1 + h_2 p_2 + h_3 t_1 + h_4 t_2$$

where

$$
\begin{aligned}
h_1 &= 2s^3 - 3s^2 + 1 \\
h_2 &= -2s^3 + 3s^2 \\
h_3 &= s^3 - 2s^2 + s \\
h_4 &= s^3 - s^2
\end{aligned}
$$

for $0 \le s \le 1$

Program 20.4 shows the routine used for evaluating splines. Figure 20.14 shows the output from this function when evaluated with a starting point at one, with a tangent of zero, and an ending point of zero with a tangent of zero. The *Hermite_Spline* function was then executed with s ranging from zero to one.

Program 20.4: Evaluating a Simple Cubic Hermite Spline Section

```
float Hermite_Spline(float s)
{ float ss = s*s;
  float sss= s*ss;
  float h1 = 2*sss - 3*ss + 1;
  float h2 = -2*sss + 3*ss;
  float h3 = sss - 2*ss + s;
  float h4 = sss - ss;
  float value = h1*starting_point_location
              + h2*ending_point_location
              + h3*tangent_for_starting_point
              + h4*tangent_for_ending_point;
  return value;
}
```

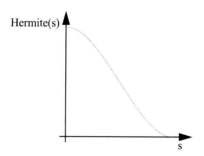

Figure 20.14: Cubic Hermite spline curve

20.7.2 Control Algorithm and Feedback

Larger, more complex curves can be achieved by concatenating a number of cubic Hermite spline sections. This results in a set of curves that are capable of expressing the control signals necessary for legged robot locomotion. The spline controller consists of a set of joined Hermite splines. The first set contains robot initialization information, to move the joints into the correct positions and enable a smooth transition from the robot's starting state to a traveling state. The second set of splines contains the cyclic information for the robot's gait. Each spline can be defined by a variable number of control points with variable degrees of freedom. Each pair of a start spline and a cyclic spline corresponds to the set of control signals required to drive one of the robot's actuators.

An example of a simple spline controller for a robot with three joints (three degrees of freedom) is shown in Figure 20.15. Each spline indicates the controller's output value for one actuator.

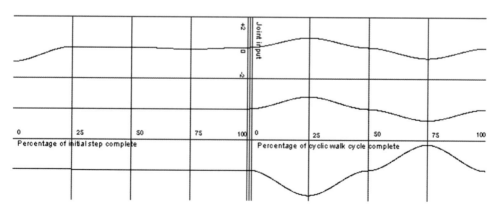

Figure 20.15: Spline joint controller by Adrian Boeing, UWA

There are a number of advantages offered by Hermite spline controllers. Since the curve passes through all control points, individual curve positions can be predetermined by a designer. This is especially useful in situations where the control signal directly corresponds to angular values, which in our case are servo positions. Program 20.5 provides a simplified code snippet for calculating the position values for a one-dimensional spline.

Program 20.5: Evaluating a Concatenated Hermite Spline

```
Hspline hs[nsec]; // A spline with nsec sections
float SplineEval(float s)
{ int sect; //what section are we in?
  float z; // how far into that section are we?
  float secpos;
  secpos=s*(nsec-1);
  sect=(int)floorf(secpos);
  z=fmodf(secpos,1);
  return hs[sect].Eval(z);
}
```

There is a large collection of evidence that supports the proposition that most gaits for both animals and legged robots feature synchronized movement (see Reeve[24]). That is, when one joint alters its direction or speed, this change is likely to be reflected in another limb. Enforcing this form of constraint is far simpler with Hermite splines than with other control methods. In order to force synchronous movement with a Hermite spline, all actuator control points must lie at the same point in cycle time. This is because the control points represent the critical points of the control signal when given default tangent values.

Most control methods require a form of feedback in order to correctly operate. Spline controllers can achieve walking patterns without the use of feedback; however, incorporating sensory information into the control system allows a more robust gait. The addition of sensory information to the spline control system will enable a bipedal robot to maneuver on uneven terrain, as shown in an experiment below.

In order to incorporate sensor feedback information into the spline controller, the controller is extended into another dimension. The extended control points specify their locations within both the gait's cycle time and the feedback value. This results in a set of control surfaces for each actuator. Extending the controller in this form significantly increases the number of control points required. Figure 20.16 shows a resulting control surface for one actuator.

---

[24]R. Reeve, *Generating walking behaviours in legged robots*, Ph.D. Thesis, University of Edinburgh, 1999.

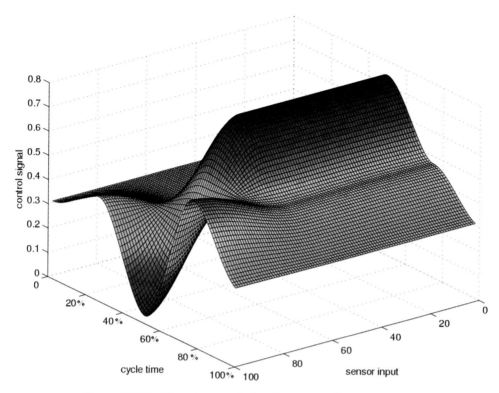

Figure 20.16: Generic extended spline controller by Adrian Boeing, UWA

The actuator evaluates the desired output value from the enhanced controller as a function of both the cycle time and the input read from the sensor. The most appropriate sensory feedback was found to be an angle reading from an inclinometer placed on the robot's central body (torso). Thus, the resultant controller is expressed in terms of the percentage cycle time, the torso angle and the output control signal.

20.7.3 Controller Evolution

The first step in building a genetic controller is to find an encoding format that can be evolved by the genetic algorithm. The parameters for the spline control system are simply the position and tangent values of the control points that are used to describe the spline. Thus, each control point has three different values that can be encoded:

- The position in the cycle time (i.e., position along the x-axis)
- The value of the control signal at that time (i.e., position along the y-axis)
- The tangent value.

461

To allow these parameters to evolve with a genetic algorithm in minimal time, a more compact format of representing the parameters is desired. This can be achieved by employing fixed point values instead of floating-point numbers.

For example, if we wanted to encode the range [0, 1] using 8-bit fixed point values, then the 8 bits can represent any integer value from 0 to 255. By simply dividing this value by 255, we can represent any number ranging from 0 to 1, with an accuracy of 0.004 (= 1/256).

The curve shown in Figure 20.15 was generated by a one-dimensional spline function, with the first control point ($s = 0$) at position 1 with tangent value of 0, and the second control point ($s = 1$) at position 0 with tangent value of 0. If an encoding which represented each value as an 8-bit fixed point number from 0 to 1 is used, then the control parameters in this case would be represented as a string of 3 bytes with values of (0, 255, 0) for the first control point's position and tangent, and (255, 0, 0) for the second control point's position and tangent.

Thus, the entire spline controller can be directly encoded using a list of integer control point values for each actuator. An example structure to represent this information is shown in Program 20.6.

Program 20.6: Direct Encoding Structure

```
struct controlpoint
{ unsigned char x, y, tangent;
};
struct splinecontroller
{ controlpoint init_spline [num_splines][num_controlpoints];
  controlpoint cyclic_spline[num_splines][num_controlpoints];
};
```

There are a number of methods for optimizing the performance of the genetic algorithm. One method for increasing the algorithm's performance is *staged evolution*. This concept is an extension to *Behavioural Memory* and was first applied to controller evolution by Lewis, Fagg, Bekey.[25] Staged evolution divides a problem task into a set of smaller, manageable challenges that can be sequentially solved. This allows an early, approximate solution to the problem to be solved. Then, incrementally increasing the complexity of the problem provides a larger solution space for the problem task and allows for further refinements of the solution. Finally, after solving all the problem's subtasks, a complete solution can be determined. Solving the sequence of subtasks is typically achieved in less time than required if the entire problem task is tackled without decomposition.

---

[25]M. Lewis, A. Fagg, G. Bekey, *Genetic Algorithms for Gait Synthesis in a Hexapod Robot*, in Recent Trends in Mobile Robots, World Scientific, New Jersey, 1994, pp. 317–331 (15).

This optimization technique can also be applied to the design of the spline controller. The evolution of the controller's parameters can be divided into the following three phases:

1. Assume that each control point is equally spaced in the cycle time. Assume the tangent values for the control points are at a default value. Then only evolve the parameters for the control points' output signal (y-axis).
2. Remove the restriction of equidistant control points and allow the control points to be located at any point within the gait time (x-axis).
3. Allow refinement of the solution by evolving the control point tangent values.

To evolve the controller in this form, a staged encoding method is required. Table 20.5 indicates the number of control points required to represent the controller in each phase. In the case of an encoding where each value is represented as an 8-bit fixed point number, the encoding complexity directly corresponds to the number of bytes required to describe the controller.

Evolution Phase	Encoding Complexity
Phase 1	$a(s + c)$
Phase 2	$2a(s + c)$
Phase 3	$3a(s + c)$

Table 20.5 Encoding complexity with a (number of actuators), s (number of initialization control points), c (number of cyclic control points)

20.7.4 Controller Assessment

In order to assign a fitness value to each controller, a method for evaluating the generated gait is required. Since many of the generated gaits result in the robot eventually falling over, it is desirable to first simulate the robot's movement in order to avoid damaging the real robot hardware. There are many different dynamic simulators available that can be employed for this purpose.

One such simulator is *DynaMechs*.[26] The simulator implements an optimized version of the Articulated Body algorithm and provides a range of integration methods with configurable step sizes. The package is free, open source and can be compiled for Windows or Linux. The simulator provides information about an actuator's location, orientation and forces at any time, and this information can be utilized to determine the fitness of a gait.

---

[26]S. McMillan, *DynaMechs (Dynamics of Mechanisms): A Multibody Dynamic Simulation Library*, online: http://dynamechs.sourceforge.net/index.html.

A number of fitness functions have been proposed to evaluate generated gaits. Reeve[27] proposed the following sets of fitness measures:

- FND (forward not down)
  The average speed the walker achieves minus the average distance of the center of gravity below the starting height.
- DFND (decay FND)
  Similar to the FND function, except it uses an exponential decay of the fitness over the simulation period.
- DFNDF (DFND or fall)
  As above, except a penalty is added for any walker whose body touches the ground.

These fitness functions do not consider the direction or path that is desired for the robot to walk along. Thus, more appropriate fitness functions can be employed by extending the simple *FND* function to include path information and including terminating conditions. The terminating conditions assign a very low fitness value to any control system which generates a gait that results in:

- A robot's central body coming too close to the ground.
  This termination condition ensures that robots do not fall down.
- A robot that moves too far from the ground.
  This removes the possibility of robots achieving high fitness values early in the simulation by propelling themselves forward through the air (*jumping*).
- A robot's head tilting too far forward.
  This ensures the robots are reasonably stable and robust.

Thus, the overall fitness function is calculated, taking into account the distance the robot moves along the desired path, plus the distance the robot deviates from the path, minus the distance the robot's center of mass has lowered over the period of the walk, as well as the three terminating conditions.

20.7.5 Evolved Walking Gaits

This system is capable of generating a wide range of gaits for a variety of robots. Figure 20.17 shows a gait for a simple bipedal robot. The robot moves forward by slowly lifting one leg by rotating the hip forward and knee backward, then places its foot further in front, straightens its leg and repeats this process. The genetic algorithm typically requires the evaluation of only 1'000 individuals to evolve an adequate forward walking pattern for a bipedal robot.

---

[27]R. Reeve, *Generating walking behaviours in legged robots*, Ph.D. Thesis, University of Edinburgh, 1999.

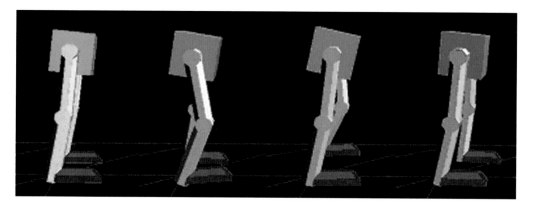

Figure 20.17: Biped gait, A. Boeing, UWA

Figure 20.18 shows a gait generated by the system for a tripod robot. The robot achieves forward motion by thrusting its rear leg toward the ground and lifting its forelimbs. The robot then gallops with its forelimbs to produce a dynamic gait. This illustrates that the system is capable of generating walking patterns for legged robots, regardless of the morphology and number of legs.

Figure 20.18: Tripod gait, A. Boeing, UWA

The spline controller also evolves complex dynamic movements. Removing the termination conditions allows for less stable and robust gaits to be evolved. Figure 20.19 shows a jumping gait evolved for an android robot. The resultant control system depicted was evolved within 60 generations and began convergence toward a unified solution within 30 generations. However, the gait was very unstable, and the android could only repeat the jump a few times before it would fall over.

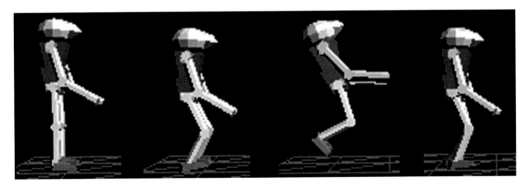

Figure 20.19: Biped jumping, A. Boeing, UWA

The spline controller shown in Figure 20.16 uses sensory information from an inclinometer located in the robot's torso. The inclinometer reading was successfully interpreted by the control system to provide an added level of feedback capable of sustaining the generated gait over non-uniform terrain. An example of the resultant gait is shown in Figure 20.20. The controller was the result of 512 generations of evaluation.

Figure 20.20: Biped walking over uneven terrain, A. Boeing, UWA

The results in Figure 20.21 demonstrate the increase in fitness value during the evolution of the extended controller. A rapid increase in fitness values can clearly be observed at around 490 generations. This corresponds to the convergence point where the optimal solution is located. The sharp increase is a result of the system managing to evolve a controller that was capable of traversing across flat, rising and lowering terrains.

This chapter presented the fundamentals of genetic algorithms including a practical application of walking robot gait generation. The control system was shown to describe complex dynamic walking gaits for robots with differing morphologies. A similar system can be employed to control any robot consisting of multiple actuators and could also be extended to evolve the robot's morphology in unison with the controller. This would enable the robot's physical design to be improved to optimally suit its desired purpose. Further extensions of this could be used to automatically construct the designed robots using 3D printing technology, removing the human completely from the robot design process (see Lipson, Pollack[28]).

---

[28]H. Lipson, J. Pollack, *Evolving Physical Creatures*, 2006, online: http://citeseer.nj.nec.com/523984.html.

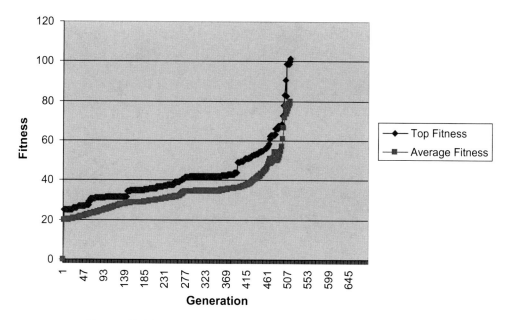

Figure 20.21: Fitness versus generation for extended spline controller

20.8 Tasks

1. Use a GA to evolve a walking gait for Starman. Use the distance covered from its starting position over a fixed amount of simulation time as the fitness value.

2. Use a GA to evolve the shortest collision-free path of a manipulator arm from A to B in the presence of multiple obstacles.

3. Use a GA to evolve the shortest driving path of a robot in a distance node graph.
Refer to the *Travelling Salesman Problem*[29].

4. Use a physics simulation engine in combination with a GA to evolve a gait for a 6-legged walking robot.

5. Use a physics simulation engine in combination with a GA to evolve a gait for a bipedal walking robot.

6. Use a physics simulation engine in combination with a GA to evolve the creature's physical structure alongside its motion algorithm.

---

[29]Wikipedia: *Travelling salesman problem*, online: https://en.wikipedia.org/wiki/Travelling_salesman_problem.

DEEP LEARNING

21

Most advances in engineering come from the availability of techno-
logically superior components. For example, researchers and engineers can spend years designing, building and perfecting a sensor
system, but once you can buy a commercially built system for a reasonable
price, this opens up a lot more applications.

There have been enormous advancements especially in robot sensor technology over the last couple of decades. Especially cameras, Lidars and IMUs
have technically matured and come down in price to a point where almost
every robotics application will want to use them.

Development of software packages has taken a similar path. Not having to
program neural networks or genetic algorithms from the ground up (like we
have done for educational purposes in the two previous chapters) will greatly
increase software productivity and outcomes. By using advanced AI and
learning tools, it is now possible to set up and train complex learning systems
in a relatively short time. Acknowledging this principle, we want to look at the
usage of advanced deep learning AI tools in this chapter, rather than trying to
code such a system from scratch.

The term *deep learning* refers to the depth of neural networks used for a
particular application. While traditional neural networks often had only three
layers (see chapter on neural networks), modern deep learning networks can
have dozens of layers. For example, Nvidia's network for end-to-end learning
for self-driving cars (see Bojarski et al.[1]) has 12 layers, Google's general-
purpose network *MobileNets*[2] has 25 layers.

---

[1]Bojarski, Testa, Dworakowski, Firner, Flepp, Goyal, Jackel, Monfort, Muller, Zhang, Zhang, Zhao, Zieba, *End to End Learning for Self-Driving Cars*, arXiv:1604.07316, 2016, pp. (9).

[2]Howard, Zhu, Chen, Kalenichenko, Wang, Weyand, Andreetto, Adam, *MobileNets: Efficient Convolutional Neural Networks for Mobile Vision Applications*, arXiv:1704.04861v1, 2017, pp. (9).

© The Author(s), under exclusive license to Springer Nature Singapore Pte Ltd. 2022
T. Bräunl, *Embedded Robotics*,
https://doi.org/10.1007/978-981-16-0804-9_21

This is an interesting repetition in the history of neural networks. The first neural networks proposed by Rosenblatt as *Perceptrons*[3] only had two layers and since it could be proven that such simple networks cannot solve any meaningful problem, neural network research did not progress for many years until three-layer networks were proposed (see, e.g., Rumelhart, McClelland[4]). After that, there was the belief that three layers are sufficient and even a proof was done that any pattern recognition problem can be solved in principle by a network with only three layers.[5] This has now been superseded by the huge success of deep neural networks in the last few years.

Figure 21.1 shows the principle of several problem-solving techniques. While in the traditional engineering approach, the complete solution is coded directly, the standard AI-neural network approach uses only a relatively shallow network. As a consequence, a significant part of the problem solution still has to be coded, e.g., in the form of data preprocessing. In the deep learning method, on the other hand, especially for end-to-end learning, a large network takes care of solving the complete problem. No additional coding is required.

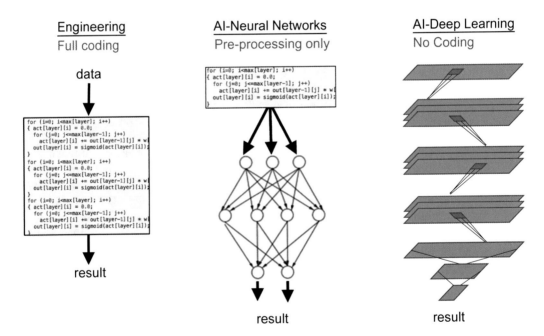

Figure 21.1: Problem-solving methods

[3]F. Rosenblatt, *Principles of Neurodynamics*. Spartan Books, Washington DC, 1962.

[4]D. Rumelhart, J. McClelland (Eds.), *Parallel Distributed Processing*, 2 vols., MIT Press, Cambridge MA, 1986.

[5]This somehow reminds me of the proof from my student days that anything computable can be executed by a Turing machine.

21.1   TensorFlow and Caffe

TensorFlow[6] presents itself as an *"end-to-end opensource machine learning platform"* and was developed as an open-source project by Google. Caffe[7] describes itself as a *"deep learning framework with expression, speed and modularity in mind"* and was developed by the Berkeley AI Research team (BAIR). Caffe uses the MPI multi-tasking communication library for multi-node applications but lacks the high-level API and Python tools offered in TensorFlow. Because of the lack of Python support, conducting training sequences in Caffe requires the use of the C++ command-line interface.

The Caffe framework has a significantly better performance of factor 1.2 to 5.0 than TensorFlow on some benchmarking[8] tests, which gives it an advantage for a use in production systems, but TensorFlow allows easier experimentation for research applications due to its easy-to-use API.

Figure 21.2 shows a screenshot from the *TensorFlow playground*.[9] This Web page allows the setting or various learning parameters and applies them directly to a training pattern. In the middle of the page, a neural net of multiple layers can be constructed. The results can be immediately seen graphically in the 2D map on the right.

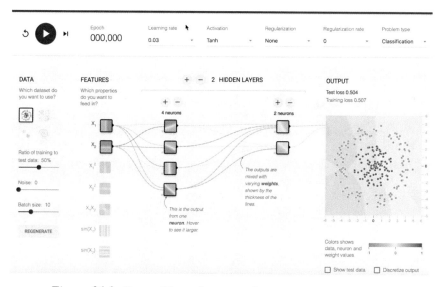

Figure 21.2: TensorFlow playground

[6]TensorFlow, *An end-to-end open source machine learning platform*, online: https://www.tensorflow.org, 2020.

[7]Caffe, *Deep learning framework*, online: https://caffe.berkeleyvision.org, 2020.

[8]TensorFlow vs Caffe, online: https://www.educba.com/tensorflow-vs-caffe/.

[9]TensorFlow Playground, online: https://playground.tensorflow.org/.

TensorFlow provides a number of tools that aid tailoring of the learning process. TensorBoards[10] are a valuable visualization tool for keeping track of a system's learning progress and to profile and fine-tune learning parameters. Figure 21.3 shows the TensorBoards for a traffic sign recognition implementation by Jordan King[11] at UWA.

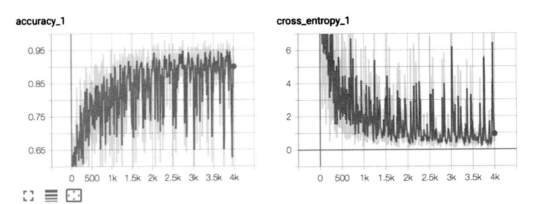

Figure 21.3: TensorBoard for traffic sign recognition problem showing accuracy (left) and cross-entropy (right) as a function of training cycles by J. King, UWA

21.2   Carolo-Cup Competition

Carolo-Cup[12] is an annual student event for autonomous model cars organized by the Technical University Braunschweig, Germany. Although the use of learning methods and AI tools is not a requirement, most teams do use various AI methods for their vehicles. Figure 21.4 shows an identical scene from our setup of the real[13] and simulated Carolo-Cup.

---

[10]TensorBoard, online: https://www.tensorflow.org/tensorboard.

[11]J. King, SIFT-like Keypoint Cluster-Based Traffic Sign Recognition with Deep Learning, Master thesis, supervised by T. Bräunl, UWA, Oct. 2019, pp. (61).

[12]Carolo-Cup, https://wiki.ifr.ing.tu-bs.de/carolocup/en/carolo-cup.

[13]The left side of our Carolo-Cup board is a recycled playing field from robot soccer (spot the goals) and the right side is a pegboard from the Micromouse competition (see the border pieces).

Figure 21.4: Scene from real (top) and simulated Carolo-Cup setup in EyeSim (bottom)

The Carolo-Cup allows the development and testing of tasks relevant for automotive research. The algorithms developed for the small robot cars can then be extended for applications in real autonomous vehicles. Areas covered in this competition are:

- Lane detection and lane keeping
- Collision avoidance
- Detection of other vehicles and pedestrians
- Traffic sign recognition
- Automated parking
- Automated overtaking
- Automated intersection control
- Automated zebra crossing detection
- Automated speed control following signage
- Vehicle-to-vehicle and vehicle-to-base-station communication

In the following sections, we are looking at just two aspects of the competition, traffic sign recognition and lane detection, both solved with deep learning methods.

21.3   Traffic Sign Recognition

One important feature of smart cars is automated traffic sign recognition. This problem has been investigated in the automotive industry for over 20 years, originally with *engineering-style* image processing algorithms, but more recently with deep learning approaches.

The traditional image processing approach for this task is to use shape (e.g., circle, square, triangle and hexagon) and color (e.g., red, blue and white) detection algorithms to find an image area that could qualify as a traffic sign, then search this area in more detail for lines, digits or symbols. As you can imagine, this is a tremendous software engineering task, especially as signs can be partially occluded or warped when seen at an angle.

Since neural networks are an ideal tool for classification, it is an obvious idea to use them for traffic sign recognition. However, only the strength of deep learning/deep neural networks has proven to deliver reliable results under realistic conditions. The end-to-end deep learning approach does away with any image preprocessing algorithms and presents the camera input directly to the deep neural network, which will then learn to detect and distinguish a set of traffic signs.

Figure 21.5: Some of the traffic signs used in Carolo-Cup

Jordan King[14] implemented a deep learning system for traffic sign recognition in the Carolo-Cup competition environment (Figure 21.5). Because of the limited processing power of the Raspberry Pi, a color-based region-of-interest (RoI) preprocessing step is used. Then, only smaller image snippets, possibly containing a traffic sign, are presented to the network. After converting the RBG image to HSI (hue, saturation, intensity), a hue histogram is generated to find possible traffic sign candidates, which are then classified by the deep learning network MobileNets.[15]

---

[14]N. Burleigh, J. King, T. Bräunl, *Deep Learning for Autonomous Driving*, Intl. Conference on Digital Image Computing: Techniques and Applications (DICTA), Dec. 2019, Perth, pp. (8).

[15]A. G. Howard and M. Zhu, *MobileNets: Open-Source Models for Efficient On-Device Vision*, 14 June 2017, online: https://ai.googleblog.com/2017/06/mobilenets-open-source-models-for. html.

Figure 21.6: Traffic sign preprocessing steps by Jordan King, UWA

In Figure 21.6, the individual steps are shown. Starting with the original RGB image (top left), we conduct an HSI thresholding for the red color hue. In the top right image, only those image pixels appear white, all others are black. A histogram of the matching red hue over the x-axis and y-axis (gray bars in the bottom left image) gives us the rough outline and size of the traffic sign candidate. Only this subimage is then classified by MobileNets (bottom right image) in a much shorter time than presenting the full image. The typical subimage size is only 128×128 pixels.

About 1'500 images from robots driving in the Carolo-Cup environment have been used to retrain the network from the ImageNet Large Visual Recognition Challenge.[16] This type of *transfer learning* is also referred to fine-tuning a model. Since only the last layers are retrained with the new dataset and all earlier layers stay as they are, this drastically reduces the amount of training time required.

---

[16]ImageNet Large scale Visual Recognition Challenge (ILSVRC), 2020, online: http://www.image-net.org/challenges/LSVRC/.

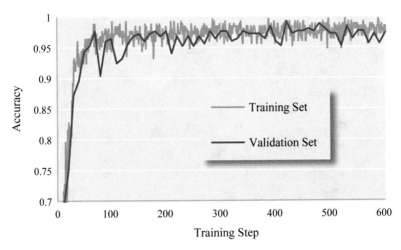

Figure 21.7: TensorFlow training and validation of traffic sign detection by J. King, UWA

Figure 21.7 shows the training process using TensorFlow's TensorBoard.[17] We can see that already after a few hundred training cycles the network achieves an accuracy above 95%. Of all recorded images, 85% were selected for the training set and 15% for the test set. The network's performance in precision and recall values demonstrates an almost perfect detection rate, as shown in Table 21.1.

Object class	True positives	True negatives	False positives	False negatives	Accuracy (%)
Background	64	161	0	1	99.56
Stop	33	192	1	0	99.56
Parking	40	184	2	0	99.12
Give way	39	187	0	0	100.00
Pedestrian Crossing	47	177	0	2	99.12

Table 21.1 Precision and recall for retrained network

---

[17]TensorBoard: TensorFlow's visualization toolkit, 2020, online: https://www.tensorflow.org/tensorboard.

Due to the small image snippet, the processing time is relatively short on a Raspberry Pi 3 controller, as shown in Table 21.2. The larger the network grows (in MB), the longer the processing time gets. While a small 2 MB network can be processed in 0.04 s (24 Hz), a large 17 MB network requires 0.27 s (3.7 Hz). This is the tradeoff between accuracy and processing speed. However, it will only get better with newer, faster embedded processor generations.

Architecture	Width multiplier	Input resolution	Model size (MB)	Mean inference time, T_i (s)
MobileNets-V1	0.25	128 × 128	1.90	0.04191
MobileNets-V1	0.50	128 × 128	5.34	0.10290
MobileNets-V1	0.75	128 × 128	10.35	0.16809
MobileNets V1	1.0	128 × 128	16.92	0.26692

Table 21.2 NN inference time on a Raspberry Pi 3B with TensorFlow Lite

Figure 21.8 shows the traffic sign detection from a related implementation for the Carolo-Cup environment by Sun et al., UWA.[18]

Figure 21.8: Traffic sign detection in Carolo-Cup environment by Sun et al., UWA

[18]S. Sun, J. Zheng, Z. Qiao, S. Liu, Z. Lin, T. Bräunl, *Architecture of a driverless robot car based on EyeBot system*, 3rd International Conference on Robotics: Design and Applications (RDA 2019), Xi'an, China, April 2019.

21.4 End-To-End Learning for Autonomous Driving

If deep neural networks can classify traffic signs and other objects, why not try them at classifying road views from the driver's perspective and let them decide on the correct steering angle? This is the idea behind deep learning for autonomous driving. As before, the less image preprocessing is required, the better. This end-to-end learning method for autonomous driving was first implemented and published by Nvidia researchers Bojarski et al.[19]

In a similar, but simplified approach, N. Burleigh[20] at UWA implemented an end-to-end deep learning method for autonomous robot driving in the Carolo-Cup environment (see Figure 21.9).

Figure 21.9: Carolo-Cup environment for autonomous driving and EyeBot robot

The deep network structure for this implementation is shown in Figure 21.10. A grayscale image of resolution 200×66 pixels is presented to the input layer of the deep neural network. Three subsequent 5×5 convolutional layers, followed by two 3×3 convolutional layers reduce the network to 18×64 neurons. Four subsequent fully connected layers reduce the net to 100, 50, 10 and finally just a single output neuron, which determines the steering angle of the autonomous vehicle.

---

[19]M. Bojarski, D. Del Testa, D. Dworakowski, B. Firner, B. Fleep, P. Goyal, L. Jackel, M. Monfort, U. Muller, J. Zhang, X. Zhang, J. Zhao, K. Zieba, *End to End Learning for Self-Driving Cars*, Nvidia Corporation, Holmdel, 2016, online: https://arxiv.org/abs/1604.07316.

[20]N. Burleigh, J. King, T. Bräunl, *Deep Learning for Autonomous Driving*, Intl. Conference on Digital Image Computing: Techniques and Applications (DICTA), Dec. 2019, Perth, pp. (8).

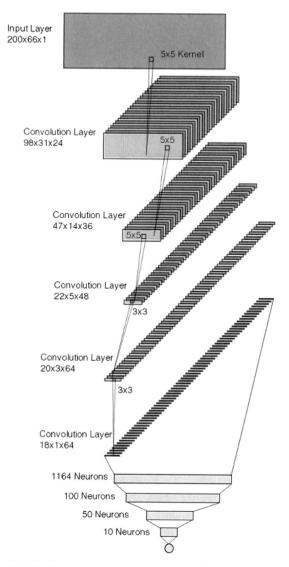

Figure 21.10: Deep network structure by N. Burleigh, UWA

To collect training data, the robot was manually driven around the track and about 1'000 images were captured together with the correct matching steering angles. Speed was reduced to 10 cm per second at a sampling rate of 5 Hz to give the human driver adequate time to select the best steering angle. Sample data collection and training was first conducted on the EyeSim simulator and then later repeated on the real EyeBot robots. In this process, every camera image will get an associated steering able (see sample images in Figure 21.11).

...   .

turn left   drive straight   turn right

Figure 21.11: Sample training images collected from a manual drive for turning left, driving straight, or turning right, respectively

Training was conducted in TensorFlow over 150'000 steps within 24 hours on a workstation with GPU coprocessors (Figure 21.12). The loss comes down to a manageable level after about 30'000 iterations.

A significant problem is the Raspberry Pi camera's limited field of view of only 62.2° vertical and 48.8° horizontal. This severely restricts the view of the lane markings, especially when driving a curve, and therefore deteriorates the learning performance for autonomous driving. Using an alternative camera module from a third-party supplier with an interchangeable board lens plus a wide fish-eye lens will improve learning rates and the overall lane detection performance.

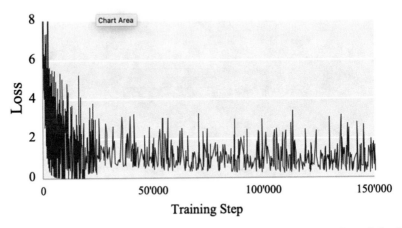

Figure 21.12: Training loss over training steps in TensorBoard, by B. Burleigh UWA

We use the following definition for an *autonomy rate (AR)*:

$$AR = \frac{N(interventions)}{N(track\_segments) \cdot N(completed\_circuits)}$$

With this, Table 21.3 shows the deep network's driving performance. Not surprisingly, driving failures (collisions) and near misses (temporary lane departures) increase at higher speeds when keeping the same processing frequency.

Speed (mm/s)	Avg. failures	Avg. near misses	Adjusted autonomy rate (%)
50	0	2	83.3
100	0	2	83.3
150	4	2	50.0

Table 21.3 Deep NN driving performance

Table 21.4 contrasts a conventional autonomous driving implementation based on a Hough line detection by N. Burleigh, UWA, with the deep neural network approach described here. Although the Hough method (itself a resource-hungry algorithm) runs faster than the deep learning method, the advantage of the neural network approach is that is does not require programming and is versatile in the sense that it can perform over a range of different driving scenarios and is more robust due to NN's inherent generalization ability.

Driving method	Frame rate (FPS)		
	Minimum	Average	Maximum
NN Driving	4	9	11
Hough Lines	13	16	20

Table 21.4 Comparison of traditional engineering implementation with deep learning

Program 21.1 shows the program section for recording training data. The robot camera's Sobel edge image is displayed as feedback for the driver. The original camera images are stored in every iteration (every 100 ms) in subdirectory *img* under continuously incremented filenames 0000.pgm, 0001.pgm and so on (see Figure 21.13). The corresponding steering angles are included as a comment in the second line of each PGM image file (see Figure 21.14).

Pressing *KEY1* and *KEY2* (either the physical buttons on the robot or a Bluetooth game controller, or the softkeys via remote desktop on a PC) will change the steering angle by $10°$ to the left or right, respectively.

Program 21.1: Recording Training Data

```
#include "eyebot.h"
#include "files.inc"
#define SAFE 300
#define HORIZON 120

int main ()
{ BYTE img[QVGA_PIXELS], sob[QVGA_PIXELS];
  char fname[20];
  int k, angle=0, frame=0;

  CAMInit(QVGA);
  LCDMenu("START", " ", " ", " ");
  KEYWait(KEY1);

  LCDMenu("LEFT", "RIGHT", " ", "END");
  do
  { CAMGetGray(img);
    IPSobel(img, sob);
    LCDImageGray(sob);
    LCDLine(0, HORIZON, 319, HORIZON, RED); // horizon
    k=KEYRead();
    if (k==KEY1) angle += 10;
        else if (k==KEY2) angle -= 10;
    LCDSetPrintf(18,0, "Curve %4d", angle);
    VWCurve(100, angle, 50); // moderate speed
    sprintf(fname, "img/%04d.pgm", frame++);
    IPWriteFileGrayCom(fname, img, angle); //write file
    OSWait(100); // wait 0.1 s
  } while (k != KEY4);
  return 0;
}
```

As the standard Carolo-Cup track shown in Figure 21.4 only has right turns, it is essential to let the robot also drive in the reverse direction when recording training data. Otherwise, the robot has never been exposed to left turns and the learning method will fail to perform correctly in a left turn.

21.4   End-To-End Learning for Autonomous Driving

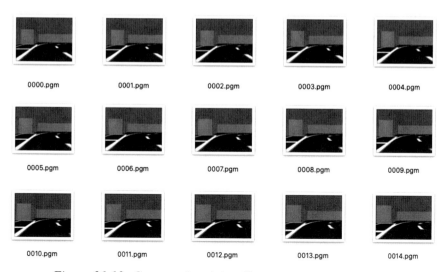

Figure 21.13: Generated training files

The ASCII header of each file (even binary grayscale files have ASCII headers) is shown in Figure 21.14. After the first line with the code for a binary PGM (*P5*) follows the second line with a comment, initiated by character '#' in which we have now written the current steering angle, in this case –20. The third line has the image dimensions in pixels and the fourth line shows the maximum grayscale value (255). After that follows the image data as a sequence of binary grayscale values (here omitted).

Any program that analyses this image file (as in our case the learning program) can now directly read the matching angle of the training pattern directly from the same file.

```
P5
#  -20
320  240
255
...
```

Figure 21.14: Sample header of generated image file (steering angle –20)

In Figure 21.15 we can now finally see the process of recording training data. The human driver needs to steer the robot as good as possible to get the best datasets for the subsequent learning phase. Program 21.1 immediately stores each camera image frame together with the current steering angle for the whole driving procedure. If driving errors are made, then these files can be deleted from the sequence before starting the learning procedure.

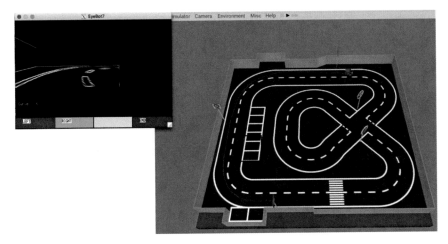

Figure 21.15: Collecting training data in EyeSim

21.5 Tasks

1. Make yourself familiar with TensorFlow by using the online TensorFlow Playground.
   Try different NN settings and observe the outcome in the detection field.

2. Use MobileNets to detect a limited number of office objects: phone, mug, mouse, monitor, laptop, keyboard, table, chair. Test your learning outcome by presenting various objects from these categories to the system.

3. Use TensorFlow for implementing a traffic sign recognition system as described in this chapter.

4. Write a program to semi-automatically collect image training data for Carolo Cup. A human driver can steer the robot along a near-optimal path.

5. Use TensorFlow for implementing a lane keeping system for autonomous driving in the Carolo Cup competition as described in this chapter.

6. Implement s system that automatically detects training scenarios that are underrepresented (e.g. more right curves than left curves). This can be used to expand the training set in a second step and thereby improve the learning outcome.

OUTLOOK

In this book, we have presented general and applied embedded systems, especially for mobile robot systems. We looked at hardware and CPU design, interfacing of sensors and actuators, feedback control, operating system functions, device drivers, multitasking and system tools. On the robot design side, we have presented driving, walking, swimming, diving and flying robots. And on the robot application side, we examined kinematics, localization, navigation and various AI techniques including neural networks and genetic algorithms. A number of detailed programming examples were presented to aid the understanding of this very practical subject area.

Of course, time does not stand still. When we started the EyeBot robots design, we developed several generations of our own EyeBot controller, originally based on a Motorola 68332, later based on a Gumstix chip set and a Xilinx FPGA. With the arrival of Arduino, Raspberry Pi and similar systems, we finally decided to make use of this cheap ubiquitous hardware and benefit from almost annually renewed and more powerful versions.

Arduino, Raspberry and similar are still experimental platforms and must not be used for industrial applications, especially not for anything safety-critical. Commercial embedded systems have to meet additional requirements such as an extended temperature range and electromagnetic compatibility (EMC), which will make them more reliable, but it also makes them more expensive for the same performance. These systems must be able to function in a harsh environment, at cold or hot temperatures, and in the presence of electromagnetic noise, while their own level of electromagnetic emission is strictly limited.

With such rapid development in processor and image sensor chips, the advances in electric motors, gearboxes and battery technology seem slow. However, one should not forget that improvements in processor speed and image sensor resolution are mainly a consequence of miniaturization—a technique that cannot easily be applied to other components.

T. Bräunl, *Embedded Robotics*,
https://doi.org/10.1007/978-981-16-0804-9_22

The largest development effort still remains in software. Several person-years are required for a project like the RoBIOS operating system, including cross-compiler adaptations, operating system routines, system tools, simulation systems and application programs.

While it has been estimated that up to 99% of all produced CPUs are being used to build embedded systems, most of us own over 100 embedded systems today. Even small and unassuming electric household devices are now being fitted with embedded controllers. With robot vacuums, robot lawn mowers and smart vehicles, the first autonomous robots have already entered our homes. Although it will still be a long road until we can delegate general household chores to a robot assistant, more and more embedded systems and smart devices are invading our daily lives—without us even noticing.

APPENDICES

APPENDIX A: RoBIOS LIBRARY

Version 7.2, Jan. 2021—RoBIOS is the operating system for the EyeBot controller.

The following libraries are available for programming the EyeBot controller in C/C++ or Python. Unless noted otherwise, return codes are 0 when successful and non-zero if an error has occurred.

In application source files include:

```
#include "eyebot.h".
```

Compile application to include RoBIOS library:

```
$gccarm myfile.c -o myfile.o
```

- LCD output (Sect. A.1)
- Key input (Sect. A.2)
- Camera (Sect. A.3)
- Image processing (Sect. A.4)
- System functions (Sect. A.5)
- Timer (Sect. A.6)
- USB/serial (Sect. A.7)
- Audio (Sect. A.8)
- Distance sensors (Sect. A.9)
- Servos and motors (Sect. A.10)
- V-omega driving interface (Sect. A.11)
- Digital and analog I/O (Sect. A.12)
- IR remote control (Sect. A.13)

Appendix A: RoBIOS Library

- Radio communication (Sect. A.14)
- Multitasking (Sect. A.15)
- Simulation (Sect. A.16).

A.1 LCD Output

```
int LCDPrintf(const char *format, ...);                              // Print string and arguments on LCD
int LCDSetPrintf(int row, int column, const char *format, ...);      // Printf from given position
int LCDClear(void);                                                  // Clear the LCD display and display buffers
int LCDSetPos(int row, int column);                                  // Set cursor position in pixels
int LCDGetPos(int *row, int *column);                                // Read current cursor position
int LCDSetColor(COLOR fg, COLOR bg);                                 // Set color for subsequent printf
int LCDSetFont(int font, int variation);                             // Set font for subsequent print operation
int LCDSetFontSize(int fontsize);                                    // Set font-size (7..18)
int LCDSetMode(int mode);                                            // Set LCD Mode (0=default)
int LCDMenu(char *st1, char *st2, char *st3, char *st4);             // Set menu entries for soft buttons
int LCDMenuI(int pos, char *string, COLOR fg, COLOR bg);             // Set menu for i-th entry with color
int LCDGetSize(int *x, int *y);                                      // Get LCD resolution in pixels
int LCDPixel(int x, int y, COLOR col);                               // Set one pixel on LCD
COLOR LCDGetPixel (int x, int y);                                    // Read pixel value from LCD
int LCDLine(int x1, int y1, int x2, int y2, COLOR col);              // Draw line
int LCDArea(int x1, int y1, int x2, int y2, COLOR col, int fill);    // Draw filled/hollow rectangle
int LCDCircle(int x1, int y1, int size, COLOR col, int fill);        // Draw filled/hollow circle
int LCDImageSize(int t);                                             // Define image type for LCD
int LCDImageStart(int x, int y, int xs, int ys);                     // Define image start position and size
int LCDImage(BYTE *img);                                             // Print color image
int LCDImageGray(BYTE *g);                                           // Print gray image [0..255] black..white
int LCDImageBinary(BYTE *b);                                         // Print binary image [0..1] white..black
int LCDRefresh(void);                                                // Refresh LCD output
```

Font Names and Variations:

HELVETICA (default), TIMES, COURIER

NORMAL (default), BOLD

Color Constants (COLOR is data type "int" in RGB order):

RED (0xFF0000), GREEN (0x00FF00), BLUE (0x0000FF), WHITE (0xFFFFFF), GRAY (0x808080), BLACK (0)

ORANGE, SILVER, LIGHTGRAY, DARKGRAY, NAVY, CYAN, TEAL, MAGENTA, PURPLE, MAROON, YELLOW, OLIVE

LCD Modes:

LCD_BGCOL_TRANSPARENT, LCD_BGCOL_NOTRANSPARENT, LCD_BGCO-L_INVERSE, LCD_BGCOL_NOINVERSE, LCD_FGCOL_INVERSE,

LCD_FGCOL_NOINVERSE, LCD_AUTOREFRESH, LCD_NOAUTOREFRESH, LCD_SCROLLING, LCD_NOSCROLLING, LCD_LINEFEED,

LCD_NOLINEFEED, LCD_SHOWMENU, LCD_HIDEMENU, LCD_LISTMENU, LCD_CLASSICMENU, LCD_FB_ROTATE, LCD_FB_NOROTATION

Appendix A: RoBIOS Library

A.2 Keys

```
int KEYGet(void);              // Blocking read (and wait) for key press (returns KEY1..KEY4)
int KEYRead(void);             // Non-blocking read of key press (returns NOKEY=0 if no key)
int KEYWait(int key);          // Wait until specified key has been pressed (use ANYKEY for any key)
int KEYGetXY (int *x, int *y); // Blocking read for touch at any position, returns coordinates
int KEYReadXY(int *x, int *y); // Non-blocking read for touch at any position, returns coordinates
```

Key Constants:
KEY1..KEY4, ANYKEY, NOKEY

A.3 Camera

```
int CAMInit(int resolution);   // Change camera resolution (will also set IP resolution)
int CAMRelease(void);          // Stops camera stream
int CAMGet(BYTE *buf);         // Read one color camera image
int CAMGetGray(BYTE *buf);     // Read gray scale camera image
```

For the following functions, the Python API differs slightly as indicated.

```
def CAMGet     () -> POINTER(c_byte):
def CAMGetGray() -> POINTER(c_byte):
```

Resolution Settings:
QQVGA(160x120), QVGA(320x240), VGA(640x480), CAM1MP(1296x730), CAMHD(1920x1080), CAM5MP(2592x1944), CUSTOM (LCD only)
Variables CAMWIDTH, CAMHEIGHT, CAMPIXELS (=width*height) and CAMSIZE (=3*CAMPIXELS) will be automatically set,
(BYTE is data type "unsigned char").

Constant sizes in bytes for color images and number of pixels:
QQVGA_SIZE, QVGA_SIZE, VGA_SIZE, CAM1MP_SIZE, CAMHD_SIZE, CAM5MP_SIZE
QQVGA_PIXELS, QVGA_PIXELS, VGA_PIXELS, CAM1MP_PIXELS, CAMHD_PIX-ELS, CAM5MP_PIXELS

Appendix A: RoBIOS Library

Data Types:

```
typedef QQVGAcol BYTE [115][5][3]; typedef QQVGAgray BYTE [115][5];
typedef QVGAcol BYTE [224][302][3]; typedef QVGAgray BYTE [224][302];
typedef VGAcol BYTE [480][640][3]; typedef VGAgray BYTE [480][640];
typedef CAM1MPcol BYTE [730][1296][3]; typedef CAM1MPgray BYTE [730][1296];
typedef CAMHDcol BYTE[1080][1920][3]; typedef CAMHDgray BYTE[1080][1920];
typedef CAM5MPcol BYTE[1944][2592][3]; typedef CAM5MPgray BYTE[1944][2592];
```

A.4 Image Processing

Basic image processing functions using the previously set camera resolution are included in the RoBIOS library. For more complex functions see the OpenCV library.

```
int    IPSetSize(int resolution);                                    // Set IP resolution using CAM constants
int    IPReadFile(char *filename, BYTE* img);                        // Read PNM file
int    IPWriteFile(char *filename, BYTE* img);                       // Write color PNM file
int    IPWriteFileGray(char *filename, BYTE* gray);                  // Write gray scale PGM file
void   IPLaplace(BYTE* grayIn, BYTE* grayOut);                       // Laplace edge detection on gray image
void   IPSobel(BYTE* grayIn, BYTE* grayOut);                         // Sobel edge detection on gray image
void   IPCol2Gray(BYTE* imgIn, BYTE* grayOut);                       // Transfer color to gray
void   IPGray2Col(BYTE* imgIn, BYTE* colOut);                        // Transfer gray to color
void   IPRGB2Col (BYTE* r, BYTE* g, BYTE* b, BYTE* imgOut);          // Transform 3*gray to color
void   IPCol2HSI (BYTE* img, BYTE* h, BYTE* s, BYTE* i);             // Transform RGB image to HSI
void   IPOverlay(BYTE* c1, BYTE* c2, BYTE* cOut);                    // Overlay c2 onto c1, all color images
void   IPOverlayGray(BYTE* g1, BYTE* g2, COLOR col, BYTE* cOut);     // Overlay gray image g2 onto g1, using col
COLOR  IPPRGB2Col(BYTE* r, BYTE g, BYTE b);                          // PIXEL: RGB to color
void   IPPCol2RGB(COLOR col, BYTE* r, BYTE* g, BYTE* b);             // PIXEL: color to RGB
void   IPPCol2HSI(COLOR c, BYTE* h, BYTE* s, BYTE* i);              // PIXEL: RGB to HSI for pixel
BYTE   IPPRGB2Hue(BYTE r, BYTE g, BYTE b);                           // PIXEL: Convert RGB to hue
void   IPPRGB2HSI(BYTE r, BYTE g, BYTE b, BYTE* h, BYTE* s, BYTE* i); // PIXEL: Convert RGB to hue, sat, int;
```

For the following functions, the Python API differs slightly as indicated.

```
from typing import List
from ctypes import c_int, c_byte, POINTER

def IPLaplace     (grayIn: POINTER(c_byte)) -> POINTER(c_byte):
def IPSobel       (grayIn: POINTER(c_byte)) -> POINTER(c_byte):
def IPCol2Gray    (img: POINTER(c_byte))    -> POINTER(c_byte):
def IPCol2HSI     (img: POINTER(c_byte))    -> POINTER(c_byte) -> List[c_byte, c_byte, c_byte]:
def IPOverlay     (c1: POINTER(c_byte), c2: POINTER(c_byte))   -> POINTER(c_byte):
def IPOverlayGray (g1: POINTER(c_byte), g2: POINTER(c_byte))   -> POINTER(c_byte):
```

A.5 System Functions

```
char * OSExecute(char* command);              // Execute Linux program in background
int OSVersion(char* buf);                     // RoBIOS Version
int OSVersionIO(char* buf);                   // RoBIOS-IO Board Version
int OSMachineSpeed(void);                     // Speed in MHz
int OSMachineType(void);                      // Machine type
int OSMachineName(char* buf);                 // Machine name
int OSMachineID(void);                        // Machine ID derived from MAC address
```

A.6 Timer

```
int    OSWait(int n);                              // Wait for n/1000 sec
TIMER OSAttachTimer(int scale, void (*fct)(void)); // Add fct to 1000Hz/scale timer
int    OSDetachTimer(TIMER t);                     // Remove fct from 1000Hz/scale timer
int OSGetTime(int *hrs,int *mins,int *secs,int *ticks); // Get system time (ticks in 1/1000 sec)
int OSGetCount(void);                              // Count in 1/1000 sec since system start
```

A.7 USB/Serial Communication

```
int    SERInit(int interface, int baud,int handshake); // Init communication HDT file
int    SERSendChar(int interface, char ch);     // Send single character
int    SERSend(int interface, char *buf);       // Send string (Null terminated)
char SERReceiveChar(int interface);             // Receive single character
int    SERReceive(int interface, char *buf, int size); // Receive String
int    SERFlush(int interface);                 // Flush interface buffers
int    SERClose(int interface);                 // Close Interface
```

Communication Parameters:

Baudrate: 50 .. 230400

Handshake: NONE, RTSCTS

Interface: 0 (serial port), 1..20 (USB devices, names are assigned via HDT entries)

A.8 Audio

```
int AUBeep(void);                      // Play beep sound
int AUPlay(char* filename);            // Play audio sample in background (mp3 or wave)
int AUDone(void);                      // Check if AUPlay has finished
int AUMicrophone(void);                // Return microphone A-to-D sample value
```

Use analog data functions to record microphone sounds (channel 8).

A.9 Distance Sensors

Position sensitive devices (PSDs) are using infrared beams to measure distance and need to be calibrated in HDT (https://robotics.ee.uwa.edu.au/ eyebot/HDT7.html) to get correct distance readings.

Light detection and ranging (LIDAR) is a single-axis rotating laser scanner.

```
int PSDGet(int psd);                         // Read distance value in mm from PSD
int PSDGetRaw(int psd);                       // Read raw value from PSD sensor [1..6]
int LIDARGet(int distance[]);                 // Measure distances in [mm];
int LIDARSet(int range, int tilt, int points); // range [1..360°], tilt angle down
```

```
PSD Constants:
PSD_FRONT, PSD_LEFT, PSD_RIGHT, PSD_BACK
assuming PSD sensors in these directions are connected to ports 1, 2,
3, 4.


LIDAR Constants:
LIDAR_POINTS Total number of points returned
LIDAR_RANGE Angular range covered, e.g., 180°
```

A.10 Servos and Motors

Motor and servo positions can be calibrated through HDT (https://robotics. ee.uwa.edu.au/eyebot/HDT7.html) entries.

```
int SERVOSet(int servo, int angle);          // Set servo [1..14] position
int SERVOSetRaw (int servo, int angle);      // Set servo [1..14] position bypassing HDT
int SERVORange(int servo, int low, int high); // Set servo [1..14] limits in 1/100 sec
int MOTORDrive(int motor, int speed);        // Set motor [1..4] speed in percent
int MOTORDriveRaw(int motor, int speed);     // Set motor [1..4] speed bypassing HDT
int MOTORPID(int motor, int p, int i, int d); // Set motor [1..4] PID controller
int MOTORPIDOff(int motor);                  // Stop PID control loop
int MOTORSpeed(int motor, int ticks);        // Set controlled motor speed
int ENCODERRead(int quad);                   // Read quadrature encoder [1..4]
int ENCODERReset(int quad);                  // Set encoder value to 0 [1..4]
```

A.11 Omega Driving Interface

This is a high-level wheel control for differential driving. It always uses motor 1 (left) and motor 2 (right).

Motor spinning directions, motor gearing and vehicle width are set in the HDT (https://robotics.ee.uwa.edu.au/eyebot/HDT7.html) file.

```
int VWSetSpeed(int linSpeed, int angSpeed);          // Set fixed linSpeed  [mm/s] and [deg/s]
int VWGetSpeed(int *linSspeed, int *angSpeed);       // Read current speeds [mm/s] and [deg/s]
int VWSetPosition(int x, int y, int phi);            // Set robot position to x, y [mm], phi [deg]
int VWGetPosition(int *x, int *y, int *phi);         // Get robot position as x, y [mm], phi [deg]
int VWStraight(int dist, int lin_speed);             // Drive straight, dist, lin. speed
int VWTurn(int angle, int ang_speed);                // Turn on spot, angle, ang. speed
int VWCurve(int dist, int angle, int lin_speed);     // Drive Curve, dist, angle, lin. speed
int VWDrive(int dx, int dy, int lin_speed);          // Drive x[mm] straight and y[mm] left,
int VWRemain(void);                                  // Return remaining drive distance in [mm]
int VWDone(void);                                    // Non-blocking check it drive is finished
int VWWait(void);                                    // Suspend current thread until finished
int VWStalled(void);                                 // Returns number of stalled motor
```

All VW functions return 0 if OK and 1 if error (e.g., destination unreachable)

A.12 Digital and Analog Input/Output

```
int DIGITALSetup(int io, char direction);            // Set IO line [1..16]
int DIGITALRead(int io);                             // Read and return individual input line
int DIGITALReadAll(void);                            // Read and return all 16 io lines
int DIGITALWrite(int io, int state);                 // Write individual output [1..16] to 0 or 1
int ANALOGRead(int channel);                         // Read analog channel [1..8]
int ANALOGVoltage(void);                             // Read analog supply voltage in [0.01 Volt]
int ANALOGRecord(int channel, int iterations);       // Record analog data
int ANALOGTransfer(BYTE* buffer);                    // Transfer previously recorded data
```

Default for digital lines: [1..8] are input with pull-up, [9..16] are output

Default for analog lines: [0..8] with 0: supply-voltage and 8: microphone

IO settings are: i: input, o: output, I: input with pull-up res., J: input with pull-down res.

A.13 IR Remote Control

These commands allow sending commands to an EyeBot via a standard infrared TV remote (IRTV). IRTV models can be enabled or disabled via a HDT (https://robotics.ee.uwa.edu.au/eyebot/HDT7.html) entry.

Supported IRTV models include: Chunghop L960E Learn Remote.

```
int IRTVGet(void);                                   // Blocking read of IRTV command
int IRTVRead(void);                                  // Non-blocking read, return 0 if nothing
int IRTVFlush(void);                                 // Empty IRTV buffers
int IRTVGetStatus(void);                             // Checks to see if IRTV is activated
```

Defined constants for IRTV buttons are

IRTV_0 .. IRTV_9, IRTV_RED, IRTV_GREEN, IRTV_YELLOW, IRTV_BLUE, IRTV_LEFT, IRTV_RIGHT, IRTV_UP, IRTV_DOWN, IRTV_OK, IRTV_POWER

A.14 Radio Communication

These functions require a WiFi modules for each robot, one of them (or an external router) as master, all others in slave mode.

Radio can be activated/deactivated via a HDT (https://robotics.ee.uwa.edu.au/eyebot/HDT7.html) entry. The names of all participating nodes in a network can also be stored in the HDT file.

```
int RADIOInit(void);                              // Start radio communication
int RADIOGetID(void);                             // Get own radio ID
int RADIOSend(int id, char* buf);                 // Send string (Null terminated) to ID
int RADIOReceive(int *id_no, char* buf, int size); // Wait for message
int RADIOCheck(void);                             // Check if message is waiting
int RADIOStatus(int IDlist[]);                    // Returns number of robots (incl. self)
int RADIORelease(void);                           // Terminate radio communication
```

```
ID numbers match last byte of robots' IP addresses.
```

```
def RADIOReceive(int size) -> [id_no, buf]  # max 1024 Bytes
def RADIOStatus()           -> int IDList[]  # max  256 entries
```

A.15 Multitasking

For multitasking, simply use the pthread functions.

A number of multitasking sample programs are included in the demo/MULTI directory.

A.16 Simulation *only*

These functions will **only** be available when run in a simulation environment, in order to get ground truth information and to repeat experiments with identical setup.

```
void SIMGetRobot (int id, int *x, int *y, int *z, int *phi);
void SIMSetRobot (int id, int  x, int  y, int  z, int  phi);
void SIMGetObject(int id, int *x, int *y, int *z, int *phi);
void SIMSetObject(int id, int  x, int  y, int  z, int  phi);
int  SIMGetRobotCount()
int  SIMGetObjectCount()
```

```
id=0 means own robot; id numbers run from 1..n
```

Thomas Bräunl, Remi Keat, Marcus Pham, 1996–2021.

Appendix B: EyeBot-IO7 Interface

Connect the IO7 board via the micro-USB connectors to an embedded controller, Mac, Windows or Linux PC. If you want to use the motor or servo drivers, you have to connect an external battery or power supply (do not power from USB in that case).

- Use system file EyeBot_M7.inf (https://robotics.ee.uwa.edu.au/eyebot/EyeBot-IO/EyeBot_M7.inf)
- Mac connect with: source /dev/tty.usbmodem…
- Windows connect with putty (https://www.putty.org/)
- Windows may require device tools lib-usb (https://sourceforge.net/projects/libusb-win32/) or Zadig (https://zadig.akeo.ie)
- The system will start with a greeting message after booting:
  ```
  EyeBot IO7 ver. <xx.x>, <date>
  ```
- The prompt before every command in is '>'.
  If there is an error in binary mode, the prompt gets changed to '#'.
- Commands can be given in ASCII mode (see below) or binary mode (command char. + 0x80).
  Return values are in ASCII/binary matching the given command.
- Unless otherwise noted, each command and each parameter uses 1 byte, and for multiple return bytes, the highest byte is sent first.

Appendix B: EyeBot-IO7 Interface

List of interface commands

Cmd.		Description	Parameters
h	help	print list of commands	none
O	Output	text output in verbose mode - verbose [0 ...1] Ex. O 0 (turn off verbose mode)	O \<verbose>
E	demo	will drive robot along a path and print relevant sensor data	none
m	motor	set motor speed (uncontrolled) - motor [1..4] - speed [-100 .. 100] Ex.: m 1 50, m 2 -100	m \<motor> \<speed>
M	motor	set motor speed (controlled) - motor [1..4] - ticks per 1/100 sec. [-128 .. +127] will automatically enable PID control	M \<motor> \<ticks>
d	PID	set PID parameters - motor [1..4] - p, i, d [0..255] PID parameters Ex.: p 3 4 1 1	d \<motor> \<p> \<i> \<d>
s	servo	set servo position - servo [1..14] - position [0..255] Ex.: s 2 200, s 10 0	s \<servo> \<position>
s	servo	set servo position - servo [1..14] - position [0..255] Ex.: s 2 200, s 10 0	s \<servo> \<position>
S	servo	set servo limits - servo [1..14] - lower, upper [0..255] in ms/100 lower & upper specify servo signal uptimes position values 0 (lower) and 255 (upper)	S \<servo> \<lower> \<Upper>
p	PSD	read PSD distance sensor sensor [1..6] Ex.: p 5 (returns 2 Byte value, analog)	p \<sensor>
e	encoder	print encoder value encoder [1..4] Return: 2 Byte value Ex. e 4	e \<encoder>
v	version	print version number Return: 4 Bytes (characters)	none

Appendix B: EyeBot-IO7 Interface

Cmd.		Description	Parameters
V	voltage	print supply voltage level Return: 2 Byte value [V/100]	none
i	input	read digital input port port [0..15] Ex.: i 5	i <port>
I	input	read ALL digital input ports Return: 2 Bytes	none
o	output	set digital output port port [0..15], bit [0..1] Ex.: o 7 1	o <port> <bit>
c	config	set i/o port configuration port [0..15], state [i, o, I, J] i=input, o=output, I=in+pull_up, J=in +pull_down default i/o [0..7] = I, [8..15] = o Ex.: c 14 i; c 2 o	c <port> <state>
a	analog	read analog input port [0..8] a0 is supply voltage, a8 is microphone input a1..a7 are free pins on board. Max. 2.7V Return: 2 Byte value [V/100] Ex.: a 5	a <port>
r	record	record analog input port [0..14] 0 is supply volt., 1..7 pins, 8 is microphone, 9..14 are PSD1..PSD6 iterations [1..2048] at 1kHz Ex.: r 8 1000 record for 1sec microphone input	r <port> <iterations>
R	read	read recorded analog values transfers previously recorded byte values Ex.: R	none
l	led	set led status led [1..4], bit [0..1] Ex.: l 2 1	l <led> <bit>
t	time	print 100Hz timer counter Return: 4 Byte value	none
T	timer	set timer frequency timer [1..100] set timer in [10Hz] Ex.: T 100 for 1'000Hz	T <timer>
u	upload	set board into USB-upload mode for software upgrade	none

Appendix B: EyeBot-IO7 Interface

Cmd.		Description	Parameters
w	**v-omega**	init v-omega driving always uses motors M1(left and M2(right) - ticks per m [1..65'535] (2 Bytes) - base width in mm [1..65'535] (2 Bytes) - max speed in cm/s [1..65'535] (2 Bytes) - dir [1..4]: 1= M1-counter-clock & M2-clockwise; 2=clock&counter; 3=clock&clock; 4=counter&counter Ex.: w 20 75 200 1 (for 20 ticks/dm, 75mm base, 2m/s max, cnt/clock)	w \<tick\> \<base\> \<max\> \<dir\>
W	**v-omega**	stop v-omega driving	none
A	**v-omega**	change v-omega parameters - vv, tv, vw, tw [0.255] proportional/ integral parameters for v-omega controller, v typically around 7 (parameter 70) [1/10], t typically around 0.2 (param. 20) [1/100] Ex.: A 70 30 70 10 (for vv=7, tw=0.3 vw=7, tw=0.1)	A \<vv\> \<tv\> \<vw\> \<tw\>
x	**driveV**	set vehicle speed - linear speed in cm/s [-32'768..+32767] (2 Bytes) - angular speed in rad/10s [-128..127] (1 Byte) Ex.: x 100 0 for straight 1m/s x 50 1 for curve to forward left 5.7°/s	x \<linear\> \<angular\>
X	**driveV**	get vehicle speed print linear speed [cm/s] (2 Bytes) and angular speed in rad/10s (1 Byte) Return: 3 Byte value	none
q	**pose**	get vehicle pose print x, y, phi in [mm] and [100rad] (2 Bytes each) initially all set to 0 Return: 6 Bytes	none
Q	**pose**	set vehicle pose - x,y in mm [-32768 .. 32767] (2 Bytes) - phi in 100rad [-314 .. 314] (2 Bytes) Ex.: Q 200 100 157 (x=0.2m y=0.1m, phi=90°)	Q \<x\> \<y\> \<phi\>
y	**driveS**	drive straight - speed in mm/s [0..65535] (2 Bytes) - distance in mm [-32768 .. 32767] (neg. for backwards, 2 Bytes) Ex.: y 100 -1000 for 1m backwards y 1000 1000 for 1 m fast forward	y \<speed\> \<distance\>

Appendix B: EyeBot-IO7 Interface

Cmd.		Description	Parameters
Y	driveS	drive turn - rot_speed in 0.1*rad/s [0..65535] (2 Bytes) - angle in 0.1*rad [-32768 .. 32767] (positive=counter-clock, 2 Bytes) Ex.: Y 1000 -6284 full clockwise rotation (2*Pi)	y \<rot_speed \<angle>
C	driveC	drive curve - speed in mm/s [0..65535] mm (2 Bytes) - distance in mm [-32768 .. 32767] (neg. for backwards, 2 Bytes) - angle in 0.1*rad [-32768 .. 32767] (positive=counter-clock, 2 Bytes) Ex.: C 1000 2071 1000 1m forward with 90 deg (Pi/4) counter-clock	C \<speed> \<distance> \<angle>
z	driveR	drive remaining return remaining distance to goal (0 if reached) in mm or radians Return: 2 Byte value	none
Z	driveD	drive done or stalled return 1 if previous drive is finished, otherwise 0	none
L	stalled	check if a motor has stalled return 0 if no motor has stalled return stall-bit-values for Motors[1..4], e.g., 0b1111 = all motors stalled	none

Thomas Bräunl, Franco Hidalgo, Remi Keat 2021.

APPENDIX C: HARDWARE DESCRIPTION TABLE

Each EyeBot controller contains a hardware description table (HDT) as a file stored in location:

`/home/pi/hdt.txt`

The HDT files contain important entries, which allow changing robot characteristics without having to change the robot's program. Important HDT entries are

- Robot name (for WiFI SSID broadcast)
- Robot group names (for radio communication)
- Select WiFI master or slave mode
- Select LCD screen orientation
- Motor and servo calibration
- PSD sensor calibration
- Enabling infrared remote control
- Robot dimensions and motor arrangement for differential drive
- Port names of attached USB devices.

Supported infrared remote is Chunghop L960E (code 786):

On remote press SET + TV buttons for 3 s. (light on), press 7 8 6, then SET.

Appendix C: Hardware Description Table

Example HDT file:

```
# EyeBot Controller Name: EYEBOT Name
EYEBOT myrobot

# Network Settings: SSID Network_Name |0 (MASTER=Hotspot) or 1
(SLAVE=join network) | Password (SSID)
# Default is creating hotspot with SSID derived from PI's serial
number, password raspberry
SSID robonet 0 secret

# RoBIOS run at startup (0/1)
ROBIOS 1

# LCD orientation (0/1)
LCD 0

# Infrared Remote enabled (0/1)
IRTV 1

# USB Devices: USB Number | DeviceName
USB 1 EyeBot7
USB 2 GPS

# Servo Number [1..14] |Min-time (us) | Max-time (us) | TableName (opt.)
SERVO 1 1000 2000
SERVO 2 700 1500 Servo_Table

# ENCODER Number [1..4] | Clicks per meter
ENCODER 1 10000
ENCODER 2 10000

# PSD Number [1..6] | TableName
PSD 1 PSD_TableA
PSD 2 PSD_TableA
PSD 3 PSD_TableB

# DRIVE Wheel distance (mm) | Max motor speed (cm/s) | Motor1/left
dir. |Motor2/right dir. (1=clockwise)
DRIVE 100 200 0 1
```

Appendix C: Hardware Description Table

```
# IRTV Name | Type | Length | tog_mask | inv_mask | mode | bufsize | delay
# Currently support remote is Chunghop L960E, code 786
# On remote press SET+TV buttons for 3 sec. (light on), press 7 8 6, then
SET.
IRTV "IRTV0" 0 4 0 0 0 4 20


# Path to demo and software folders
DEMOPATH /home/pi/eyebot/demo
SOFTWAREPATH /home/pi/usr/software


# ———————————————— TABLES (Optional) ————————————————
# Motor Linearisation Table 101 values: 0 .. 100
TABLE Motor_Table
0    1 2 3   4  5  6  7  8  9 10 11 12 13 14 15 16 17 18 19
20   21 22 23 24 25 26 27 28 29 30 31 32 33 34 35 36 37 38 39
40   41 42 43 44 45 46 47 48 49 50 51 52 53 54 55 56 57 58 59
60   61 62 63 64 65 66 67 68 69 70 71 72 73 74 75 76 77 78 79
80   81 82 83 84 85 86 87 88 89 90 91 92 93 94 95 96 97 98 99
100
END TABLE


# Thomas Bräunl, Remi Keat, Marcus Pham 2021.
```

APPENDIX D: ROBOT PROGRAMMING PROJECTS

D.1 Games

Experiment 1. Etch-a-Sketch

This program implements the *Etch-a-Sketch* children's game. Four buttons are used in a consistent way for moving the drawing pen left/right and up/down.

```
#include <eyebot.h>
#define X 320
#define Y 240

int main()
{ int k;
  int x=0, y=0, sign=1;

  LCDMenu("X","Y","+/-","END");
  while(KEY4 != (k=KEYRead()))
  { LCDPixel(x,y, WHITE);
    switch (k)
    { case KEY1: x = (x + sign + X) % X; break;
      case KEY2: y = (y + sign + Y) % Y; break;
      case KEY3: sign = -sign; break;
    }
```

Appendix D: Robot Programming Projects

```
  LCDSetPrintf(18,0, "%3d:%3d (x,y)", x,y);
  }
  return 0;
}
```

Experiment 2. Reaction Test Game

This program implements a reaction test game. After pressing the *GO* button, it waits a random amount of time before starting the timer. Early presses are detected and reported as cheating.

```
#include <eyebot.h>

int main()
{ float time;
  struct timeval stop, start;

  while (1)
  { LCDPrintf("Reaction Test\n");
    LCDMenu("GO", "GO", "GO", "END");
    if (KEYGet()==KEY4) return 0;
    time = 1000.0 + 7000.0 * drand48(); // 1..8 sec
    LCDMenu(" ", " ", " ", " ");
    usleep(1000 * (int) time);

    LCDMenu("HIT", "HIT", "HIT", "HIT");
    if (KEYRead()) LCDPrintf("No cheating !!\n\n");
    else
    { gettimeofday(&start, NULL);
      KEYWait(ANYKEY);
      gettimeofday(&stop, NULL);
      LCDPrintf("time: %lu micro sec\n\n",
        (stop.tv_sec - start.tv_sec)*1000000 +
        (stop.tv_usec - start.tv_usec));
    }
  }
}
```

Appendix D: Robot Programming Projects

D.2 Driving

Experiment 3. Drive a Fixed Distance and Return

The easiest driving solution uses the *VWStraight* and *VWTurn* functions.

```
#include <eyebot.h>

int main()
{ VWStraight(500, 150); VWWait(); // Sraight
  VWTurn(180, 60);        VWWait(); // Turn
  VWStraight(500, 150); VWWait(); // Straight back
}
```

Experiment 4. Drive in a Square

Driving a square requires a loop of four iterations of driving straight and turning.

```
#include "eyebot.h"

int main()
{ for (int i=0; i<3*4; i++)              // run 3 squares
  { VWStraight(400, 300); VWWait(); // wait until completed;
    VWTurn(90, 90); VWWait();              // wait until completed
  }
}
```

Experiment 5. Drive in a Circle

Either *VWDriveCurve* or the low-level *VWSetSpeed* can be used for driving in a circle. Shown here is the simplest version in Python.

```
#!/usr/bin/env python3
from eye import *

VWSetSpeed(300, 15)
```

Appendix D: Robot Programming Projects

Experiment 6. Drive and Steer an Ackermann Vehicle

```
#include "eyebot.h"

void Mdrive(char* txt, int drive, int steer)
/* Print txt and drive motors and encoders for 1.5s */
{ LCDPrintf("%s\n", txt);
  MOTORDrive(1, drive);
  SERVOSet (1, steer);
}

int main ()
{ LCDPrintf("Ackermann Steering\n");
  Mdrive("Forward",      20, 128); OSWait(1000);
  Mdrive("Backward",    -20, 128); OSWait(1000);
  Mdrive("Left Curve",   10,   0); OSWait(2000);
  Mdrive("Right Curve", 10, 255); OSWait(2000);
  Mdrive("Stop",              0,  0);
  return 0;
}
```

Experiment 7. Drive and Steer a Mecanum Vehicle

```
#include "eyebot.h"
#define S 25 // Speed

void Mdrive(char* txt, int Fleft, int Fright, int Bleft, int
Bright)
/* Print txt and drive motors and encoders for 1.5s */
{ LCDPrintf("%s\n", txt);
  MOTORDrive(1, Fleft);
  MOTORDrive(2, Fright);
  MOTORDrive(3, Bleft);
  MOTORDrive(4, Bright);
  OSWait(1500);
  LCDPrintf("Enc.%5d %5d\n %5d %5d\n",
    ENCODERRead(1), ENCODERRead(2), ENCODERRead(3), ENCODERRead(4));
}

int main ()
{ LCDPrintf('Mecanum Wheels\n');

  Mdrive('Forward',     S,  S,  S,  S);
  Mdrive('Backward',   -S, -S, -S, -S);
  Mdrive("Left",       -S,  S,  S, -S);
```

```
Mdrive('Right',       S,-S,-S, S);
Mdrive('Left45',      0, S, S, 0);
Mdrive('Right45',     S, 0, 0, S);
Mdrive('Turn Spot L', -S, S,-S, S);
Mdrive('Turn Spot R', S,-S, S,-S);
Mdrive('Stop',        0, 0, 0, 0);
return 0;
}
```

D.3 Sensors

Experiment 8. Straight and Back with Distance Sensor

In this experiment the robot is to drive forward until an obstacle or a wall is encountered, then stop, turn 180° and finally drive back.

```
#include <eyebot.h>

int main()
{ int x, y, p;
  for (int i=0; i<2; i++) // run twice
  { VWStraight(500, 50); // 0.5m in ca. 5s
    while (!VWDone())
    { VWGetPosition(&x, &y, &p);
      LCDPrintf("X:%4d; Y:%4d; P:%4d\n", x, y, p);
      if (PSDGet(2) < 100) VWSetSpeed(0,0); // STOP if obstacle
    }
    VWTurn(180, 60);   // half turn (180 deg) in ca. 3s
    VWWait();     // wait until completed
  }
}
```

Appendix D: Robot Programming Projects

Experiment 9. Drive While Displaying the Camera Image

In this program the robot drives and turns while continuously reading and displaying the camera image.

```
#include <eyebot.h>

void cam_display()
{ QVGAcol img;
  while (!VWDone())
  {CAMGet(img);
   LCDImage(img);
  }
}

int main()
{ CAMInit(QVGA);
  for (int i=0; i<2; i++)   // run twice
  { VWStraight(500, 50);    // 0.5m in ca. 5s
    cam_display();
    VWTurn(180, 60);        // half turn (180 deg) in ca. 3s
    cam_display();
  }
}
```

Experiment 10. Use the Pan–Tilt Movement of the Camera

```
#include "eyebot.h"

void checkpos(int pan, int tilt)
{ BYTE img[QVGA_SIZE];

  SERVOSet(1, pan);
  SERVOSet(2, tilt);
  LCDSetPrintf(0,0,"PanTlt: %3d %3d\n", pan, tilt);
  CAMGet(img);
  LCDImage(img);
}

int main()
{ int pos;

  CAMInit(QVGA);

  for (pos=128; pos<255; pos++) checkpos(pos, 128);
  for (pos=255; pos>0;   pos- ) checkpos(pos, 128);
  for (pos=0;   pos<128; pos++) checkpos(pos, 128);
```

Appendix D: Robot Programming Projects

```
    for (pos=128; pos<255; pos++) checkpos(128, pos);
    for (pos=255; pos>0;   pos- ) checkpos(128, pos);
    for (pos=0;   pos<128; pos++) checkpos(128, pos);

    return 0;
}
```

Experiment 11. Motion Detection with Camera

By subtracting the pixel values of two subsequent grayscale images, motion can be detected. This program only detects the presence of motion, but does not determine its location in the image.

```
#include "eyebot.h"
#define RES QVGA
#define SIZE QVGA_SIZE

void image_diff(BYTE i1[SIZE], BYTE i2[SIZE], BYTE d[SIZE])
{ for (int i=0; i<SIZE; i++)
    d[i] = abs(i1[i] - i2[i]);
}

int avg(BYTE d[SIZE])
{ int i, sum=0;
  for (i=0; i<SIZE; i++)
    sum += d[i];
  return sum / SIZE;
}

int main()
{ BYTE image1[SIZE], image2[SIZE], diff[SIZE];
  int avg_diff, delay, i=0;

  CAMInit(RES);
  LCDMenu(" ", " ", " ", "END");
  CAMGetGray(image1);

  while (KEYRead() != KEY4)
  { i++;
    if (i%2=0) CAMGetGray(image1);
         else  CAMGetGray(image2);

    image_diff(image1, image2, diff);
    LCDImageGray(diff);
    avg_diff = avg(diff);
    LCDSetPrintf(0,50, "Avg = %3d", avg_diff);
```

```
    if (avg_diff > 15) // Alarm threshold
    { LCDSetPrintf(2,50, "ALARM!!!");
      delay = 10;
    }
    if (delay) delay-;
       else LCDSetPrintf(2,50, " "); // clear text after delay
  }
}
```

Experiment 12. Display the Lidar Sensor Data

```
#include "eyebot.h"

int main ()
{ int i, scan[360];

  while (1)
  { LCDMenu("SCAN", " ", " ", "END");
    if (KEYGet() == KEY4) return 0;
    LCDClear();
    LIDARGet(scan);
    for (i=0; i<360; i++)
      LCDLine(i,250-scan[i]/10, i,250, BLUE);
    LCDLine(180,0, 180,250, RED);    // straight (0°)
    LCDLine( 90,0,  90,250, GREEN); // left (-90°)
    LCDLine(270,0, 270,250, GREEN); // right (+90°)
    LCDSetPrintf(19,0, " -90 0 +90");
  }
}
```

D.4 Navigation

Experiment 13. Dead Reckoning

This program lets the robot drive a sequence of straight segments and rotations, depending on button presses. Starting with initial pose (0, 0, 0), it continuously updates and displays its current pose using dead reckoning. This program can be used to test how quickly a robot's internal position and orientation errors accumulate.

```
#include <eyebot.h>

int main()
{ int x, y, phi;

  LCDPrintf("Dead Reckoning Experiment\n");
  LCDMenu("STRAIGHT", "LEFT", "RIGHT", "END");
  while(1)
  { VWGetPosition(&x,&y,&phi);
    LCDPrintf("Position: (%d, %d) Angle: %d\n", x, y, phi);
    switch(KEYGet())
    { case KEY1: VWStraight(250, 125); VWWait(); break;
      case KEY2: VWTurn( 90, 45); VWWait(); break;
      case KEY3: VWTurn(-90, 45); VWWait(); break;
      case KEY4: return 0;
    }
  }
}
```

D.5 Robot Groups

Experiment 14. Following a Leading Robot

The following program implements a predator behavior. This robot will follow another robot (prey), trying to catch it. It detects the prey robot by using a Lidar scan, so this requires that no other robot or obstacle is in its field of view.

```
#include "eyebot.h"

int main ()
{ int i, min_pos, scan[360];

  while (KEYRead()!=KEY4)
  { LCDClear();
    LCDMenu(" ", " ", " ", "END");
    LIDARGet(scan);
    min_pos = 0;
    for (i=0; i<360; i++)
    { if (scan[i] < scan[min_pos]) min_pos = i;
      LCDLine(i,250-scan[i]/50, i,250, BLUE);
    }
```

```
    VWSetSpeed(300, 180-min_pos);
    LCDSetPrintf(20,0,"minimum at %d (%d)\n",
     min_pos, scan[min_pos]);
    if (scan[min_pos] > 5000) VWSetSpeed(0,0); // stop
    OSWait(100); // 0.1 sec
  }
}
```

D.6 Submarines

Experiment 15. Let a Submarine Dive by the Push of a Button

```
#include "eyebot.h"

// Thruster IDs.
#define LEFT 1
#define FRONT 2
#define RIGHT 3
#define BACK 4

#define PSD_DOWN 6 // PSD direction

void dive(int speed)
{ MOTORDrive(FRONT, speed);
  MOTORDrive(BACK, speed);
}

int main()
{ BYTE img[QVGA_SIZE];
  char key;

  LCDMenu("DIVE", "STOP", "UP", "END");
  CAMInit(QVGA);
  do { LCDSetPrintf(19,0, "Dist to Ground:%6d\n",
        PSDGet(PSD_DOWN));
       CAMGet(img);
       LCDImage(img);

       switch(key=KEYRead())
       { case KEY1: dive(-100); break;
         case KEY2: dive(   0); break;
         case KEY3: dive(+100); break;
```

Appendix D: Robot Programming Projects

```
        }
    } while (key != KEY4);
  return 0;
}
```

Experiment 16. Drive a Submarine on the Surface

```c
#include "eyebot.h"

// Thruster IDs.
#define LEFT 1
#define FRONT 2
#define RIGHT 3
#define BACK 4

// PSD direction
#define PSD_FRONT 1
#define PSD_FRONT_LEFT 2
#define PSD_FRONT_RIGHT 3
#define PSD_BACK_LEFT 4
#define PSD_BACK_RIGHT 5
#define PSD_DOWN 6

void drive(int l_speed, int r_speed)
{ MOTORDrive(LEFT, l_speed);
  MOTORDrive(RIGHT, r_speed);
}

void PSD_All(int psd[])
{ int i;
  for (i=1; i<=6; i++) psd[i] = PSDGet(i);
}

int main()
{ BYTE img[QVGA_SIZE];
  int psd[7];
  char key;

  LCDMenu("FORWARD", "LEFT", "RIGHT", "END");
  CAMInit(QVGA);
  do { CAMGet(img);
       LCDImage(img);
       PSD_All(psd);
       LCDSetPrintf(19,0, "Dist %6d\n", psd[PSD_FRONT]);
```

```
    switch(key=KEYRead())
    {case KEY1: drive( 100, 100); break;
     case KEY2: drive(-100, 100); break;
     case KEY3: drive( 100, -100); break;
    }
  } while (key != KEY4);
 return 0;
}
```

D.7 Manipulators

Experiment 17. Move a Manipulator Repeatedly between Positions

```
#include "eyebot.h"

int main ()
{ while(1)
  { // pos. 1
    SERVOSet(1, 200);
    SERVOSet(2,  10);
    SERVOSet(3,  65);
    SERVOSet(4, 240);
    SERVOSet(5,  60);
    SERVOSet(6,  10);
    OSWait(4000);

    // pos. 2
    SERVOSet(2, 245);
    SERVOSet(3,   0);
    OSWait(2000);

    // pos. 2
    SERVOSet(1,   0);
    OSWait(2000);
  }
}
```

Appendix D: Robot Programming Projects

Experiment 18. Move a Manipulator one Joint at a Time

```
#include "eyebot.h"

int main ()
{ int angles[6] = {0,0,0,0,0,0};
  int joint = 0, done = false;

  LCDMenu("JOINT", "+", "-", "END");
  while (!done)
  { LCDSetPrintf(0,0, "Joint %d - Angles %3d %3d %3d %3d %3d %3d",
    joint+1, angles[0],angles[1],angles[2],
             angles[3],angles[4],angles[5]);
    switch (KEYGet())
    { case KEY1: joint = (joint+1)%6; break;
      case KEY2: angles[joint] +=10; break;
      case KEY3: angles[joint] -=10; break;
      case KEY4: done = true;
    }
    SERVOSet(joint+1, angles[joint]); // drive to new angle
  }
}
```

Printed in the United States
by Baker & Taylor Publisher Services